DISTRIBUTION-FREE STATISTICAL TESTS

JAMES V. BRADLEY

BEHAVIOR RESEARCH LABORATORY

ANTIOCH COLLEGE

PRENTICE-HALL, INC.

Englewood Cliffs, New Jersey

Current printing (last digit):
10 9 8 7 6 5

Library of Congress Catalog Card Number 68–22803

Printed in the United States of America

PREFACE

Unlike parametric tests, many of the best known distribution-free tests rest upon a mathematical foundation no more sophisticated than elementary probability theory, and their derivation is well within the grasp of anyone not frightened by high school algebra. This convenient feature obviates the uncomprehending "cookbook" approach to statistics which has characterized the explication of classical parametric tests in textbooks aimed at users innocent of higher mathematics. Thus while the nonmathematical scientist may have to take the validity and appropriateness of parametric tests on faith, most distribution-free tests subject him to no such humiliatingly unscientific procedure. Their derivation is well within his comprehension, placing in his hands a tool made scientific by his understanding of its mechanism. This is the case provided, of course, that he learns distribution-free tests from a source that gives their derivation, rather than from a cookbook.

This book is intended as the antithesis of a cookbook. Although examples of application are given, the major emphasis is placed upon the logical and mathematical rationale of the tests on the theory that if the rationale is really understood the proper application of the tests becomes virtually self-evident. The level of mathematics required is as follows. Formulas and concepts from elementary probability theory are used throughout. For the most part these are restricted to formulas for permutations, combinations, and the probability fraction, which are developed in Chapter 3, using only algebra. Algebra alone is used in deriving the finite-sample distribution of the test statistic. The *limiting* distribution is either simply stated without proof or partially derived, still using only algebra, but sometimes employing more sophisticated information from probability theory, which is stated but not proved. There are two lapses into concepts from calculus, both of which occur in a very casual way: once when discussing the discredited "Normal Law of Error" in Chapter 1, and once in discussing asymptotic relative efficiency in Chapter 3. Neither is critical to an understanding of the discussion. Thus all that is absolutely necessary so far as mathematics per se is concerned is a knowledge of algebra. However, some knowledge of elementary probability would be helpful, as would an inkling of what calculus is about. The level of *statistical* sophistication required is more difficult to state. Although most other statistical concepts used are at least briefly explained when introduced, it is assumed that the reader already understands what a statistical test is and knows the meaning of basic statistical terms such as correlation and variance. Furthermore, although not absolutely necessary, some knowledge of parametric statistics is desirable since comparisons are frequently made between distribution-free and parametric tests, e.g., a parametric test is often the comparison test for the relative efficiency of a distribution-free test. Much more important than specific prerequisites, however, is a general facility in statistical modes of thought. Thus while perhaps not absolutely indispensible, a thorough knowledge of elementary statistical theory is highly desirable and some knowledge of parametric tests would be helpful.

The common practice of grouping tests according to the function served is eschewed in this book. Instead, for the sake of conceptual and pedagogical economy, they are grouped into families having a common type of derivation. Thus all tests whose test statistic is binomially distributed, i.e., having a "binomial" derivation, are treated in the same chapter. Having grasped the derivation of one such test, the reader need exert little additional effort to comprehend the remainder. Important interrelationships between tests are often illuminated by such grouping.

The first three chapters serve to introduce and lay the groundwork for the subject. Chapters 4 to 6 concern tests that are based upon (i.e., use as input information) the *relative magnitude* of observations, such as their ranks. Chapters 7 to 10 concern tests based upon *frequencies* with which observations fall into certain categories. And, in the sense that the observations are not differentially weighted on the basis of their magnitudes but only counted as present or absent, the remaining chapters are also based upon frequencies.

The "relative magnitude" tests tend to be more efficient and their derivations less complex (since only the denominator of the probability fraction is mathematically derived, the numerator being obtained by brute-force enumeration). However, the "frequency" tests of Chapters 7 to 10 tend to be more specific about what *particular* aspect of the population distribution is being tested. The book was written in such a way that either the "relative magnitude" chapters from 4 to 6 or the "frequency" chapters from 7 to 13 may be taken up following Chapter 3.

The writer is indebted to the many statisticians who have encouraged and assisted him in this undertaking, especially those who granted permission to republish their tables, and particularly to the late Frank Wilcoxon, who made helpful comments upon an earlier version of the book. For permission to publish the tables appearing in the Appendix, the writer is grateful to B. M. Bennett. Z. W. Birnbaum, Frieda Swed Cohn, M. Csorgo, E. S. Edgington, C. Eisenhart, D. J. Finney, M. Friedman, G. J. Glasser, I. Guttman, R. A. Hall, P. Hsu, L. Kaarsemaker, S. K. Katti, J. H. Klotz, R. Latscha, W. J. MacKinnon, L. H. Miller, D. B. Owen, D. Teichroew, L. R. Verdooren, A. van Wijngaarden, Roberta A. Wilcox, F. Wilcoxon, R. F. Winter, Addison-Wesley Publishing Company, American Cyanamid Company, *Annals of Mathematical Statistics*, *Biometrika*, Cambridge University Press, The Florida State University, *Journal of the American Statistical Association*, the Mathematical Centre at Amsterdam, Sandia Corporation, *Statistica Neerlandica*, *Technometrics*, and the United States Atomic Energy Commission. The figures in Chapters 1 and 2 are taken from the first, fourth, sixth, and seventh in a series of Aerospace Medical Research Laboratory Technical Reports entitled *Studies in Research Methodology*, authored by the writer (see references 2–4 in Chapter 2). Figure 1-2 first appeared in the writer's Ph.D. thesis, *An Empirical Investigation of the Central Limit Theorem Applied to Time Scores*, Psychology Department, Purdue University, January 1962, Major Professor: B. J. Winer. The difficult task of typing a manuscript heavily laden with mathematical derivations and statistical symbols was performed with great skill by Mrs. Marie A. Payton. To all of the above the author expresses his gratitude.

James V. Bradley

CONTENTS

1

INTRODUCTION

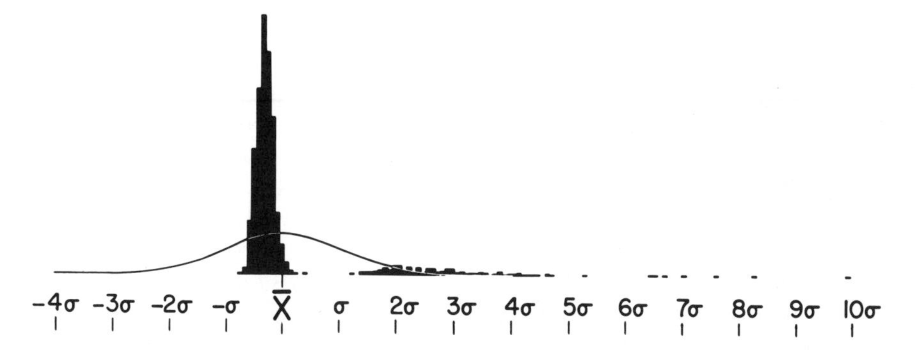

Fig. 1-1. Radically nonnormal distribution obtained in a routine experiment by the author. (Histogram is based upon 2520 observations, i.e., time scores for the operation of a push-button; smooth curve is normal distribution with same mean, variance, and area as histogram.)

Distribution-free statistical tests can be traced at least as far back as 1710, when John Arbuthnott[1] used the statistical Sign test in an attempt to prove the wisdom of Divine Providence. However, the combinatorial mathematics upon which such tests are based began to develop much earlier. During and after the Renaissance, the odds for success at games of chance were known only empirically from personal estimates based upon large numbers of observations. Certain nobles, cavaliers, and gentlemen of leisure suspected that their mathematician friends (or sometimes protégés) could calculate the odds exactly, thereby giving them the edge over their opponents. Under their impatient and excited urgings mathematicians and scientists as famous as Galileo, Pascal, and Fermat began to develop the mathematics for combinations of events. The fundamental principle which evolved is that if an event

[1] John Arbuthnott, "An Argument for Divine Providence, taken from the constant Regularity observ'd in the Births of both Sexes," *Philosophical Transactions*, **27** (1710), 186–190.

can occur in N different, equally likely ways and if n of these are regarded as successes, the rest as failures, then the probability of success is n/N. Numerator and denominator of the probability fraction were both finite and were obtained by combinatorial formulas or, failing that, by simple enumeration of the respective numbers of "ways." The probabilities so obtained were exact and the methods simple. Probabilities for most modern distribution-free statistical tests are derived in essentially the same way and are characterized by the same qualities of simplicity, exactitude, and finiteness of the number of "ways" an event can occur.

Toward the end of this period, Bernoulli developed the binomial theorem, which is the basis for many distribution-free tests (including Arbuthnott's Sign test), and it was published in 1713 in *Ars Conjectandi.*[2] In 1733, de Moivre found the limiting form of a binomial distribution when the number of trials becomes infinite. This asymptotic distribution was what is now called a normal distribution. This first discovery of the normal distribution received little attention and it was discovered a second time by Laplace about a half-century later, and a third time, still later, by Gauss. Laplace and Gauss are two of the most illustrious names in the field of mathematics; both men enjoyed tremendous professional prestige in their own time as well as thereafter. Both applied their talents to problems of astronomy, and both derived the normal distribution as a mathematical function intended to describe the distribution of errors of astronomical observations. Presumably many factors contribute to such an error of observation: variations in temperature and density of successive layers of atmosphere, vibration of the telescope, slippage in the meshing of the numerous cogwheels which position the telescope and hold it in place, the myriad fluctuations in the physiological condition of the observer, etc. If each such cause contributes an "elementary error" which is relatively small and independent of all others, and if the error of observation is the algebraic sum of such elementary errors, then as the number of elementary errors approaches infinity, the distribution of errors of observation approaches the normal distribution. This is the necessary rationale for derivations, and the basis of a school of thought, starting with Laplace and continuing through the astronomer-mathematicians Hagen[3] and Bessel to the astronomer-social scientist Quetelet.[4] Mathematically, these derivations are special forms or special cases of the Central Limit Theorem; the physical rationale is different, but analogous. As was the case with de Moivre's derivation, the normal distribution was obtained by Laplace and

[2] That is, *The Art of Conjecture*, sometimes translated as *The Art of Guessing*.

[3] Hagen's derivation of the normal distribution from the hypothesis of elementary errors is a very simple and straightforward one. It is recapitulated in Don Lewis, *Quantitative Methods in Psychology* (New York: McGraw-Hill Book Company, 1960), pp. 205–209.

[4] A key figure in the history of the Normal Law of Error. See ref. [4].

his followers as the limiting form or asymptotic distribution approached by some other distribution when one of its parameters becomes infinite. An entirely different approach to the Error Law is based upon a combination of intuition and fact as evidenced by empirical frequency distributions of errors of observation in astronomy. It seemed reasonable that errors of observation should be as likely to be positive as to be negative and should become less and less frequent as they increased in absolute magnitude. And, in fact, the empirical error distributions appeared to be unimodal and symmetric about the true value, on either side of which they seemed to decrease monotonically. On the basis of these considerations one can specify that the slope, dy/dx, of the error curve is zero (i.e., the curve is horizontal) both when the size x of the errors is zero (at which point the curve reaches a maximum so that the second derivative is negative) and when the relative frequency y of the errors is zero. One of the simplest equations which embodies these characteristics is $dy/dx = -Cxy$. This can be rewritten $dy/y = -Cx\,dx$, and integrated to give

$$\ln y = \frac{-Cx^2}{2} + K$$

or

$$y = e^{-(Cx^2/2)+K} = ke^{-Cx^2/2}$$

the essential equation for a normal curve. Thus one in effect decides what the general shape of the curve should be on the basis of both a priori and empirical considerations, and then determines a simple mathematical function which supplies the required characteristics. Gauss's derivation of the error function was in the spirit of this approach. (Still another entirely different derivation, based on geometrical considerations, was presented by Herschel.) Because of recourse to the calculus in their derivation, these Error Functions are continuous, and therefore infinite, distributions. (Also x must be able to assume infinite values.) Therefore, while the normal distribution is not obtained as an asymptotic distribution in these types of derivation, infinity is still involved.

Empirical distributions of actual errors of astronomical observation were found to be rather well fitted by the Error distribution, independently derived by Laplace and Gauss. The Belgian astronomer Quetelet was greatly interested in social statistics, and discovered that the Error distribution which worked so well in astronomy was also a reasonably good fit to empirical distributions of anthropological measurements upon military personnel and to empirical distributions in other areas. Furthermore, a close analogy to the doctrine of "elementary errors" was often available. The yield of an acre of wheat is obviously the sum of the yields of the individual wheat shoots. A city's consumption of water or gas is the sum of those of its individual users. A man's height may be regarded as the sum of the lengths of various bones of the leg, spinal vertebra, and skull. Or, taking another approach, the height of an organism might be regarded as the sum of the effects of a multitude of individual impulses to growth arising from various causes during the maturation of the organism. The Error distribution began to fit, and to explain, almost everything and soon it was regarded as a population archetype; that which was Gaussian was considered normal and that which was non-Gaussian was regarded as abnormal. In the words of Arne Fisher,[5]

> It was the great Belgian astronomer and statistician, Quetelet, who first introduced exact measurements in the study of biological and anthropological phenomena and showed that a number of collected statistical data on heights, weights and chest measurements of recruits exhibited a close conformity to the Gaussian law of error, although the variation among the individual objects as measured could not be considered solely as errors in the original sense of the word.
>
> Investigations along this line were greatly accelerated by the discoveries of Quetelet. All sorts of measurements were taken and the rapidly growing collections of statistical data relating to economic and social conditions as recorded by various government statistical bureaus furnished material for further investigations. But unfortunately in all these investigations the Gaussian error law came to act as a veritable Procrustean bed to which all possible measurements should be made to fit. The belief in authority so typical of modern German learning and which had also spread to America was too great to question the supposed generality of the law discovered by the great Gauss. Statisticians could not conciliate themselves with the thought of the possible presence of "skew" frequency curves, although numerous data offered complete defiance to the Gaussian dogma and exhibited a markedly skew frequency distribution. Supposedly great authorities argued naively that the reason the data did not fit the curve of Gauss was that the observations were not numerous enough to eliminate the presence of skewness. In other words, skewness was regarded as a by-product of sampling and was believed could be made to disappear completely if we could take an infinite number of observations.

The *Zeitgeist* is further described, most colorfully, by van Dantzig and Hemelrijk:[6]

> ... Laplace's discovery in 1778 that the normal law of errors results from a large number of independent elementary errors, whatever their individual "laws" (assumed to be identical) may be. [and] Its somewhat more elementary treatment by Poisson and, in particular, its imbedding in the formalism of least squares by Gauss (1809) rapidly worked as the Mephistophelian drink: "Mit diesem Trank im Leibe siehst eine Helena

[5] *The Mathematical Theory of Probabilities*, trans. Charlotte Dickson and William Bonynge, 2nd ed. (New York: The Macmillan Company, 1923), Vol. 1, p. 181.

[6] D. van Dantzig, and J. Hemelrijk, "Statistical Methods Based on Few Assumptions," *Bulletin of the International Statistical Institute*, **34** (1954), Part II, 239–267.

> in jedem Weibe":[7] many statisticians soon believed to find the normal distribution almost always and everywhere.

Indeed, in some quarters the normal distribution seems to have been regarded as embodying metaphysical and awe-inspiring properties suggestive of Divine Intervention. The chaos of unpredictable, in fact often unknown, individual effects ("elementary errors") seemed to have been transmuted into complete lawfulness (the Normal Law) simply by taking their aggregate. The scientific predictability of mass phenomena, e.g., the number of murders committed per year, appeared to have been reconciled, via the Normal Law, with the theological doctrine of free will, i.e., indeterminism of individual actions. If Nature had brought order out of chaos by a distribution of values symmetrically and unimodally dispersed about their mean, then perhaps the mean was that one constant (and therefore even more "orderly") value toward which Nature was striving. Thus in the Normal Mystique[8] the mean appears to have been regarded as the truest value of all. If a distribution is known to be normal, its mean and variance are all the additional information needed to specify the distribution completely. Therefore, all too frequently interest centered upon the identification of these two parameters, normality being taken for granted. (Other distributions were known and even developed during this period. However, their applicability appeared to be highly particularistic in contrast to the near universality of the normal. The normal distribution held center stage.)

While causes may be numerous, however, they cannot really be infinite. Some may predominate, so that the resulting effect is *not* relatively small. And it is unlikely that their effects are entirely independent and summative. Neighboring wheat shoots are not influenced by independent soil fertilities. Changes in the weather which influence the water or gas consumption of one dwelling are likely to exert similar influence upon that of another, and a single factory may account for most of the water consumption of a small town. If the individual bones are among the "elementary errors" contributing to the height of a man, then fewer elementary errors contribute to the length of a single bone. Elderton and Woo [2] "reach the suggestive conception that the simpler the organ measured, the less likely we are to meet with normal distributions." Nor is it likely that the effect of a growth impulse fails to interact with that of the immediately preceding impulse and simply sums with it. Although the normal distribution extends to minus infinity, it is impossible (not simply improbable) for the height of a man to assume negative values. Finally, the square or cube of a normally distributed variate is not normally distributed. So if the length of a bone or organ is normally distributed, its cross-sectional area or volume should not be. (To digress for a moment, the

[7] "With this drink in the belly one sees Helen of Troy in every woman."

[8] Called the "Quetelet Mystique" by Lancelot Hogben [5].

Laplacean Error Law based on the hypothesis of elementary errors implies that the greater the number of sources of error the more nearly normal a distribution will be. This implies that in laboratory experiments high degrees of experimental control are incompatible with normality; the greater the amount of control, the less approximately normal the distribution, until finally the measured variable's distribution is the result of a single manipulated cause. Experiments discussed and referenced by Guilford [3] strongly suggest that this is in fact the case. Thus we have the paradox that the large statistical power of normality-assuming tests is indirectly dependent upon a relative absence of power in the experiment to which they are applied.) The fact was that both Laplace's and Gauss's derivations were simply mathematically aided deductions from unproven postulates which they found intuitively acceptable [6]. It is a tribute to their astuteness as astronomers that their intuitions produced an Error Law so well accommodated to the data in that area (although even here objections have been raised against their postulates and the optimality of their Error Law has been questioned [1]). However it is a long step from astronomy to anthropology, and intuitions which prove excellent in one field may be entirely inappropriate to the other [5]. It was their less sophisticated followers, not Laplace and Gauss, who took such steps, but their prestige accompanied the misapplication of their methods.

The logical discrepancies between mathematical model and reality began to be realized by scientists in general (the more astute had been protesting all along) and, as larger and larger quantities of data became available, the nonnormality of more and more empirical distributions became apparent. However, while it was realized that, for purely logical reasons, exact normality could not be expected, a surprisingly large number of empirical distributions appeared to be quasi-normal. The fit to normality was often quite good over the central 80 to 90 per cent of the distribution, and there was seldom enough data to tell much about the goodness of fit at the tails, which usually are, by nature, improbable. Thus while quasi-universal normality was no longer an article of faith, it was replaced by a belief in quasi-universal quasi-normality. Much of the evidence upon which this belief was based came from the areas of agriculture and anthropometry, and there is no denying that many (although by no means all) of the populations dealt with in these areas are remarkably close to normal.

Against this background the best-known, classical parametric tests began to emerge in the early part of the twentieth century. They were developed by mathematical statisticians associated with the fields of agriculture, anthropology, and, in one case, the brewing industry. It is not surprising therefore that, although aware of the nonnormality of some empirical distributions, Fisher, Pearson, and "Student" developed tests and statistics which presupposed a normally distributed population. The mathematics was formidable, at least to the layman. The normal distribution was used in the derivation of the test statistic. Furthermore, in cases involving more than one population, it was found that the derivation and the resulting test statistic could be greatly simplified by setting the variances of the various distributions equal to one another, thus introducing a second precondition—that population variances be homogeneous. The validity of the test depended upon the validity of the preconditions. Thus, in tests for means, the test statistic had the derived distribution provided that all populations were (a) normally distributed, with (b) a common variance (when more than one population was sampled), and with (c) a specified mean, in the one-sample case, or, in the multisample case, means differing by specified amounts (usually zero). Since its mean and variance are the only additional information required to specify a normal distribution completely, it is obvious that the tests for equal means were in fact tests for identical normal populations. Only if one knows that conditions (a) and (b) are true does the test become a test of the validity of condition (c), and only if they are exactly true does the test become an exact test for condition (c). However, because they believed conditions (a) and (b) to be approximately true in the vast majority of cases, the test developers presented them as tests for condition (c). Whatever nervousness they may have felt about this was soon dispelled by the fact that the tests (usually conducted at the .05 level of significance) worked remarkably well in the research areas with which their developers were associated (thus lending empirical support to the supposition that a close approximation to normality and homogeneity was all that was needed and that the tests were most sensitive to the location parameter). It appeared that agricultural and anthropological experimenters had a powerful research technique, and research workers in other areas, particularly the behavioral sciences and engineering, were eager to adopt it. Mathematicians and mathematical statisticians, associated with no particular area of application, were also attracted by the new techniques, often not so much because of their practical value, as because of the power and elegance of the calculations and the interesting mathematical properties of the resulting statistics. If agricultural and anthropological statisticians were unconcerned about the preconditions of normality and homogeneity because they believed them to be approximately true, mathematicians were even less concerned about their validity since their discipline permits them to postulate whatever conditions they wish. Their attitude has been described by Peters:[9]

> Once making the assumptions, the mathematics is simple and exact and fascinatingly beautiful; and mathematicians will frankly say that it is our concern as researchers, not theirs, whether the assumptions are legitimate in the particular research situations with which we work. It happens that in most of the research in our field the assumptions are so far-fetched as to abort the results for careful work.

[9] C. C. Peters, "Misuses of the Fisher Statistics," *Journal of Educational Research*, **36** (1943), 546–549.

Of course, the end result of the mathematical calculations is qualified by the preconditions, but the mathematicians felt little obligation to stress this fact verbally, since the qualifications were implicit in the derivation. When reference to the qualifications was made in words, the restrictive preconditions were tragically labeled "assumptions." To the layman unable to follow the derivation but ambitious enough to read the words, it sounded as if the mathematician had esoteric *mathematical* reasons for believing in at least quasi-universal quasi-normality and quasi-universal quasi-homogeneity. The situation was reminiscent of that described by Lippmann in a much-quoted remark to Poincaré:[10] "*Tout le monde y* [*la loi des erreurs*] *croit cependant, me disait un jour M. Lippmann, car les expérimentateurs s'imaginent que c'est un théorème de mathématiques, et les mathématiciens que c'est un fait expérimental.*"[11]

Thus, in contrast to statistics at the time of Bernoulli, classical parametric tests were based upon asymptotic or "infinite" population distributions, were inexact, and were mathematically complex. So complex, in fact, was the mathematics that few save professional mathematicians could understand it. It was so far above the highest level comprehended by the typical research worker that virtually no attempt was made to explain a test's derivation to the experimenter who must use it. Instead, research workers were instructed in the use of these tests by means of manuals which presented procedures of application in minute detail while grossly neglecting the mathematical logic upon which the tests were based. Some did not even explain application in general terms, but resorted rather to endless examples which the user was to follow as paradigms. Only the most casual reference was made to the underlying assumptions, if indeed they were not ignored completely. "Assume" was practically synonymous with "take for granted." Such textbooks became known as "cookbooks." At best they resulted in the acquisition of skill unaccompanied by understanding. At worst, they laid the groundwork for outlandishly irrational statistical gaucheries and blunders. Bizarre misapplications of statistical methods could be traced and attributed to vague or ambiguous passages in one of the cookbooks. Statistical disputes among nonmathematicians often took the form of citing Authorities back and forth, the Authority, in each case, being the author of the cookbook from which the respective disputants had acquired their concepts. Statistics became the *bête noire* of the graduate student seeking to understand what he is taught and the Achilles' heel of the practicing research worker. The weakest link in the scientific chain was often the statistical test at which point scientific logic was replaced by blind faith in the wisdom of the Authority and the absence of typographical errors in the cookbook.

[10] H. Poincaré, *Calcul des Probabilités*, 2nd Ed. (Paris: Gauthier-Villars, 1912), p. 171.

[11] "Everyone believes in it [the law of errors] however, said Monsieur Lippmann to me one day, for the experimenters fancy that it is a theorem in mathematics and the mathematicians that it is an experimental fact."

Among professional statisticians, however, there remained a lingering doubt. It does not follow logically that approximate normality and homogeneity insure approximate validity of a test which assumes exact normality and exact homogeneity. A number of sampling studies were conducted to check the extent to which the tests remained approximately valid when their assumptions were violated in various ways. This was often done empirically by drawing a large number of samples from an assumption-violating population (or populations), calculating the test statistic from each sample (or set of samples) and constructing the frequency distribution of these empirical values of the test statistic. This sampling distribution of the test statistic under assumption-violating conditions was then compared with the theoretical distribution of the test statistic when all assumptions are met. Sometimes the sampling distribution under violation of assumptions was obtained mathematically, the procedure thereafter being the same. In the mathematical studies the populations to be sampled appear to have been chosen more for their mathematical convenience than for their prevalence in practice. Such studies generally investigated populations represented by various geometric shapes such as a rectangle or triangle or by mathematical functions such as the various types of Pearson curve. In the empirical studies when populations of real data were used, they tended to come from fields such as agriculture, anthropology, or meteorology and to be nearly normal in shape (see Fig. 1–2). However, many of the empirical studies sampled from the same kinds of artificial populations investigated mathematically, perhaps because highly nonnormal populations of real data were unavailable in these fields. In any case the more nonnormal of the artificial populations appear to have been regarded in some quarters as incorporating a "degree" of nonnormality as extreme as any likely to be encountered in practice, and the results of studies based upon them appear to have been regarded as "bounds" for the effects of nonnormality upon the test statistic.[12] Most of the studies were crude by present standards. The empirical studies almost never drew enough samples to attain high precision of estimate and were hagridden with chance effects. The mathematical studies often resorted to still further assumptions in obtaining an "approximation" to the true sampling distribution of the test statistic under assumption-violating conditions. Discrepancies from the normal-theory distribution of the statistic were often attributed largely to chance or to the mathematical approximations which had been used. Often the fit was excellent over the central 80 per cent of the distribution but increasingly poor at increasingly remote tail regions, and often much was made of the former while the latter was depreciated and attributed to the insufficient number of data points for good precision of estimate at the tails. Frequently the rule against small expected frequencies in chi-square tests of fit was invoked as

[12] Occasionally voices were raised in explicit denial of such a conclusion, but the *Zeitgeist* was against them, and they went largely unheeded by the authors of the cookbooks.

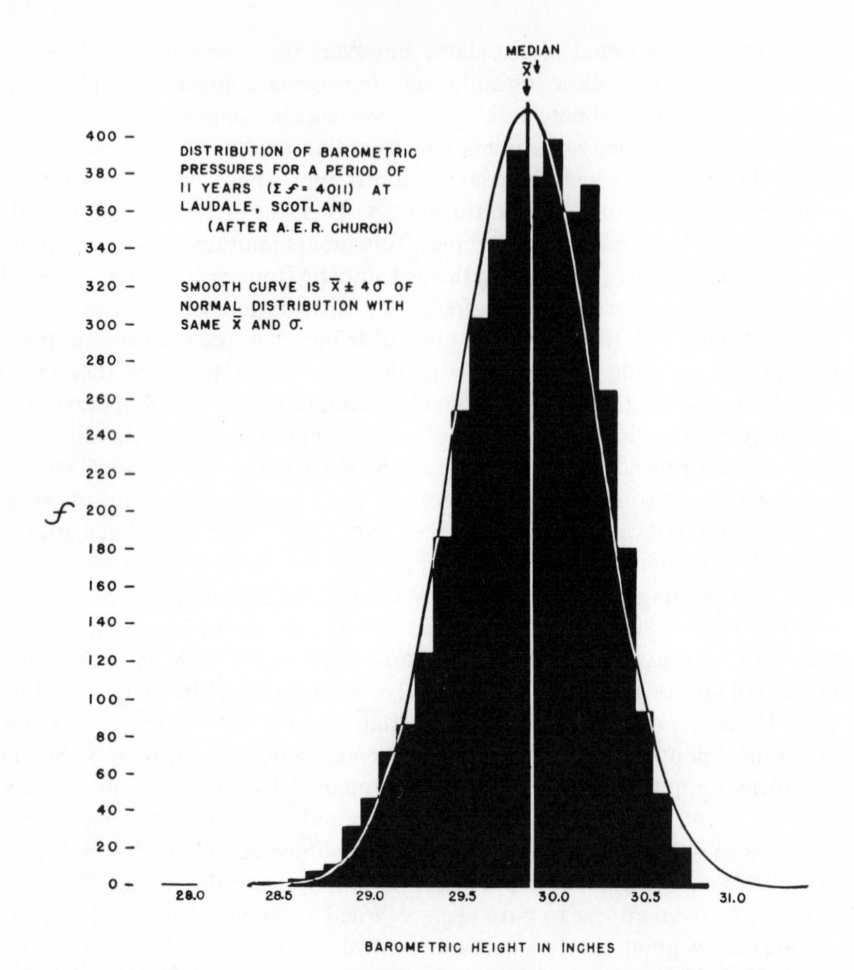

Fig. 1-2. Population sampled in an early study on the effect of nonnormality.

justification for the pooling of the tail frequencies, thereby conveniently obscuring their increasingly poor fit with increasingly remote tail regions. In short, methodological shortcomings were often compounded by prejudicial treatment or interpretation of the data by the investigator. This frame of mind and the general historical context in which it appeared have been eloquently described by Geary:[13]

> Our historian will find a significant change of attitude about a quarter-century ago following on the brilliant work of R. A. Fisher who showed that, when universal normality could be assumed, inferences of the widest

[13] Quoted by permission of author and of the editor and trustees of *Biometrika* from R. C. Geary, "Testing for Normality," *Biometrika*, **34** (1947), 209–242.

> practical usefulness could be drawn from samples of any size. Prejudice in favor of normality returned in full force and interest in non-normality receded to the background (though one of the finest contributions to non-normal theory was made during the period by R. A. Fisher himself), and the importance of the underlying assumptions was almost forgotten. Even the few workers in the field (amongst them the present writer) seemed concerned to show that "universal non-normality doesn't matter": we so wanted to find the theory as good as it was beautiful. References (when there were any at all) in the text-books to the basic assumptions were perfunctory in the extreme. Amends might be made in the interest of the new generation of students by printing in leaded type in future editions of existing text-books and in all new text-books: *Normality is a myth; there never was, and never will be, a normal distribution.* This is an over-statement from the practical point of view, but it represents a safer initial mental attitude than any in fashion during the past two decades.

The reason for the type of personal bias mentioned is not hard to find. At that time there was no general alternative to normality-assuming statistics. Accordingly, the investigator was highly motivated to find in their favor unless overwhelmed by contrary evidence (which was often prevented by methodological inadequacies). Still, the evidence sometimes was overwhelming and some investigators resisted the prejudicial pressures better than others. It became apparent that under violation of assumptions normal-theory statistics are in fact approximate tests, that in a surprising number of cases the approximation is excellent, in many cases the approximation is fair-to-good, but that in some perfectly realistic cases it is so bad as to be unacceptable. Furthermore, although formal presentation of evidence was generally lacking, it was becoming increasingly apparent in the behavioral sciences that quasi-normality and quasi-homogeneity are not nearly so likely to be encountered there as they are in agriculture and physical anthropology. Indeed, behavioral scientists often found themselves in the predicament of having virtually no inkling as to the shape of a distribution, save that which could be inferred from the same small sample to be used in conducting the statistical test. The time was ripe for development of tests which presuppose little about the sampled population's shape and whose validity does not depend upon it.

I. R. Savage[14] places the "true beginning" of distribution-free statistics in 1936. As already mentioned, Arbuthnott used a distribution-free test in 1710. Furthermore, other distribution-free tests were developed long before 1936. However, the important quality which appeared at that time was a complete self-consciousness on the part of the test developers that their statistics were not dependent upon the shape of the underlying distribution of magnitudes and therefore were the much-needed remedy for the ills of classical statistics.

[14] I. R. Savage, "Bibliography of Nonparametric Statistics and Related Topics," *Journal of the American Statistical Association*, **48** (1953), 844–906.

Furthermore, it is only from 1936 on that distribution-free statistics began to take the form of a separate discipline. In this new discipline, no attempt was made to specify or identify the exact form of the sampled population of magnitudes. Instead, interest focused upon quantities and characteristics of the *sample* whose distributions were sensitive to the tested hypothesis and could be expressed by exact combinatorial formulas. Thus after two hundred years of wandering through a statistical wilderness in which actual populations were equated with (i.e., replaced by) infinite or asymptotic distributions and probabilities were obtained by integrating over mathematical density functions, statistics entered an era in which populations could be whatever they were and probabilities, as in the time of Bernoulli, were the ratio of the number of successful outcomes of an event to a finite number of possible outcomes.

However, the influence of past attitudes was not easily overcome. Classical statistics had used the mean and variance as measures, i.e., indices, of location and dispersion. This had been done mainly because it was mathematically convenient to do so, rather than because the mean and variance are any truer measures of location and dispersion than many others. A good (although debatable) case can be made for means and variances as indices of location and dispersion when the normality assumption holds; the argument loses much of its force, however, when the assumption fails. The medians, interquartile ranges, etc. used by the new statistics have certain unique advantages, i.e., are superior in certain ways, even under normality. And they may be far more appropriate indices of location and dispersion when the sampled population is nonnormal. Nevertheless, even the developers of distribution-free statistics felt somewhat shamefaced about them, regarding them as approximate, "quick and dirty," or makeshift, because they were not direct tests of the location and dispersion parameters dealt with by classical tests. This feeling was reinforced by the discovery of the apparent inefficiency of the new statistics. Partly because it was mathematically convenient (i.e., easier) and partly perhaps because of bias, the earliest efficiency figures were obtained by comparing the distribution-free test with a parametric test under common conditions meeting all the assumptions of the latter. Thus the parametric test was permitted both to hurl the challenge and to choose the weapons. Under these loaded conditions the best parametric test was found to be more efficient than (or, at worst, equally efficient to) its distribution-free competitor. Actually the difference in efficiency was often surprisingly small; however, more was made of the unfavorable direction of the difference than of its often favorably small extent. The new statistics carried still another imaginary burden of guilt. The magnitudes of observation values were generally not used as such by the tests, which, instead, used frequencies or order relationships, such as ranks. It seemed as though these magnitudes contained "information" which was used by the parametric test but was "thrown away" by its distribution-free counterpart. Actually, the utilization of the additional sample information is made possible by the additional population "information" embodied in the parametric test's assumptions. Therefore, the distribution-free test is discarding information only if the parametric test's assumptions are known to be true. Or, in other words, if one knows that all parametric assumptions are met, the parametric test is the proper choice.

If the professionals had a somewhat distorted perspective on distribution-free statistics, the situation among the laity was pure chaos. A fantastic folklore sprang up among research workers in the various areas of application. On the one hand, distribution-free tests were regarded as second-class statistics, hopelessly inferior to parametric statistics, which were considered to represent a sort of ideal or state of perfection to which they could never attain; the latter were pure and powerful, the former dirty, inefficient, and wasteful of information, regardless of whether the original data existed in the form of magnitudes, ranks, or frequencies and irrespective of the meeting of assumptions or the type of hypothesis tested. But while definitely inferior to classical tests insofar as power and accuracy were concerned, they were imbued with properties nothing short of magical in respect to applicability. Thus, on the other hand, they were widely believed to make no assumptions whatever, and, in effect, to presuppose nothing in regard to careful experimentation, accuracy in collecting data, etc. In some there was an unshakable conviction that if an experiment had been so badly designed or executed that no trustworthy conclusions could be drawn from the data by parametric tests (for example, when data have been collected under an inadvertently confounded design or with an apparatus known to have malfunctioned at unidentifiable points in the sequence), somehow the rules of logical inference would suspend themselves out of courtesy to distribution-free methods, which would therefore remain applicable. Thus it was widely believed that the new statistics embodied a perfect antidote for experimental gaucheries. Frequently one encountered both types of fallacy in the same person—a belief in the supernatural powers of a dirty, inefficient statistic. Such fallacies reveal a total lack of comprehension of just what it is that the statistical test is doing and show that cookbook learning is just as dangerous (although less justified) when distribution-free tests are involved as it was with their parametric predecessors.

At present the professionals have long since developed a realistic perspective toward the new statistics. The efficiency of distribution-free relative to classical tests has been investigated under common *non*parametric conditions, i.e., nonnormal populations and/or heterogeneous variances, and the new statistics have often proven superior, sometimes infinitely so. There are no delusions about having discarded information and few apologies for testing medians rather than means. There appears however to be a cultural lag between professionals and laity. The latter have grown more sophisticated, but have not entirely lost their delusions, perhaps because of the intellectual ravages of a cookbook education, perhaps because they are still in the Gaussian grip of the Normal Mystique.

REFERENCES

1. Campbell, N. R., *Foundations of Science: The Philosophy of Theory and Experiment* (formerly entitled *Physics: The Elements*). New York: Dover Publications, Inc., 1957, pp. 457–521, especially 477–485.
2. Elderton, E. M., and T. L. Woo, "On the Normality or Want of Normality in the Frequency Distributions of Cranial Measurements," *Biometrika*, **24** (1932), 45–54.
3. Guilford, J. P., *Psychometric Methods.* New York: McGraw-Hill Book Company, 1954, pp. 142–144.
4. Hankins, F. H., "Adolphe Quetelet as Statistician," *Studies in History, Economics and Public Law*, **31** (1908), No. 4, 1–133.
5. Hogben, L., *Statistical Theory: The Relationship of Probability, Credibility and Error.* London: George Allen & Unwin, Ltd., 1957, pp. 159–181.
6. Jevons, W. S., *The Principles of Science: A Treatise on Logic and Scientific Method.* New York: Dover Publications, Inc., 1958, pp. 374–398.

2

DISTRIBUTION-FREE VERSUS CLASSICAL TESTS

2.1 DEFINITIONS

The terms **nonparametric** and **distribution-free** are not synonymous, and neither term provides an entirely satisfactory description of the class of statistics to which they are intended to refer. This is discussed at length in [9]. Popular usage, however, has equated the terms and they will be used interchangeably throughout this book. Roughly speaking, a nonparametric test is one which makes no hypothesis about the value of a parameter in a statistical density function, whereas a distribution-free test is one which makes no assumptions about the precise form of the sampled population. The definitions are not mutually exclusive and a test can be both distribution-free and parametric. (For example the Sign test, which does not assume any exact shape for the sampled population of variate values, tests the hypothesis that the parameter p of a binomial distribution has the value .5.) Of the two terms, *distribution-free* comes closer to describing the quality that makes the tests desirable. In order to be entirely clear about what is meant by distribution-free it is necessary to distinguish between three distributions: (a) that of the sampled population; (b) that of the observation-characteristic actually used by the test; and (c) that of the test statistic. The distribution from which the tests are "free" is that of (a), the sampled population. And the freedom that they enjoy is usually relative. Frequently the assumption is made that the sampled population is continuously distributed and sometimes more elaborate assumptions are made, such as the assumption that the sampled populations have symmetrical distributions or identical shapes. However, the assumptions are never so elaborate as to imply a population whose distribution is completely specified. The reason that no elaborate assumptions are made about the population distribution of variate magnitudes is very simple: the magnitudes are not used as such in the test, nor is any other strongly population-linked attribute of the variate. Instead, *sample*-linked characteristics of the *obtained observations*, i.e., ordinal or categorical relationships within the set of observations defined by the actually obtained samples, such as ranks, positions in sequence, or frequencies, provide the information used by the test statistic. And of course the distribution of the characteristic used must be known just as must the population distribution of magnitudes in the parametric case. However, whereas the latter is generally an infinite distribution whose exact form is therefore inexactly known at best, the former is generally a discrete finite distribution which, if the assumptions hold, can be specified completely and exactly on the basis of a priori considerations. The rank of an observation's magnitude or its position in sequence among other observations, or the frequency of appearance of a certain class of observation, are all properties of the observation with reference to the *sample*, rather than to the sampled population, in the sense that they cannot be specified from an isolated observation and can only be known after the entire sample, or at least the preceding observations, have been drawn and examined. Therefore, the use of ordinal or categorical observation characteristics, rather than continuous observation magnitudes, renders distribution (a) somewhat irrelevant, and tends to make distribution (b) exactly specifiable and discrete. This and the finiteness of the sample make (b) finite and permit exact combinatorial formulas to be used to obtain distribution (c) exactly. Thus, while both parametric and nonparametric tests require that the form of *a* distribution, associated with observations, be fully known, that knowledge, in the parametric case, is generally not forthcoming and the required distribution of *magnitudes* must therefore be "assumed" or inferred on the basis of approximate or incomplete information. In the nonparametric case, on the other hand, the distribution of the observation *characteristic* is usually known precisely from a priori considerations and need not, therefore, be "assumed." The difference, then, is not one of requirement but rather of what is required and of certainty that the requirement will be met.

Because they do not use magnitudes as such, distribution-free tests do not test for parameters computed from them in the same sense that classical tests test for equal means, say, or for identical variances. Instead, the analogous distribution-free tests might test for equal medians or identical interquartile ranges, i.e., values which can be computed from nonmagnitudinal observation characteristics such as frequency, or position in rank order. Of course, a distribution-free test may be indirectly a test for parameters based on magnitudes; for example, if populations are known to be symmetrical, a test for equal medians becomes, in addition, a test for equal means; likewise, a test which shows two populations to be identical also shows them to have equal variances.

2.2 COMPARISON OF DISTRIBUTION-FREE AND CLASSICAL TESTS WITH REGARD TO SPECIFIC ATTRIBUTES

Both distribution-free and classical tests have points of superiority, and which type of test should be used depends upon a number of specific conditions as well as upon the sophistication of the user. The comparison, however, is generally quite favorable to distribution-free tests. Some advantages and disadvantages of distribution-free relative to parametric tests are outlined in the following paragraphs.

2.2.1 Simplicity of Derivation

Most distribution-free tests can be derived using simple combinatorial formulas, whereas the derivation of classical tests requires a level of mathematics far above the highest level attained by the typical research worker. However, the logic and appropriateness of a test's application, the assumptions it makes, and its sensitivity to assumption violation all hinge upon its derivation. If the research worker understands the derivation, he can deduce or infer much of this necessary information for almost any application he may contemplate, thus operating with a maximum of comprehension and flexibility. If he does not understand it, he is reduced to performing tests according to the uncomprehending "cookbook" procedure of following a paradigm while obeying certain highly overgeneralized rules of thumb. In the opinion of the writer, this simplicity of derivation is by far the most important advantage of distribution-free statistics since, for research workers ignorant of higher mathematics, it replaces a mystery-cloaked ritual with a truly scientific procedure.

2.2.2 Ease of Application

The mathematical operations required in computing the test statistic from sample observations are generally much less involved for distribution-free than for parametric statistics. Frequently all that is required is counting, or adding, subtracting, and ranking. This simplicity of application is obviously an economic advantage, permitting lower-paid, mathematically naïve personnel to be employed to reduce data and perform computations.

2.2.3 Speed of Application

When samples are of small or moderate size, distribution-free methods are generally faster than parametric techniques. This saving in computation time may be used to obtain more data, thus frequently canceling any advantage the parametric test may have in terms of statistical efficiency. When samples are large (say, $n \geqq 30$), distribution-free tests involving simple counting are generally faster, whereas those involving ranking may prove considerably more time-consuming, than standard classical tests.

2.2.4 Statistical Efficiency

As implied in the preceding paragraphs, when judged by the practical criterion of the total amount of human effort required to conduct an experiment and analyze its results, distribution-free tests are frequently, if not generally, more efficient than their parametric counterparts. (And this is especially true if one "weights" the various component efforts by the amount of sophistication demanded by the task.) When judged by the mathematical criterion of statistical efficiency, distribution-free tests are often superior to their most efficient parametric counterparts when both tests are applied under "nonparametric" conditions, i.e., conditions meeting all assumptions of the distribution-free test, but failing to meet some of the assumptions of the parametric test. When both tests are applied under "parametric" conditions, i.e., conditions meeting all assumptions of the parametric test, and therefore of both tests, distribution-free tests are usually very slightly less efficient (i.e., have relative efficiencies a shade less than 1.00) at extremely small sample sizes, becoming increasingly less efficient as sample size increases. When sample size becomes infinite, distribution-free tests generally have their lowest efficiencies relative to the most efficient, comparable parametric test under parametric conditions. This efficiency value may be as high as 1.00 or as low as zero, depending on the test.

2.2.5 Choice of Significance Levels

The distribution of the test statistic, when the null hypothesis is true, is usually continuous for classical tests and discrete for distribution-free tests. This means that, for any designated significance level α, a value of the classical statistic can be found whose cumulative probability is exactly α, whereas for the distribution-free test, such a value of the test statistic usually does not exist. Thus when using a classical test the research worker may choose any significance level he wishes, whereas when using a distribution-free test, he must either accept one of the discrete cumulative probabilities of the test statistic as his significance level or he must apply the test inexactly, using as nominal significance level a cumulative probability which the test statistic cannot actually assume and rejecting whenever it is found to have a smaller cumulative probability. The latter choice is often forced upon him by inexact tables of probabilities which list values of the test statistic which are "significant" at the standard significance levels, .05, .01, and .001. The discrepancy between the true cumulative probability of the "critical value" listed and the nominal significance level claimed for it is quite variable, but tends to increase with diminishing sample size and with diminishing nominal significance level, becoming quite appreciable at the smallest sample sizes and most extreme α's. Some tables list as critical value that value of the statistic whose true cumulative probability is closest (irrespective of whether it is larger or smaller) to the nominal significance level. More often, however, the closest cumulative probability *smaller* than the nominal significance level determines the critical value to be listed. In such cases, the user of the table may be employing a far smaller significance level (and therefore a far more stringent test) than he is able to claim in reporting the results of his test.

An increasingly prevalent modification of such tables avoids the above criticism. Critical values of the test statistic are listed, as cell entries, under a column heading which is a standard nominal significance level, just as before. But, in parentheses, under or beside each critical value is listed its exact cumulative probability. An entirely different type of table makes no mention whatever of standard significance levels and simply gives the exact cumulative probability distribution of the test statistic, thereby also obviating the criticism. In fact, some form of exact table can be found for most of the better known distribution-free tests. Exact tables, however, tend to be longer than the more concise critical-value tables. The former, therefore, tend to exist mainly in professional journals or in books of tables, while statistical textbooks are more likely to contain the latter.

One defect remains as a result of the discrete cumulative probability levels of the test statistic. Experiments performed under dissimilar sampling conditions or analyzed by different distribution-free tests cannot usually be analyzed at identical exact significance levels without resorting to unusual procedures. This is a disadvantage if one wishes a series of experiments to be statistically comparable in the sense of having been tested at a common (exact) significance level, or if one wishes to obtain an overall significance level for several experiments by combining their test probabilities by the binomial method (Fisher's method is disqualified by discreteness).

2.2.6 Logical Validity of Rejection Region

The distribution of a classical test statistic is usually continuous, increasing or decreasing smoothly, without fluctuation, except for a possible change of direction at a single mode. Unfortunately, the point probability of a nonparametric test statistic does not necessarily always increase as the test statistic approaches its most probable value. It may level off or even dip before resuming its climb. This characteristic, when it exists, may be decidedly embarrassing when the rejection region for a distribution-free test is selected on an intuitive basis. Should the rejection region be chosen as the cumulative probability for those values of the test statistic which are *least likely*, or for those which are *most distant* from the expected value of the test statistic under the null hypothesis?

2.2.7 Types of Statistics Testable

Statistics defined in terms of arithmetical operations upon observation magnitudes can be tested by classical techniques, whereas those defined by order relationships (rank) or category-frequencies, etc. can be tested by distribution-free methods. Means and variances are examples of the former and medians and interquartile ranges, of the latter.

Perhaps because of the lingering effects of the Normal Mystique, the mean and variance appear to be widely regarded as ideal statistics, such statistics as medians and interquartile ranges being considered definitely inferior. The extent to which means and variances are "superior" depends upon whether it is their mathematical versatility (i.e., their properties in mathematical manipulations, theorems, etc.), their inferential qualities as sample estimates of population statistics, or their descriptive properties as indices of location and dispersion which is in question. Their mathematical properties are truly remarkable and are extremely useful both in constructing tests, establishing theorems, etc., and in practical applications. In this sphere, the claim of superiority seems to be largely justified. Their superiority as estimators of a corresponding population statistic rests largely upon one's definition of a good estimator, and their virtues as descriptive population statistics seem to depend largely upon an implicit assumption of normality or near-normality. A good estimator is (among other things) unbiased, i.e., the sampling distribution of the sample estimator has as its location parameter the population parameter being estimated. The estimator is mean-unbiased or median-unbiased, respectively, if the mean or median of its sampling distribution coincides with the population parameter being estimated. The sample mean is a mean-unbiased estimator of the population mean, i.e., the mean of the sample means is the population mean; but, on the other hand, the sample median is a median-unbiased estimator of the population median, i.e., the median of the distribution of sample medians is the population median. Thus the "superiority" of the sample mean as an estimator rests largely upon a preference for estimation of means and for mean-unbiasedness of estimators. (The sample mean is not generally a median-unbiased estimator of the population mean unless the population is symmetric.) For symmetric populations, mean and median are identical and the sample mean and median are both mean-unbiased and median-unbiased estimators of the population mean. The better estimator in this case may be regarded as the one whose sampling distribution has the smaller dispersion. For an entire class of symmetrical distributions whose kurtosis exceeds a certain value, the sample median is a better estimator of the population mean (and also median) than is the sample mean. When kurtosis reaches its largest value, the mean is no better estimator than is a single observation, whereas, on the other hand, the median is still a highly effective estimator. (Below the critical kurtotic value, the sample mean becomes increasingly superior to the median as an estimator of the population mean.) Finally, whatever lingering prestige the sample mean enjoys by virtue of its "superiority" in point estimation is still further shaken when confidence limits are considered. The type of population distribution (or, at least the value of σ^2 when n is very large) must be known in order to construct a confidence interval for the population mean, about the sample mean, but not to establish a confidence interval for the population median, about the sample median. Some of the advantages of the median, especially relative to the mean, are entertainingly discussed in [12]. Clearly, in the sphere of estimation the good properties of means and variances have been highly overgeneralized. But it is as descriptive indices of location and dispersion that mean and variance leave the most to be desired. The median, as the fiftieth percentile, immediately yields an intuitively meaningful index of the population's center. It is an index which weights all scores by their occurrence, and therefore equally, rather than by their magnitudes; and in that sense it is maximally representative of the population as a whole. Likewise, the interquartile range (or perhaps, better, the range between the fifth and ninety-fifth percentiles) yields an intuitively meaningful index of dispersion which weights each score equally and is, in that sense, maximally representative. The mean, on the other hand, is not the center of the scores but only their center of gravity; scores are weighted by their magnitudes. The variance is not a typical value; it is an average value. And it is not an average deviation; it is an average of squared deviations. Deviations are weighted by the squares of their magnitudes. Ninety-nine per cent of the units in a population may lie just below the mean, tightly clustered within a range of only $.01\sigma$, and "balanced" by a hundredth of the population consisting of remote units lying far on the other side of the mean, covering a range of many σ's, and contributing the overwhelming proportion of the variance. There is no "information" in either the mean or the variance, or in both in combination, which distinguishes this case either from the mirror-image case or from the case in which the mean is the fiftieth percentile of a symmetrical, bell-shaped population. It seems exceedingly likely that the belief in the good descriptive qualities of mean and variance as indices of location and dispersion depends almost entirely upon an implicit assumption of normality. When the population *is* normal, the mean is also median and mode and acquires all of their good properties as indices of location; the variance, properly operated upon, yields all of the interquartile and interpercentile information one wishes and therefore implicitly acquires their good properties as indices of dispersion. The good descriptive properties of the mean and variance, therefore, appear really to belong to other statistics being acquired by the mean only when the population is symmetric and by the variance only when the population distribution is known and extensively tabled.

2.2.8 Testability of Higher-Order Interactions

Higher-order interactions can be tested with ease by classical methods. However, while distribution-free tests for them are not wanting, they tend to be complicated, awkward, and limited in application. Furthermore, many of them are inexact, their derivations being based upon the limiting case of infinite sample sizes and involving "asymptotic" formulas for the test statistic. Thus they lack many of the virtues possessed by distribution-free tests for "main effects" or first-order interactions.

2.2.9 Influence of Sample Size

The size of the sample upon which they are to be used is an extremely important factor in determining the relative merits of distribution-free and classical tests. When samples are small (say $n \leq 10$) distribution-free tests are easier, quicker, and only slightly less efficient, even if all assumptions of the parametric test have been met. At these sample sizes, violations of parametric assumptions generally have their most devastating effect, yet are most unlikely to be detected. Therefore, unless the experimenter has a priori knowledge that all parametric assumptions have been met, the wiser choice would generally appear to be a distribution-free test. When samples are large (say $n > 30$), some distribution-free tests still compare favorably with their parametric counterparts. Others, however, will have become more laborious and time-consuming, and, in contrast to parametric tests whose assumptions are met, their calculated or tabled probabilities may be only approximate. Finally, their efficiency relative to a parametric test whose assumptions are all true may have dropped to an appreciably low level. On the other hand, appreciable violations of parametric assumptions will have become more readily detectable and, in some cases, their effect may have become negligible due in part to the effect described by the Central Limit Theorem. At large sample sizes, therefore, either type of test may be superior; however, circumstances are much more favorable to parametric tests than is the case when samples are small.

2.2.10 Scope of Application

Because they are based on fewer and less elaborate assumptions than classical tests, distribution-free tests can be legitimately applied to a much larger class of populations.

2.2.11 Susceptibility to Violations of Assumptions

Two general types of assumptions may be distinguished: those which concern the method of sampling, and those which concern the nature of the sampled population. Both distribution-free and parametric tests generally require the "sampling assumptions" that sample observations have been drawn randomly and independently from their parent populations, and both are highly vulnerable to violation of this type of assumption. Unlike population assumptions, however, sampling assumptions can generally be met by rigidly adhering to certain prescribed procedures and are, in this sense, under the control of the experimenter. Obviously, the more elaborate the "population assumptions" the fewer the number of situations which meet (or nearly meet) them, and, in this sense, parametric assumptions are the more susceptible to violation. For example, the parametric assumption of normality requires that in addition to (a) being continuously distibuted and (b) being symmetrically distributed (as might be assumed by nonparametric tests), the population must also (c) be bell-shaped and (d) have infinite range, since these are all features of a Gaussian distribution. The best-known and most frequently used parametric tests generally assume that all populations are normally distributed and (when more than one population is sampled and homogeneity of variance is not what is being tested) that all populations have the same variance. The most common population assumption for nonparametric tests is that the population is continuously distributed; occasionally symmetry is also assumed, but far less frequently. But in some cases there are, in effect, *no* population assumptions, i.e., the nature of the situation is such that the existence of all required conditions is guaranteed, and, in that sense, therefore, need not be "assumed." Furthermore, the continuity assumption, in the nonparametric case, is generally a sufficient, rather than a necessary, condition, covering what are often more modest and easily satisfied, necessary assumptions, which are sometimes highly insusceptible to violation. This is discussed at length in the following chapter.

2.2.12 Detectability of Violations of Assumptions

When the nonparametric assumption of continuous distributions is violated, both the fact and the degree of the violation tend to be readily apparent from the existence of tied scores (or zero differences) in the obtained data. No such obvious indication advises the experimenter that a parametric assumption has been violated. Of course he may apply time-consuming tests for normality or homogeneity to the obtained data, but such tests are rather unsatisfactory. They are unlikely to detect any but the most extreme violations when samples are small, and they are generally almost certain to detect even the most trivially slight violations when samples are very large. There is an interesting exception, however. A particularly insidious situation occurs when a distribution is (essentially) normal except for overly thick extreme tails. At remote tail regions, the distribution can have ordinates vastly exceeding the analogous normal ordinates but still representing very small relative probability of occurrence (i.e., relative density) for the corresponding abscissas. Even when sample size is quite large the sample may be unlikely to contain any observations from the excessive remote tail. If it contains none, its frequency distribution will contain no clue as to the nonnormality of the sampled population; rather it will appear to confirm the hypothesis of normality. And if the sample contains one or even two observations from the remote tail, the rest of the sample frequency distribution appearing nicely normal, many would be tempted to dismiss the "outliers" as spurious. This type of population is not uncommon to statistical practice [2, 13] and the effect of its violation of the normality assumption upon classical statistical tests [2, 3, 4] and estimates [13] can be devastating. In the words of Tukey [13], "Nearly imperceptible non-normalities may make conventional relative efficiencies of estimates of scale and location entirely useless."

2.2.13 Effect of Violation of Assumptions

The effect of violating a test's necessary assumptions is to render the test inexact. When it is the nonparametric assumption of continuity which is violated, the sample data themselves may provide the information with which to construct bounds for the resulting inexactitude. This is discussed in the next chapter. There is no analogous method of gauging the inexactitude of the *test* (as contrasted, say, with degree of nonnormality) from sample data in the parametric case (although the unknown degree of inexactitude can sometimes be mitigated, to a usually unknown degree, by transformations, the proper choice of which presupposes considerable knowledge about the shape of the sampled populations). However, the belief is widespread that no such bounds (or transformations) are needed in (wide and easily defined classes of) specific cases because of a highly general insensitivity of parametric tests to violation of their assumptions. The validity of this belief is examined in the next section.

2.3 ROBUSTNESS OF PARAMETRIC TESTS

Rivaled only by the Normal Mystique in vast overgeneralization of specific effects is the Myth of Robustness. As was the case a century ago with the Myth of Normality, a kernel of truth has been magnified into a mountain of error.

Practically any violation of a parametric test's assumptions alters the distribution of the test statistic and changes the probabilities of Type I and Type II errors. The test is said to be robust against violation of a certain assumption if its probabilities of Type I or Type II errors (usually the former) are not appreciably affected by the violation.[1] The kernel of truth mentioned above is that for most (but not all) parametric tests there are conditions under which a fairly "large" violation of an assumption produces impressively little distortion in the distribution of the test statistic. For example, the one-sample t test (as conducted in practice) tends to be affected less and less by nonnormality as sample size increases and becomes completely impervious to it when sample size is infinite (provided that the sampled population is one for which the Central Limit Theorem holds and has a finite fourth moment). Thus (for this statistic and this assumption) any objective criterion of robustness, however stringent, can be met by taking a sample which is sufficiently large. (However, the reader will presently see that a fairly lax criterion of robustness may necessitate an $n > 4096$.) As a second example, the two-sample t test is remarkably insensitive to heterogeneity of variance provided that no other assumptions are violated and that the two samples are of exactly equal size. And when $n_1 = n_2 = \infty$, the test becomes perfectly robust against violation of the homogeneity assumption. However, the conditions which render a test "robust" tend to be highly idiosyncratic. Thus while the robustness of either the one-sample or two-sample t test against nonnormality tends to improve with increasing absolute sample size and becomes perfect when sample sizes become infinite, the same cannot be said for either (a) the robustness of the one-sample χ^2 test (the common parametric test for an hypothesized population variance) against nonnormality, or (b) the robustness of the two-sample t test against heterogeneity of variance when $n_1 \neq n_2$, i.e., for either the robustness of a different test against the same assumption or for the robustness of the same test against a different assumption. Furthermore, although the two-tailed, two-sample t test is a special case of the analysis-of-variance F test, the perfect robustness of the former against heterogeneity of variance when sample sizes are equal and infinite does not extend to the latter in the general case.

The mountain of error consists in heroic generalizations transcending qualifications and unfettered by definitions. The Myth of Robustness against which objections are raised here is encountered in its most blatant form among the statistical laity. It is presumably attributable to a none-too-careful reading of robustness studies by statistical practitioners and to an incautious summarization of results by some of their authors and by the authors of statistical textbooks. It often takes the form of a statement that "the ——— test is

[1] Actually any statistical entity is "robust" to the extent that its behavior, i.e., value or sampling distribution, is not appreciably affected by departure from normal-theory conditions.

robust against the ——— assumption," or, worse, that "the ——— test is robust." Such a statement represents pure semantic chaos. There is no commonly accepted definition of what constitutes robustness, no agreed-upon criterion which distinguishes between a condition of robustness and one of nonrobustness. Therefore while the quoted statement *may* mean something to the utterer, provided he has an implicit subjective criterion of robustness, it can be (validly) meaningful to the hearer only if he is a mind-reader. Nor does the statement specify the extent of the violation against which the test is alleged to be robust. But the "amount of robustness" tends to depend (among many other things—see below) upon the "amount of violation." Thus the only relevant variable mentioned in the statement is left unquantified. But (although there are highly qualified exceptions) "extreme" degrees of violation tend to produce "drastic" nonrobustness. Yet if we assume that the statement refers to "ordinary" degrees of violation we are back at the same semantic impasse encountered with the term "robust," for there is no commonly accepted criterion differentiating between "ordinary" and "extraordinary" degrees of violation. (To a reasearch worker in whose field they are common, violations of large degree are still "ordinary.")

The most insidious thing about the Myth of Robustness, however, is that the "degree" of a test's robustness against violation of a given assumption is *strongly* dependent upon factors which are not involved in the statement of a test's assumptions, which are often not required in a complete description of the assumption's violation, and which are not mentioned in the usual allegation of robustness, as quoted above. These factors cause no distortion of Type I or Type II errors when all assumptions are met, but greatly influence the distortion occurring under a given violation of assumptions—i.e., the factors interact with whatever violation occurs. Examples of such factors are the following:

I. Factors concerned exclusively with testing.
   1. Location of rejection region, i.e., whether left-tailed, right-tailed, or two-tailed.
   2. Size of significance level, e.g., whether .05, .01, or .001.

II. Factors concerned primarily with sampling (and exclusively with sampling and testing).
   3. Minimum sample size, i.e., size of smallest sample contributing data to the test statistic.
   4. Absolute sizes of the other samples.
   5. Relative sample sizes, i.e., size of each sample relative to each other sample.
   6. Total number of samples upon which test is based.
   7. Number of samples of each absolute (and relative) size.

III. Factors involving populations but not assumptions.

8. *Which* populations yielded *what* samples (more specifically, which populations contributed what absolute and relative numbers of samples of what absolute and relative sizes).
9. Relative variances of sampled populations, when homogeneity of variance is not an assumption (as, for example, in the two-sample Z test for equal population means when population variances are exactly known).

In addition to factors such as the above, which are involved neither in statements of the assumptions nor in descriptions of their violation, robustness is also strongly influenced by factors whose very statement implies violation of an assumption but whose effect upon robustness may not be strongly related to the *degree* of the violation. Examples of such factors are the following:

10. Relative shapes of the sampled populations (and, perhaps somewhat more dependent upon the degree of violation, absolute shapes).
11. Relative amounts of correlation between sample mean and sample variance for the various samples (and, perhaps more dependent upon degree of violation, absolute amount of correlation between sample mean and sample variance, especially for the sample of smallest size).

These lists are illustrative rather than exhaustive.

Not only do these factors tend to interact with violations of assumptions, but they also display a strong tendency to interact with each other, i.e., if the "degree" of violation is held constant and the factors are varied, "degree" of robustness tends to vary. Thus interactions tend to be of high order. In fact, for a given violation of assumptions, not one of the factors listed appears to exert its influence upon probability levels independently of all other factors. The interrelationships are complex so that the net effect is difficult to anticipate on a priori grounds. When several factors are called into play by a violation, the factors may operate in different directions (some may tend to cause the true significance level to exceed the nominal one, while others may tend to cause the reverse) and may vary in relative potency as some other factor, such as absolute sample size, increases (one factor perhaps being prepotent at one sample size, giving way to another factor at a larger sample size. But, because of the high order of the causal interactions, the net effect is perhaps better thought of in terms of augmentations and detractions than of algebraic summations and deductions.) Depending upon the particular combination of factors involved and depending upon the particular levels or values assumed by the individual factors, a given violation of assumptions occurring in a specified degree or extent may have negligible effect or devastating effect upon the probability levels of the test statistic. Furthermore, because of the complexity of the interactions involved, the consequences of assumption violation are often not only unpredictable, but sometimes run counter to naïve statistical intuition. For example, in a very large, i.e., statistically reliable, and carefully controlled empirical sampling study by the writer [3], the following paradoxes were observed, in every case all conditions other than those mentioned being held constant: (a) the robustness of a critical ratio, or Z, test was much greater for a population badly fitted by a normal distribution with the same mean, variance, and area than for a population closely fitted by such a normal distribution; (b) the robustness of a two-sample t test was enormously improved by reducing absolute sample size while holding relative sample sizes constant (in other cases the robustness of a two-sample t test was greatly improved by decreasing the size of one sample while holding the other sample size constant; in *both* types of case robustness was increased by discarding data, i.e., by "throwing away information"); (c) the robustness of a two (equal-sized) sample t test under a given violation of the normality assumption was greatly increased by also violating the assumption of homogeneous variances. None of these paradoxes was "sought" or contrived; rather they simply happened. And the list is not exhaustive. For example, the improvement in robustness with decreasing sample size was found under certain conditions for the analysis-of-variance F test, the F test for equal population variances, and the χ^2 test for an hypothesized population variance as well as for the two-sample t test.

Although they are often difficult or impossible to predict in specific instances, robustness (and nonrobustness) effects are, of course, not without mathematical rationale. This rationale is helpful in understanding the general nature of "robustness," but it is no predictive panacea in situations realistic to applied statistics. Indeed, even with the aid of mathematics and with a complete knowledge of all population, sampling, and testing information, it may be, in complex situations, impossible even to predict the direction of nonrobustness, i.e., whether the true significance level will be greater or less than the nominal one! In the following paragraphs some of this rationale will be described verbally. (Much of this rationale can be derived using only algebra and the rules for taking expected values. This has been done in [3].) Since the "sampling" assumptions of random and independent sampling are under the control of the experimenter in a way in which the "population" assumptions of normality and homogeneity are not, in what follows the "sampling" assumptions will be considered as met, and only violations of the "population" assumptions of normality and homogeneity will be considered.

The formulas for many of the most common parametric test statistics can be written in such a way as to include no individual sample observations as such, the only variable quantities being sample means, variances, and sizes, i.e., n's. The sample mean is relatively robust against nonnormality because the first two moments of its distribution are independent of the normality or nonnormality of the sampled population and because its higher moments tend to approach their normal-theory values, as sample size increases, due to the effect described by the Central Limit Theorem. Thus neither the mean nor the variance, but only the shape of the distribution of the sample mean is affected by nonnormality and that shape becomes increasingly normal with increasing sample size (if the Central Limit Theorem holds).

The Central Limit Theorem is, in fact, a major theoretical basis for robustness effects. The Central Limit Theorem, as usually presented, states that if $\bar{X}$ is the mean of n observations drawn randomly (and with replacements if the population is finite) from a population with mean μ and finite variance σ^2, then as n approaches infinity the cumulative distribution of

$$\frac{\bar{X} - \mu}{\sigma/\sqrt{n}}$$

approaches the cumulative normal distribution with zero mean and unit variance. (So the cumulative distribution of $\bar{X}$ "approaches" the cumulative normal distribution with mean μ and a variance σ^2/n, which changes with sample size.) Actually the theorem applies only within a region $\mu \pm d$, where d is a function of n [6, 7]. Of course, the theorem applies strictly only to $n = \infty$, in which case $d = \infty$ also, and if n is *really* "infinite" the matter becomes academic. However, in that case the theorem itself is academic. If by "infinite" we simply mean "large" (or even "extremely large") the problem reappears, manifesting itself in the practical sense that the fit between the true distribution of $\bar{X}$ and the approached normal distribution tends to worsen rapidly, at a given n, at increasingly remote tail regions. Unfortunately, this qualification is rarely mentioned in the statistical literature where extravagant claims about the rapidity of approach to the normal distribution are commonplace. An antidote to such claims and an illustration of the worsening of fit at increasingly remote tail regions is provided by Fig. 2-2.[2] The sampled population, Fig. 2-1, for this and all subsequent figures in this chapter involving

[2] If the distribution of $\bar{X}$ were normal, a proportion $\alpha = .05$ of the sampling distribution of $\bar{X}$ at a given n would fall below $\mu - (1.644853637\,\sigma/\sqrt{n})$. Actually, when 50,000 samples of 32 observations each were drawn from the X population, only 1,471 of their $\bar{X}$'s, or a proportion $\rho = .02942$ of the empirical sampling distribution of $\bar{X}$ at $n = 32$, fell below $\mu - (1.644853637\,\sigma/\sqrt{n})$. So, corresponding to a left-tail α of .05, $\rho = .02942$ and $\rho/\alpha = .5884$, which is the ordinate of the curve labeled $\alpha = .05$ at $n = 32$ on the "left-tail" graph. The construction of all subsequent figures is analogous (except that ρ was obtained mathematically for Figs. 2-6, 2-7, and 2-8), e.g., 29,617 of the 50,000 t's obtained at $n = 4$ fell below -2.353399992, the t value with 3 degrees of freedom that is nominally significant, under the normality assumption, at the .05 level for a left-tail test, so $\rho = .59234$, $\alpha = .05$, and $\alpha/\rho = .0844$, which is the ordinate of the curve labeled $\alpha = .05$ at $n = 4$ on the "left-tail" side of Fig. 2-4. In all cases, if the test or statistic were perfectly robust all ρ/α and α/ρ curves would tend, but for sampling error, to be horizontal lines through an ordinate of 1.00, bisecting the graph. Further details can be found in the sampling studies [2, 3, 4] from which all figures in this chapter are taken.

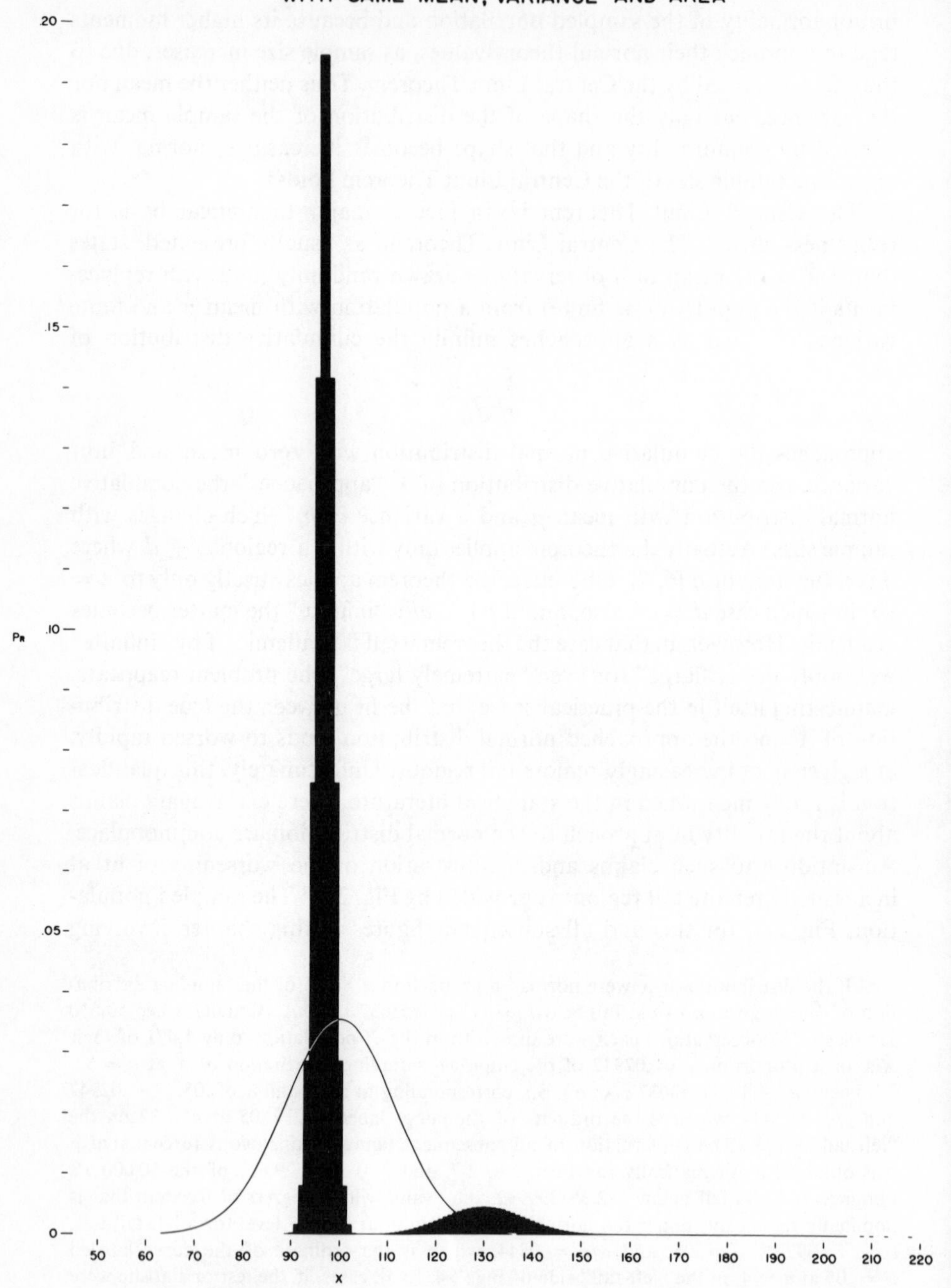

Fig. 2-1. Nonnormal population sampled to obtain the sampling distributions upon which Figs. 2-2, 2-3, 2-4, 2-5, and 2-9 are based. Population was constructed to approximate the empirical population of actual data shown in Fig. 1-1.

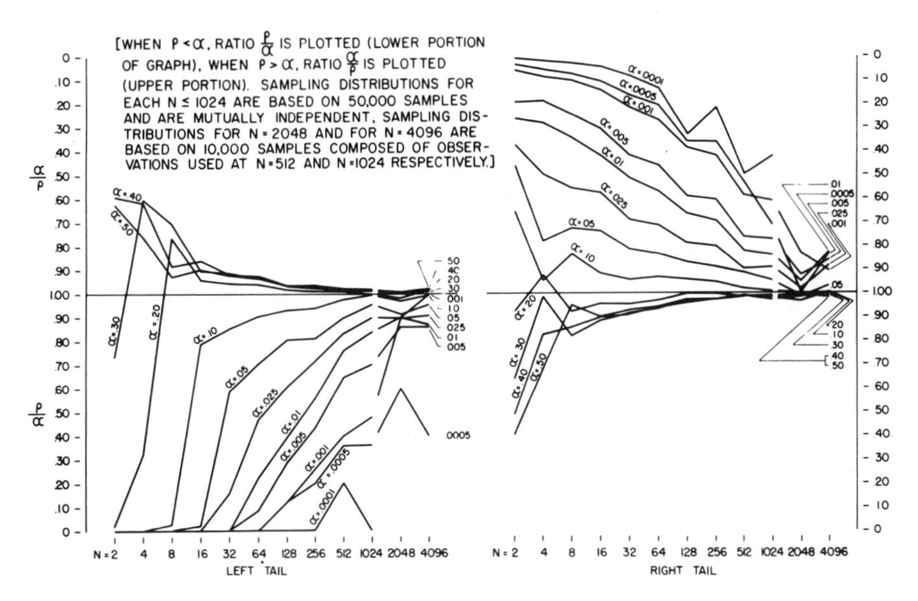

Fig. 2-2. Robustness of sample mean, or, equivalently of one-sample Z test for an hypothesized population mean, when samples are drawn from population shown in Fig. 2-1. (ρ/α may be regarded as plotted in all cases, but against a linear scale when it is less than one and against a reciprocal scale when it exceeds one.)

nonnormal populations is a population very highly similar to that shown in Fig. 1-1, on page 1. Since the one-sample Z test for an hypothesized population mean uses, when H_0 is true, a test statistic

$$Z = \frac{\bar{X} - \mu}{\sigma/\sqrt{n}}$$

which (at a given n) is simply a linear transformation on $\bar{X}$, Fig. 2-2 shows equally the robustness of the one-sample Z test for samples drawn from the same population. That test is widely regarded as one of the most robust against nonnormality!

The sample variance is quite *non*robust against nonnormality, since only the mean of its sampling distribution is independent of whether or not the sampled population is normal. (The variance of its sampling distribution depends upon the fourth, as well as the second, central moment of the sampled population.) Furthermore, there is no "Central Limit effect" for sample variances; although their distribution at $n = \infty$ has the "right" shape, it has the "wrong" variance. The variance of the sampling distribution of the sample variance *diminishes* as n increases, but its *ratio* to the corresponding "normal-theory" value does not tend to improve much, since the latter differs from it and diminishes at about the same rate with increasing n (at least above a fairly modest n value). But while robustness does not change much with changing n, it does worsen as one moves from the perfectly robust mean of the sampling distribution toward either tail. These points are illustrated by Fig. 2-3. (At $n = \infty$ the curves would have about the same ordinates as at $n = 1024$.) Since the one-sample χ^2 test for an hypothesized population variance uses, when H_0 is true, a test statistic

ROBUSTNESS OF THE SAMPLE VARIANCE FOR SAMPLES CONSISTING OF N OBSERVATIONS DRAWN FROM THE X POPULATION

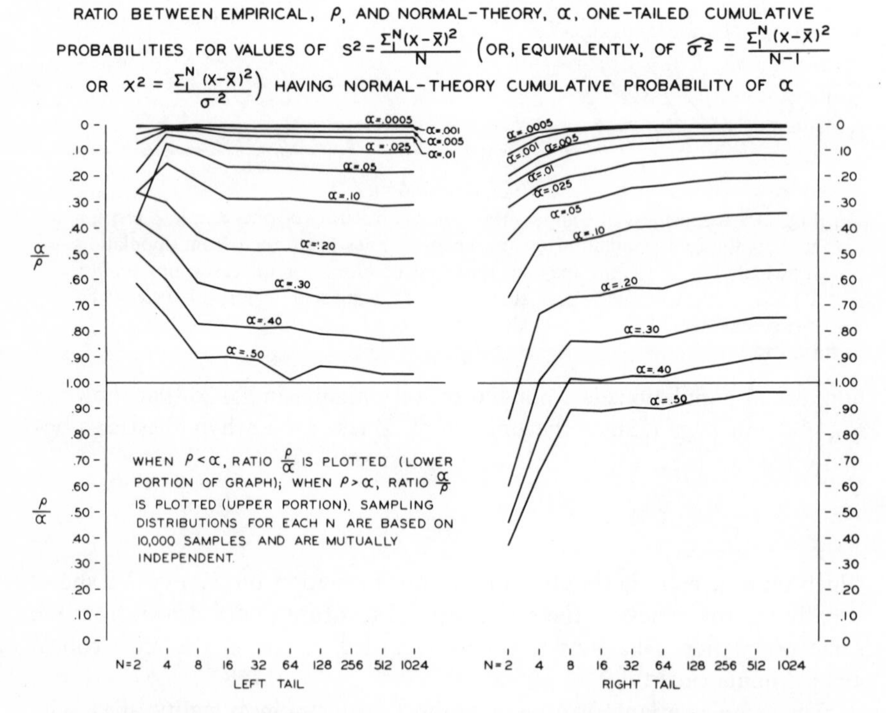

Fig. 2-3. Robustness of sample variance, or equivalently of one-sample χ^2 test for an hypothesized population variance, when samples are drawn from population shown in Fig. 2-1.

$$\chi^2 = \frac{ns^2}{\sigma^2}$$

which is simply a linear transformation upon the sample variance, the figure applies equally to the robustness of that test. The figure is quite similar to one obtained (and shown in [3]) for the two-sample F test for equal population variance

$$F = \frac{\hat{\sigma}_1^2}{\hat{\sigma}_2^2}$$

for samples drawn from the same population, as might be expected from the above reasoning. Fortunately for statistics, these two tests are generally pronounced "nonrobust."

The one- and two-sample t tests and the analysis-of-variance F test all contain sample means only in their numerators and sample variances only in their denominators. As absolute sample size increases (while relative sample sizes are held constant), the variance of the denominator shrinks relative to that of the numerator until, at $n = \infty$, its relative variance becomes zero. At that point the denominator behaves like a constant, its own expected value, i.e., the value it would have if all of the sample variances in it were replaced by *their* respective expected values, which are the same for nonnormality as for normality. Thus, as absolute sample size increases, the adverse effect of nonnormality upon the second and higher moments of the distribution of the sample variance becomes decreasingly important due to the relatively diminishing variance of the denominator as a whole, while the one moment unaffected by nonnormality becomes increasingly important. At the same time the numerator is becoming increasingly robust due to the Central Limit effect and increasingly "dominant" in the sense that its variance increasingly "accounts for" the variance of the test statistic. The net result is that the test statistic becomes perfectly robust (at standard, but not at infinitely small, significance levels [8]) against nonnormality when absolute sample size (i.e., the size of the smallest sample contributing to the test statistic) becomes infinite. We have already seen that infinity is a long way off in the case of the Z statistic (whose robustness is identical to that of the numerator of the corresponding t statistic). Infinity is still farther off in the case of the t statistic, not so much because of the nonrobustness of its denominator, per se, as because of the correlation between numerator and denominator due to the inevitable correlation between mean and variance of samples drawn from a nonnormal population. Some of these points are illustrated by Fig. 2-4. Surprising as it may be in the light of Fig. 2-4, the t test is widely regarded as highly robust against nonnormality. At this point, it is appropriate to quote Hotelling [8]:

> Central limit theorems have usually dealt only with linear functions of a large number of variates, and under various assumptions have proved

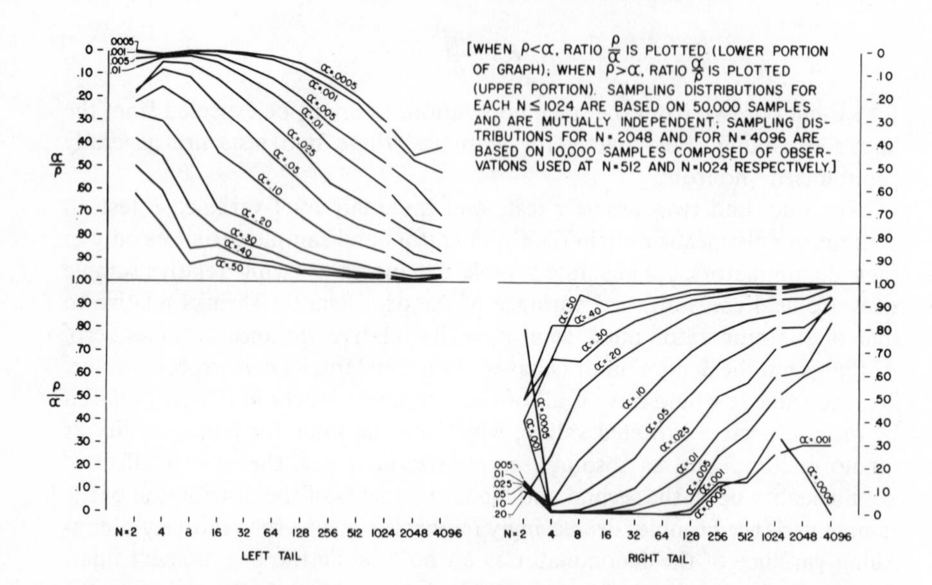

Fig. 2-4. Robustness of one-sample *t*-test for an hypothesized population mean when samples are drawn from population shown in Fig. 2-1.

> convergence to normality as the number increases. For a large but fixed number the approximation of the distribution to normality is typically close within a restricted portion of its range, but bad in the tails. Yet it is the tails that are used in tests of significance, and the statistic used for a test is seldom a linear function of the observations. The nonlinear statistics *t*, *F*, and *r* in common use can be shown . . . under certain wide sets of assumptions to have distributions approaching the normal form, though slowly in the tails. But in the absence of a normal distribution in the basic population, with independence, there is a lack of convincing evidence that the errors in these approximations are less than in the nineteenth-century methods now supplanted, on grounds of greater accuracy, by the new statistics.

Hotelling, of course, is talking about poorness-of-fit to normal-theory *limiting* distributions at *large n*; the analogous poorness-of-fit to the normal-theory distribution *itself* at *small n* can be inferred from Fig. 2-5.

We have already seen that it may take a great deal of infinity to produce

EFFECT OF X POPULATION'S NONNORMALITY UPON SIGNIFICANCE LEVELS OF SIX COMMON PARAMETRIC TEST STATISTICS AT N=8 (OR $N_1 = N_2 = 8$)

ρ IS THE EMPIRICAL SIGNIFICANCE LEVEL, IN A SAMPLING DISTRIBUTION OF K VALUES OF THE TEST STATISTIC, OF THAT VALUE OF THE TEST STATISTIC WHOSE NORMAL-THEORY SIGNIFICANCE LEVEL IS α, SO:

α = NOMINAL SIGNIFICANCE LEVEL BASED UPON FALSE ASSUMPTION OF NORMALITY
ρ = ESTIMATED TRUE SIGNIFICANCE LEVEL CORRESPONDING TO NOMINAL SIGNIFICANCE LEVEL OF α
K = SIZE OF SAMPLING DISTRIBUTION UPON WHICH ESTIMATE IS BASED

(ABSCISSA SCALE IS SIXTH ROOT)

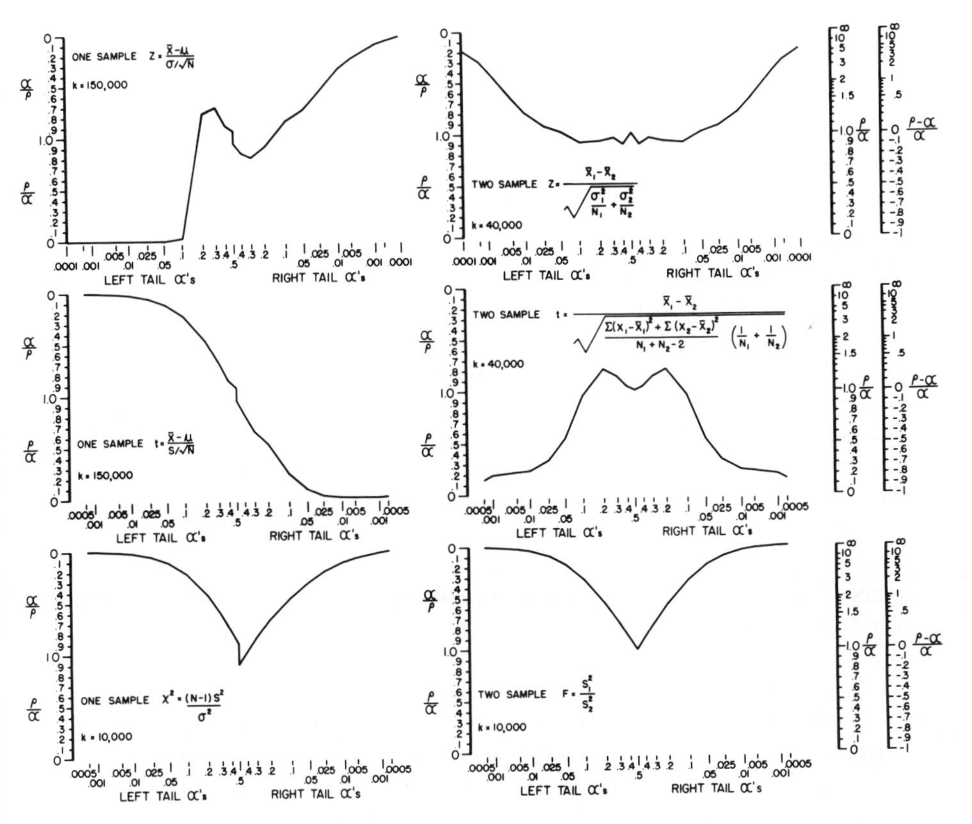

Fig. 2-5. The critical influence of significance level upon robustness of Z, t, and the variance test, in both the one- and two-sample cases, when samples contain 8 observations drawn from the population shown in Fig. 2-1. (In this figure s^2 stands for $\sum(X - \bar{X})^2/(N - 1)$, rather than $\sum(X - \bar{X})^2/N$ as in previous figures and discussion.)

excellent robustness against nonnormality for tests based on sample means or on both means and variances, but that at really "infinite" infinity perfect robustness may be attained. However, no amount of infinity will necessarily produce perfect robustness against nonnormality for the $\chi^2 = ns^2/\sigma^2$ or $F = \hat{\sigma}_1^2/\hat{\sigma}_2^2$ tests, since they are based exclusively on sample variances, nor against heterogeneity of variance for the two-sample t and analysis-of-variance F tests which assume homogeneity (although there exist highly specific exceptions). Irrespective of nonnormality or heterogeneity of variance, the mean of the sampling distribution of the sample variance

$$s^2 = \frac{\sum (X - \bar{X})^2}{n}$$

is $(n - 1)\sigma^2/n$ (alternatively the mean of the sampling distribution of the sample estimate of population variance

$$\hat{\sigma}^2 = \frac{\sum (X - \bar{X})^2}{n - 1}$$

is σ^2), where σ^2 is the *actual* variance of the particular population from which the sample was drawn. Therefore, when population variances are heterogeneous, the *locations* of the sampling distributions of the various $\hat{\sigma}^2$ are different, whereas the homogeneity assumption requires that they be the same. This difference in locations creates a bias in the denominator of F or of two-sample t, and the effect of this bias upon the robustness of the test tends to remain constant with increasing n so that "going to infinity" does not solve the problem. Figures 2-6, 2-7, and 2-8 show the effect of heterogeneity alone upon the robustness of the two-sample t test, when sample sizes are infinite. The case of $n_1 = n_2$ is famous in the literature on robustness. In that case, despite the unequal locations of $\hat{\sigma}_1^2$ and $\hat{\sigma}_2^2$, the entire denominator has its normal-theory location and the two-sample t test is impressively robust against heterogeneity alone at all sample sizes and perfectly robust when both sample sizes are infinite. This happy state of affairs (i.e., perfect robustness against heterogeneity at equal infinite sample sizes) does not extend, however, to the general case of the analysis-of-variance F test of which the two-tailed, two-sample t test is a special case.

So far, effects, while sometimes drastic, have tended to appear more or less "well-behaved" in the sense of being relatively uncomplicated. This is partly because only a single assumption was violated and partly because some of the important interacting factors were "held constant" by making them parameters for the curves or by letting them become infinite. The chaos which can result when more than one assumption is violated is illustrated in Fig. 2-9, which shows the robustness of the two-sample t test (at the testing tails, i.e., at only the standard nominal significance levels) when one sample of size $3n$ is drawn from the nonnormal population shown in Fig. 2-1, and another sample of size n is drawn from a population identical in shape to the first population, but having one-half its standard deviation. Nonnormality and heterogeneity affected robustness in the same direction for a right-tailed test and in opposite directions for a left-tailed test. For the latter, at the lower sample sizes nonnormality was dominant. As n increased the Central Limit effect reduced the relative influence of nonnormality, so that at the higher

SENSITIVITY OF t TEST TO HETEROGENEITY OF VARIANCE WHEN BOTH SAMPLES ARE OF INFINITE SIZE

RATIO BETWEEN TRUE, ρ AND NORMAL-THEORY, α, PROBABILITIES OF REJECTION WHEN TEST USES ONE-TAILED REJECTION REGION CORRESPONDING TO α = 05 AND t STATISTIC IS BASED UPON SAMPLES OF SIZE N AND γ N, RESPECTIVELY, FROM NORMAL POPULATIONS WITH EQUAL MEANS BUT VARIANCES σ^2 AND $R\sigma^2$

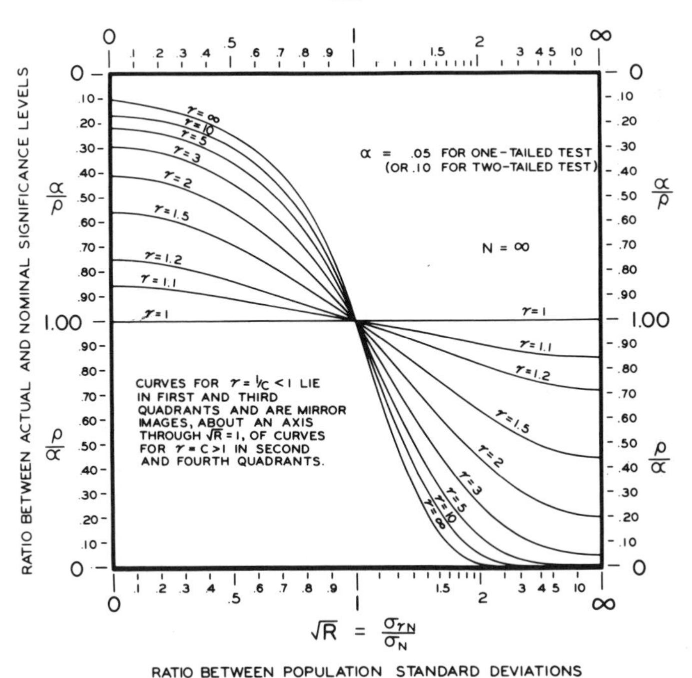

Fig. 2-6. Robustness of two-sample *t*-test against heterogeneity of variance when sample sizes are infinite, test is one-tailed, and nominal significance level is .05.

sample sizes the relatively constant influence of heterogeneity became dominant. The result is that above approximately $n = 32$ robustness diminishes with increasing sample size so that, turning the statement around, robustness can be greatly increased in this region by discarding data, i.e., by "throwing away information." Notice also that robustness does not always change monotonically with increasing n and that it does not always (at every n) worsen with

SENSITIVITY OF t TEST TO HETEROGENEITY OF VARIANCE WHEN BOTH SAMPLES ARE OF INFINITE SIZE

RATIO BETWEEN TRUE, ρ, AND NORMAL-THEORY, α, PROBABILITIES OF REJECTION WHEN TEST USES ONE-TAILED REJECTION REGION CORRESPONDING TO α = .01 AND t STATISTIC IS BASED UPON SAMPLES OF SIZE N AND γN, RESPECTIVELY, FROM NORMAL POPULATIONS WITH EQUAL MEANS BUT VARIANCES σ^2 AND $R\sigma^2$

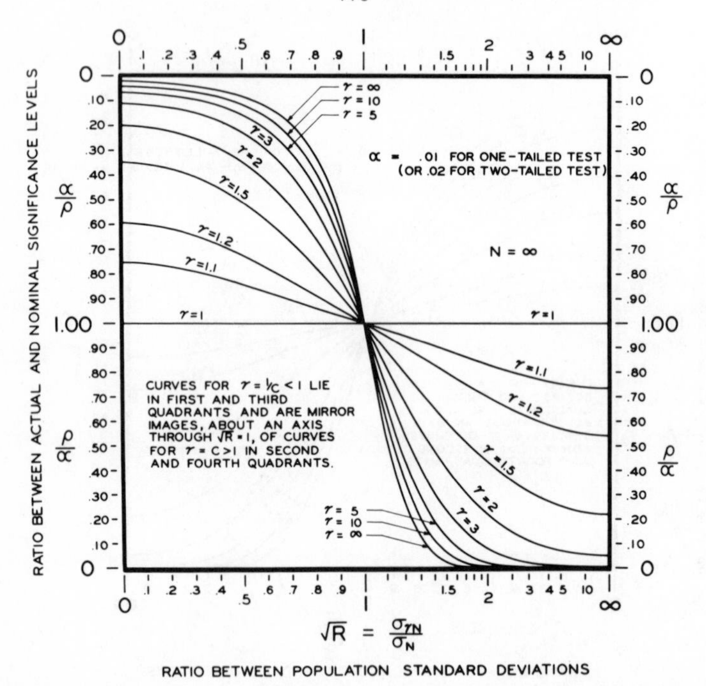

Fig. 2-7. Robustness of two-sample *t*-test against heterogeneity of variance when sample sizes are infinite, test is one-tailed, and nominal significance level is .01.

diminishing values of α. In short, whereas the general trend might be predictable from a very careful examination of theoretical considerations, not even the direction (to say nothing of the extent) of the nonrobustness is predictable in a wide range of specific cases.

The preceding figures have shown that even under the mildest of conditions parametric tests reputed to be highly robust may in fact have true

SENSITIVITY OF t TEST TO HETEROGENEITY OF VARIANCE WHEN BOTH SAMPLES ARE OF INFINITE SIZE

RATIO BETWEEN TRUE, ρ, AND NORMAL-THEORY, α, PROBABILITIES OF REJECTION WHEN TEST USES ONE-TAILED REJECTION REGION CORRESPONDING TO $\alpha = .001$ AND t STATISTIC IS BASED UPON SAMPLES OF SIZE N AND γ N, RESPECTIVELY, FROM NORMAL POPULATIONS WITH EQUAL MEANS BUT VARIANCES σ^2 AND $R\sigma^2$

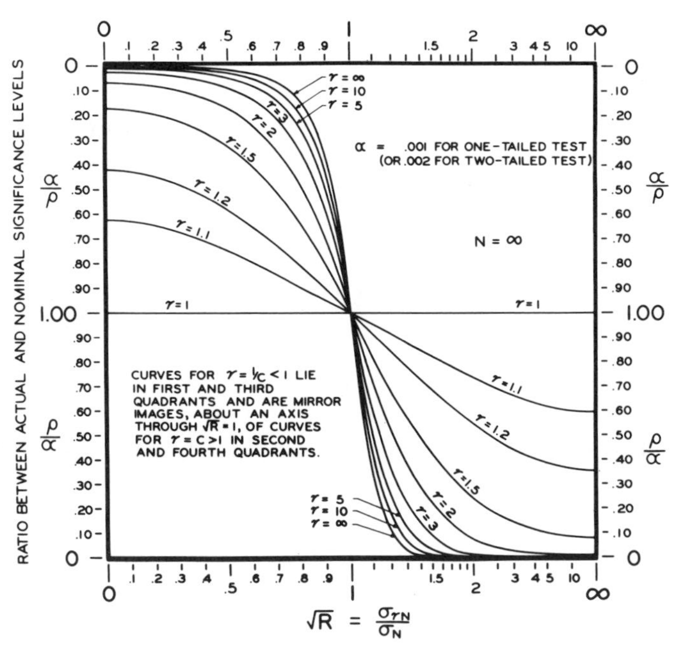

Fig. 2-8. Robustness of two-sample *t*-test against heterogeneity of variance when sample sizes are infinite, test is one-tailed, and nominal significance level is .001.

significance levels considerably different from the nominal (normal-theory) levels when samples are drawn from a population similar to that shown in Fig. 1-1. One might suppose, therefore, that the population is a rare one. If so, it is quite a coincidence that the *first* population studied by the writer (the shape of which was unknown at the beginning of the study) was of this general shape, and an even greater coincidence that a colleague, H. J. Jerison,

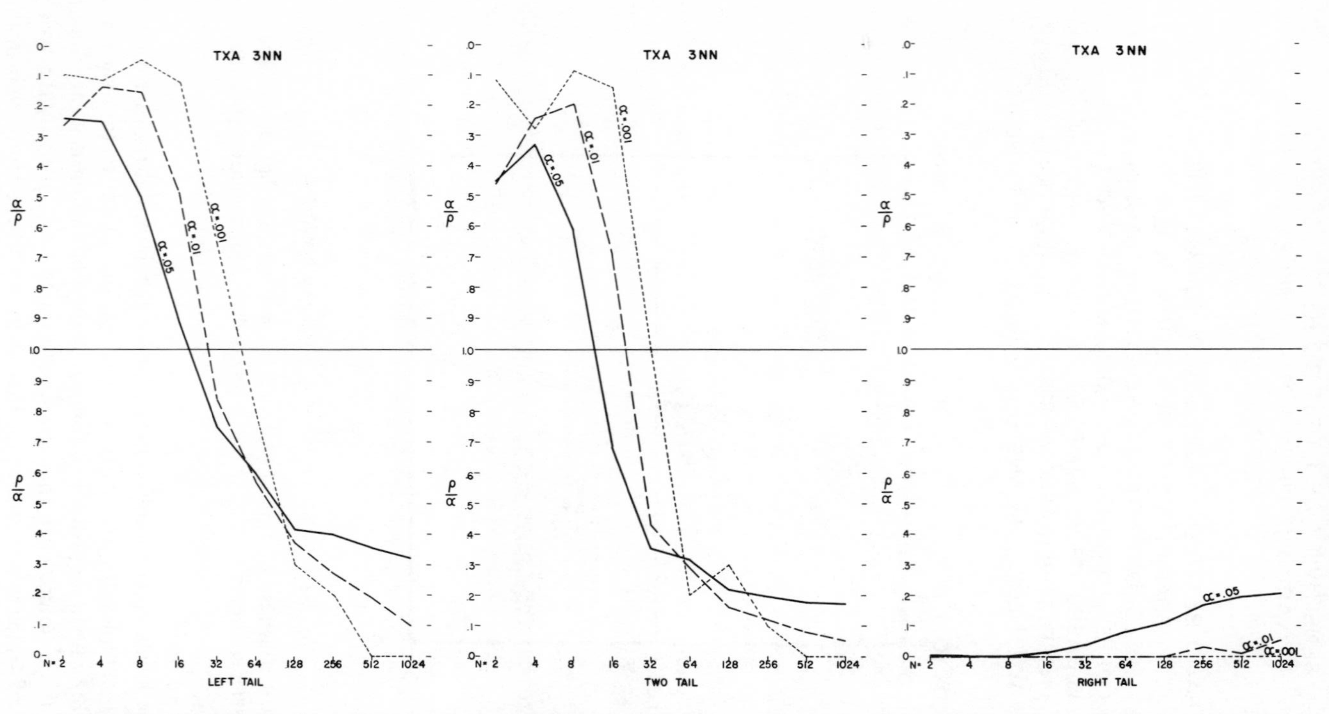

Fig. 2-9. Robustness of two-sample t-test when first sample consists of $3n$ observations drawn from population shown in Fig. 2-1 and second sample consists of n observations drawn from population identical to it except that it has one-half its standard deviation.

working in the same laboratory but entirely independently and upon an entirely different problem (and, indeed, in an entirely different research area) has obtained highly similar populations. There are, in fact, cogent theoretical reasons for believing the population to be typical for a broad general class of experimental conditions as well as for a particular common type of dependent variable. Some are discussed in [2].

Despite the myriad shortcomings of the "robustness" concept, the alleged robustness of parametric tests has sometimes been invoked as justification for regarding them as effectively distribution-free tests. Indeed, some statisticians advocate using classical *instead of* distribution-free statistics, when population shapes or relative variances are unknown, because of their allegedly superior qualities of efficiency, etc. Thus classical tests are presented as better distribution-free tests than the truly distribution-free tests themselves. Often the same people who scorn the Wilcoxon test because its asymptotic efficiency relative to Student's t test is .955, a difference of 4.5%, will cheerfully countenance a true significance level of .02 corresponding to a nominal level of .01, an error of 100%, for Student's t test under assumption violation. Not only are the two standards of rigor incompatible, but the superior efficiency of the t test does not necessarily hold when its assumptions are violated. Therefore, when its assumptions are violated, Student's t test may be inferior to the Wilcoxon test on both counts! Thus, when viewed as distribution-free tests, not only do classical tests fail to acquire many of the virtues of truly distribution-free tests, such as simplicity, but they also lose, or hazard the loss of, many of the features which made them desirable as classical tests. Not only do they become inexact—perhaps grossly so—and risk the loss of superior efficiency, but they can no longer be regarded as tests for specific population parameters. When its assumptions of normality and homogeneity are not known to be met, the t test is no longer a test for means; it is an exact test for identical normal populations. Nor is it correct to regard the test as an "inexact" test exclusively for means. The probability of rejection can exceed the nominal α value because of unequal means, unequal variances, unequal shapes, or nonnormality of shape. Therefore, if it is an "inexact" test for means it is also an "inexact" test, perhaps a less powerful one but no less valid, for variances, for normality, and for equality of shape. Thus the virtues of classical tests tend to be peculiar to the situations defined by their assumptions. The view of parametric tests as approximate distribution-free tests, therefore, appears to be barren of merit in the light of the above considerations and considering the gross variability and unpredictability of the degree of approximation.

Thus, although one would never guess it from the quoted, prototype statement, "The ——— test is robust against the ——— assumption," (a) there is no objective robustness-nonrobustness dichotomy; rather, there is a continuum of degrees of robustness, or, perhaps more logically, of nonrobustness; (b) "degree of nonrobustness" is not a simple function of "degree of violation" of an assumption, but depends instead, in a complex way, upon a multiplicity of variables. Despite these objections, the robustness concept would still have considerable practical utility if it could be shown that there is a fairly low upper bound to "degree of nonrobustness" which holds under fairly general conditions, e.g., always holds for a given test. Then the upper bound would serve, in effect, to define and dichotomize robustness versus nonrobustness. Upper bounds for nonrobustness can, in fact, sometimes be calculated, and occasionally they turn out to be low (some very low ones were found by the writer [3]). However, bounds that are sufficiently low tend to be absurdly specific [3], and bounds that are sufficiently general tend to be outlandishly high [1, 2, 3, 14], sometimes assuming the maximum possible value [3]. Nor can the results of empirical robustness studies, taken as a whole, provide highly reliable guidelines for example by indicating the probable distortion in Type I errors or the worst amount of distortion "likely to be encountered in practice." They cannot do so because, among other reasons, relatively few such studies investigated populations of real data, and artificial populations can scarcely be expected to yield results truly typical of practice. A more important reason, perhaps, is that in many specific areas of research not even the *typical* population is known. Often the only clues as to population characteristics come from the inadequately small samples taken in the experiment which raised the question in the first place. In such cases not even the practitioners know what is typical of practice, so even the test's ultimate user cannot judge the relevance of robustness findings to his specific applications. Thus the robustness concept cannot be validly invoked as a virtual carte blanche to use parametric tests with expected impunity when their assumptions are violated. Instead, the experimenter (if he wishes to make use of "robustness") must carefully sift through a sizeable literature of robustness studies until he finds one which essentially reproduced his own experimental conditions—i.e., shapes and variances of populations, sample sizes, size and location of rejection region, test used, etc.—and then be guided by its results. Of course, even this approach is not open to him unless he knows a great deal about the shapes, both absolute and relative, of his sampled populations, as is seldom the case in many areas of research, such as the behavioral sciences. And even if he is lucky enough to have detailed information about his populations, he is unlikely to find a robustness study exactly matching his experimental conditions. To the extent that they do not match (and particularly if the failure to match extends to several variables, as it almost certainly will), the experimenter must generalize (to be brutally frank, he must guess) along dimensions (i.e., across values of variables) known to have an often strong influence upon the generalized phenomenon! Thus, even if the experimenter surveyed the literature on robustness and compiled a catalogue of "degrees of nonrobustness" resulting from the various combinations of conditions which have been mathematically or empirically investigated, the catalogue would be woefully incomplete and inadequate. Indeed, in the hands of any but the most astute (and luckiest) professional statistician such a catalogue might prove the instrument of disaster.

From the above considerations it is clear that in order to convert "The ——— test is robust against the ——— assumption" into a meaningful and accurate statement about robustness, it would have to be accompanied by a quantitative definition of robustness, a complete quantitative statement of the degree or extent of the violation, and a complete specification of the exact sampling and test conditions under which the test is to be performed. If all this were done, the statement would be so particularistic as to have little general appeal, which perhaps explains why the type of statement quoted survives in its amorphous, undefined, unqualified form. It also explains why that form is so completely inaccurate and so utterly meaningless.

REFERENCES

1. Berry, A. C., "The Accuracy of the Gaussian Approximation to the Sum of Independent Variates," *Transactions of the American Mathematical Society*, **49** (1941), 122–136.

2. Bradley, J. V., *Studies in Research Methodology IV. A Sampling Study of the Central Limit Theorem and the Robustness of One-Sample Parametric Tests*,[3] AMRL Technical Documentary Report 63-29, 6570th Aerospace Medical Research Laboratories, Wright-Patterson Air Force Base, Ohio, March 1963. The body of this report (i.e., excluding extensive appendices) is essentially identical to Bradley, J. V., *An Empirical Investigation of the Central Limit Theorem Applied to Time Scores*, Ph. D. Thesis, Psychology Department, Purdue University, Lafayette, Indiana, January 1962.

3. Bradley, J. V., *Studies in Research Methodology VI. The Central Limit Effect for a Variety of Populations and the Robustness of Z, t, and F*,[3] AMRL Technical Report 64-123, Aerospace Medical Research Laboratories, Wright-Patterson Air Force Base, Ohio, December 1964.

4. Bradley, J. V., *Studies in Research Methodology VII. The Central Limit Effect for Two Dozen Populations and Its Correlation with Population Moments*,[3] AMRL Technical Report 66-242, Aerospace Medical Research Laboratories, Wright-Patterson Air Force Base, Ohio, December 1966.

5. Brown, G. W., and J. W. Tukey, "Some Distributions of Sample Means," *Annals of Mathematical Statistics*, **17** (1946), 1–12.

[3] Available from University Microfilms. See abstracts numbered 68-7446 (for Study No. IV) and 68-7447 (for Studies Nos. VI and VII) in *Dissertation Abstracts*, Part B, **28** (1968), No. 11, Monograph Section.

6. Chernoff, H., "Large-Sample Theory: Parametric Case," *Annals of Mathematical Statistics*, **27** (1956), 1–22.

7. Cramér, H., "Sur un nouveau théorème-limite de la théorie des probabilités," *Actualités scientifiques et industrielles.* Paris: Hermann and Cie., 1938, No. 736, pp. 5–23.

8. Hotelling, H., "The Behavior of Some Standard Statistical Tests under Nonstandard Conditions," in *Proceedings of the Fourth Berkeley Symposium on Mathematical Statistics and Probability*, ed. Jerzy Neyman. Berkeley and Los Angeles: University of California Press, 1961, Vol. I, pp. 319–359.

9. Kendall, M. G., and R. M. Sundrum, "Distribution-Free Methods and Order Properties," *Review of the International Statistical Institute*, **3** (1953), 124–134.

10. Scheffé, H., "The Effects of Departures from the Underlying Assumptions," *The Analysis of Variance*, New York: John Wiley & Sons, Inc., 1959, Chapter 10, pp. 331–369.

11. Siegel, S., "Nonparametric Statistics," *American Statistician*, **11** (1957), 13–19.

12. Smith, J. H., "Some Properties of the Median as an Average," *American Statistician*, **12** (1958), 24–25, 41.

13. Tukey, J. W., "A Survey of Sampling from Contaminated Distributions," in *Contributions to Probability and Statistics: Essays in Honor of Harold Hotelling*, ed. Ingram Olkin *et al.*, Stanford, Calif.: Stanford University Press, 1960, pp. 448–485.

14. Zahl, S., "Bounds for the Central Limit Theorem Error," *SIAM Journal of Applied Mathematics*, **14** (1966), 1225–1245.

3

FUNDAMENTALS

3.1 NONPARAMETRIC INFERENCE

It has already been mentioned that the information used by distribution-free tests tends to consist of nonmagnitudinal, usually sample-related characteristics of the variate values drawn from the population. The test statistic therefore is related directly and mathematically to the sample but tends to be related only indirectly and "logically" (or "procedurally") to the population, the latter relationship being largely implicit in the procedures followed in extracting from the original observations the information actually used. As a consequence of the above, what the test actually tests, in the direct, mathematical sense, is a primarily sample-related hypothesis and set of assumptions. Procedurally implicit in them are a population-related hypothesis and set of assumptions which are therefore tested indirectly by logical implication.

To be entirely explicit, let H_p and A_p be respectively the null hypothesis and set of assumptions expressed as statements about the sampled population and method of sampling. Thus H_p is the null hypothesis in which the experimenter is actually interested and A_p stands for certain assumptions he can make from facts known about the sampled population or about the intended method of sampling. Let H_s and A_s be respectively the null hypothesis and set of assumptions that the statistic actually tests, i.e., which together represent the conditions necessary to give the test statistic its null (or "tabled") distribution. One follows certain rules and procedures, R & P, in extracting from the sample the information actually used by the test statistic, in resolving ambiguities in the data, and in calculating the test statistic. Ideally these rules and procedures create a situation, i.e., establish a logical structure or context, in which if H_p and A_p are true, H_s and A_s will also be true and the test statistic will have its null distribution, whereas if H_p is false and A_p is true, H_s will be false and A_s true so that the test statistic will have a different distribution and the difference can be attributed to the falsity of H_p. Actually, however, it is not necessary that A_p and A_s always be true irrespective of the truth or falsity of the hypothesis. It is only necessary that A_p always be true when H_p is true, for if A_p can be false only when H_p is false, the test statistic will have a nonnull distribution, thereby tending to reject, only when rejection is desired.

The assumptions in A_s are necessary and indispensable. However, since it is only required that the *truth* of A_p guarantee the *truth* of A_s, the population assumptions in A_p need only be *sufficient* to imply their actually necessary counterparts in A_s. Quite a variety of different sufficient conditions may logically imply a single necessary condition, so while for a given application there is only one set of A_s, there may be a wide choice of corresponding A_p, and with certain permissible variations in the R & P the choice may be still further widened. Thus, in marked contrast to the parametric case, the experimenter may have a choice from among alternative (population) assumptions, and in some cases the alternatives may be so different as to be mutually exclusive (as when one alternative specifies a continuous population and another specifies a symmetric population which is permitted to be discrete—see Sign test for the median, and Sign test for the median difference). However, while a variety of alternative population assumptions may be logically permissible, in few cases is the experimenter likely to be aware that they are true when in fact they are. Thus very few are likely to be of any practical value. For example one such assumption (in the Sign test for the median) is that the hypothesized median of the sampled population is not one of the values comprising the population. This is logically possible if there are discontinuities in the population distribution, e.g., the population is discrete, but the experimenter is exceedingly unlikely to be aware of the truth of the assumption since the value of the median is what he is testing, and about which he is therefore largely ignorant.

For fixed R & P and for given, true A_p and A_s, H_s will be true *if and only if* H_p is true, so a test of H_s will be equally a test of H_p. It is important to understand, however, that H_p and H_s are *different* conditions which necessarily imply each other only within the context of the logical structure formed by following the proper rules and procedures and meeting the required assumptions. Thus it is quite possible for H_p to be true and H_s false if the correct R & P are not followed or if assumptions are not met. For example the H_p that M is the median of the X-population is true, but the H_s that $P(X < M) = \frac{1}{2}$ is false if 40% of the X-population lies below M, 45% above M, and 15% at M,—i.e., if the assumption in A_s that $P(X = M) = 0$ is false (see Sign test for the median). An intentional modification of any member of the logical structure may *induce* changes in other members, i.e., in order for the form of the logical relationship to hold, some of its propositions may have to be altered. Specifically, a change in the R & P may destroy the relationship between the old H_p and H_s and cause an entirely different H_p now to correspond to the same H_s. Thus a test originally designed for one purpose may, under the proper alteration in the R & P, turn out to be an excellent test for an entirely different type of application. This has been the case for many of the better-known tests, such as the Sign test, the Wilcoxon test, and Kendall's test for rank order correlation.

Clearly, the H_s and A_s are unique and basic in a sense in which H_p and A_p are not. Therefore, in the chapters that follow, when the null hypothesis and assumptions are stated formally, it will be H_s and A_s that are given. The H_p and A_p will also be stated, either directly or implicitly, usually in describing the rationale and use of the test. Generally H_p and A_p will be stated in words whereas, for maximum accuracy, H_s and A_s will sometimes be expressed in mathematical symbols and formulas. Since the distinction between H_p and H_s will be clear from context and other clues, both will be represented by the conventional H_0.

3.2 TEST ASSUMPTIONS

In common with parametric tests, distribution-free tests almost always require the "sampling" assumptions that observations are drawn *randomly* and *independently*[1] of the outcome of previous draws. Since these assumptions are concerned primarily with the *method* of sampling rather than with the nature of the sampled population, they appear in both A_p and A_s.

A large group of distribution-free tests uses only the "information" implicit in the algebraic sign of a difference-score, at least one of whose

[1] Although perhaps technically redundant from a mathematical standpoint, this phrase will be used repeatedly, since the two terms often have different connotations in the minds of experimenters. An experimenter may think of drawing subjects randomly, but he is unlikely to think of randomly "drawing" a single subject's repeated responses from an unchanging population of potential responses. Rather, he thinks of a necessity for replicated responses to be independent due to the absence of sequential effects such as learning.

members is a sample observation. Such tests require the assumption, in A_s, that the data to which the test statistic is applied can contain *no difference-scores of zero*.

A much larger group, containing most other distribution-free tests, ranks the sample observations in order of magnitude, and uses only the information implicit in the ranks. Such tests require the assumption, in A_s, that certain sets or subsets of ranks will consist of consecutive, nonrepeating integers starting with 1, i.e., that the sets can contain *no tied ranks* and no "ranks" that are not integers. (This does not necessarily mean that there must be no equal observations. Sometimes consecutive ranks can be arbitarily assigned to tied observations because any one of the possible assignments of tied-for ranks to equal observations gives the test statistic the same value. If this procedure does not impair the logical structure, it is legitimate.)

Occasionally these necessary assumptions are met simply because the nature of the situation makes zero differences or tied ranks impossible. For example, if ranks are preference ratings assigned to objects by a judge who has been instructed to guess when he is uncertain, tied ranks will not occur among that judge's ratings if the judge follows his instructions. Likewise, difference-scores of the type $X - M$ cannot be zero if the X-population does not contain the value M, and those of the type $X - Y$ cannot be zero if the X- and Y-populations contain no common value.

More often, however, undesired zero differences or equal observations can occur among the sampled "observations," but zero differences or undesirably tied ranks are prevented from occurring in the data *to which the test statistic is applied*. This is accomplished by the adoption of certain R & P which, in turn, make sense only within a context provided by certain A_p. Sometimes, of course, the necessary assumptions are simply violated, the test being used as an approximate test, as is almost always the case with parametric statistics.

3.3 THE POPULATION ASSUMPTION OF CONTINUITY

The population assumption most frequently encountered in the literature on distribution-free tests is that all sampled variates are *continuously distributed*. A continuous distribution has an infinite number of abscissas and thus contains an infinite number of different variate values any one of which has zero a priori probability of being drawn. Theoretically, therefore, samples from continuous populations will yield no difference-scores of zero and no tied observations since the first-drawn member of a group of equal observations can be considered to predesignate the remainder. In practice, however, it is possible for zero differences or tied observations to occur, despite the continuousness of the sampled variate, because no measuring instrument is capable of infinite precision, i.e., because the population of potential measurements upon the continuous variate is a discrete one. (For example, although time is continuous, an infinite number of reaction times, measured with a time clock graduated in hundredths of a second, would have a discrete distribution.) Thus the validity of the continuity assumption does not obviate zeros and ties. What it does do is to confine their "existence" to the sample (assuming, of course, that the population of interest is that of the continuous variate and not the population of potential measurements upon it). So if the continuity assumption is true and zeros or tied observations do not *happen* to occur among the sample measurements, the necessary assumptions in A_s regarding zeros and ties may be regarded as met.

3.3.1 Treatment of Zeros and Ties Under Continuity

Suppose that the variate of interest is continuously distributed and that the experimenter, being aware of this fact, adopts an A_p containing the continuity assumption, but that zero difference-scores or tied observations occur due to imprecision of measurement. In that case the zeros or ties represent variate values which actually differ but whose differences have been obscured by an artifact of measurement. Everything that needs to be known about the existence, relative frequency, location, and extent of zero or tied scores is open to the view of the experimenter; they are intrinsic to the particular obtained sample and have no existence in the population of interest. Therefore, *the "true" value of the test statistic must be its value for one of the possible resolutions of ambiguous scores*, i.e., assignments of algebraic signs to zeros or of consecutive, tied-for ranks to equal observations. *Which one* of the possible resolutions represents the true condition is inexorably indeterminable, a fact which is squarely faced only by the following method.

Method A (Recommended): Obtain Probability Bounds. The test statistic is calculated twice: once assigning algebraic signs to zero difference-scores, or assigning tied-for ranks to equal observations, in the manner most conducive to rejection of the null hypothesis; once resolving such ambiguities in the manner least conducive to rejection. Let T be the test statistic and let α_M and α_L be the respective probability levels (i.e., cumulative probabilities under the null hypothesis) associated with the two calculated values of the test statistic. Then the test outcome is simply reported in the form $\alpha_M \leqq P(T) \leqq \alpha_L$. It has been said, with considerable justification, that if both bounds fall above, or if both bounds fall below, the adopted significance level α, there is no problem; if they do not, there is no solution. The statement applies to the set of data at hand, and a "solution" which is consonant with the philosophy of this method is to keep taking more data until both bounds fall on the same side of α.

For most tests only certain types of ties are "critical," in that the value of the test statistic varies with the assignment of tied-for ranks to tied observations. Sometimes only ties between observations in different samples are critical; in other cases it is only ties between observations adjacent in sequence. When all groups of equal observations represent noncritical ties, $\alpha_M = \alpha_L$ and the inequality reduces to an equality. In such cases the occurrence of equal observations has no detrimental effect whatever upon the test. When critical zeros or ties do occur, the only detriment to the test is the imprecision inherent in the inequality used to report the probability level.

3.3.2 Treatment of Zeros and Ties Under Continuity and Additional Conditions

All but one of the methods dealt with in this section assume "equiprobability," i.e., assume that when the null hypothesis H_0 is true each of the possible resolutions of ambiguous data (i.e., each of the possible assignments of algebraic signs to zero difference-scores or of consecutive tied-for ranks to equal observations) has the same a priori probability of occurring, and therefore, if H_0 is true, is equally likely to represent the true situation. Thus, in the case of difference-scores, equiprobability demands that, when H_0 is true, difference-scores measured as zero be as likely to represent true plusses as true minuses; and this will be the case if, in the portion of the difference-score distribution whose abscissas are measurable as zeros, the area corresponding to truly negative difference-scores is equal to that corresponding to truly positive difference-scores. The equiprobability assumption and, in fact, the additional assumptions (besides continuity) of all of the methods outlined in this section will be met if the following broadly sufficient conditions are met: (1) in the case of tests based on signs of difference-scores, *the difference-score population is symmetric about zero* when H_0 is true, or, in the case of tests based on ranks, *all of the sampled populations are identical* (and unchanging) when H_0 is true; (2) although perhaps imprecise, *measurements are unbiased*, i.e., there may be variable error but there is no constant error in measurements. The null hypothesis itself may state or imply all or a portion of the circumstances necessary to insure the validity of condition (1). In what follows, in order to avoid repetitiousness of qualifications, it will be assumed that measurements are unbiased.

Symmetry of the difference-score population about zero under H_0 may sometimes be inferred from knowledge about the forms of the "contributing" populations, especially when H_0 is true. Difference-scores of the type $X - K$, where K is a constant, will be symmetric about zero if the X-population is symmetric and K is its median (as perhaps stated in H_0). Difference-scores of the type $X - Y$ will be symmetric about zero if either (1) the X- and Y-populations are both symmetric and have a common axis of symmetry (in which case for every difference $D = X - Y$ there is an equally probable difference of the same magnitude but opposite sign $-D = X' - Y'$ where X and X' are symmetrically opposite and equally probable values in the X-population, and similarly for Y and Y' in the Y-population); (2) the X- and Y-populations are identical (in which case "X" and "Y" may be regarded as entirely arbitrary, and therefore interchangeable, labels so that $P(X - Y) = P(Y - X) = P[-(X - Y)]$, and if $D = X - Y$, $P(D) = P(-D)$). Knowledge that the latter situation exists, under H_0, may occur quite naturally in practice. For example, in certain tests of randomness against trend alternatives, X and Y are observations drawn at different times from a generating "process," and if H_0 is true the "process" has not changed and the X- and Y-"populations" are therefore identical. (See Cox and Stuart's Sign test for trend.)

That the equiprobability assumption is met, for tests based on ranks, when all sampled populations are identical and unchanging can be seen as follows. Suppose that samples are drawn from two populations, X and Y, and that, due to imprecision of measurement, the first, X_1, and fifth, X_5, observations recorded in the X-sample and the fourth, Y_4, observation recorded in the Y-sample all were measured as having the value 62.5 and are therefore collectively tied for the ranks 7, 8, and 9. If the populations do not change with time, then the subscripts, indicating order in which recorded, are arbitrary labels, unrelated on an a priori basis to the *values* of the observations for which they stand. And if the X- and Y-populations are identical, this is also true of the designations "X" and "Y." So X_1, X_5, and Y_4 are arbitrary labels for observations drawn from essentially identical populations, and since they are unrelated (a priori) to observation values they are interchangeable in probability statements about observation values. Thus, considering X_1, X_5, and Y_4 to represent the unknown "true," i.e., untied, values of the observations for which they stand, $P(X_1$ has rank 7$) = P(X_5$ has rank 7$) = P(Y_4$ has rank 7$)$, and likewise $P(X_1 > X_5 > Y_4) = P(X_1 > Y_4 > X_5) = P(X_5 > X_1 > Y_4) = P(X_5 > Y_4 > X_1) = P(Y_4 > X_1 > X_5) = P(Y_4 > X_5 > X_1)$ so that $P(X_1$ has rank 7, X_5 has rank 8, and Y_4 has rank 9$)$ has the same value for all six of the possible interchanges of position among X_1, X_5, and Y_4. But this is the same as saying that each of the possible resolutions of ties is equally likely, a priori, to have occurred, which is what is assumed by "equiprobability."

None of the following methods identifies the one "true" resolution of ambiguous data, and the consequence of their inability to do so is a loss of power in the long run. Some of the methods are so intrinsically crude and makeshift as to distort the null distribution of the test statistic: when H_0 is true the method may give the test statistic a value or may report a probability level corresponding to none of the values or cumulative probabilities in the null distribution of the test statistic. This may open the way for logical absurdities (as shown in [7] for a method not discussed herein), it may completely defeat the method's own purpose [2], or it may bias the test (see [12]).

Method B (Recommended When Its Assumptions Are Met): Drop the Zeros and Reduce n Accordingly. If n_0 of the the n difference-scores are zeros, they are eliminated from the data, and the test statistic is based upon the algebraic signs and sample size of the $n - n_0$ remaining difference-scores. This method assumes that in the difference-score population the ratio of measurably positive to measurably negative difference-scores is the same as that of truly positive to truly negative difference-scores. This assumption is not quite the same as equiprobability, but is met under the same sufficient conditions of unbiased measurement and symmetry about zero for the difference-score population. When its assumptions are met, this method results in an exact and valid test upon a sample of reduced size, which reduction therefore diminishes the power of the test (at least in the long run). The loss of power under this method, however, has been found to be less than that suffered under methods C or E [3, 8].

Method C: Randomize. Randomly assign a plus or a minus to each zero difference-score or randomly assign to equal observations the ranks they would have if not tied, i.e., if differing very slightly. In the case of the zeros, the assignment of signs might be made on the basis of a coin toss, which, since $P(\text{Head}) = P(\text{Tail}) = .5$, introduces equiprobability (and, of course, for analogous reasons equiprobability is also involved in the assignment of ranks). The primary virtue of this method is that it gives the test statistic the same distribution, when the null hypothesis and the equiprobability assumption are true, that it would have if the continuity assumption had not been violated, i.e., it does not distort the null distribution of the test statistic. Its main disadvantage is that instead of using an "intuitively reasonable" estimate of the true value of the test statistic (such as the average, or most probable, of its possible values) it permits pure chance to make the determination. This is particularly unnerving when an extreme chance event (i.e., assignment) calls for rejection (or acceptance) whereas a less extreme chance event would not have.

Method D: Obtain Some Sort of Average Probability. Resolve ambiguities in every possible way and for each way calculate the test statistic and ascertain its cumulative probability. The average, median, or midrange of these probabilities is taken to be the probability level of the test. (By considering the average to be a weighted average with all weights equal to one, we see that the equiprobability assumption is implicit in this method.) If the average or median is used the method tends to be extremely tedious (and much more complicated than it sounds) unless the ambiguities are quite limited in both frequency and extent. In order to obtain the midrange probability, however, it is only necessary to proceed as in Method A and calculate $(\alpha_M + \alpha_L)/2$. Probabilities obtained by these methods may not equal the probability level corresponding to any of the possible resolutions of ambiguous data, and, in that sense, they are distorted.

Method E: Count Half the Zeros as Positive, Half as Negative. (The sign of an odd zero is decided by the flip of a coin.) The equiprobability assumption is more immediately apparent in this method than in any other. This method is based upon the supposition that its obtained probability level will approximate the average probability. It therefore seeks to accomplish roughly the same objective as Method D without resorting to the excessive labor required to accomplish it exactly.

Method F: Use Midranks. Assign to each member in a group of equal observations the average of the set of consecutive ranks for which they are collectively tied. (This introduces the equiprobability assumption for essentially the same reasons that it occurs in Method D.) Thus the members of a group of equal observations are assigned the average of the consecutive ranks the members of that group would have if finer measurements upon them disclosed that they were not intrinsically equal. This method is based upon the implicit supposition that the use of the average rank tends to give the test statistic about the average of the values it would assume under all possible resolutions of ties. Thus, as does Method E, it attempts to accomplish on the basis of an assumed relationship what Method D (using the average probability) does directly. For some tests this method completely defeats its own purpose, causing the test statistic to assume the largest or smallest, rather than something close to the average, of the values it could have under all possible resolutions of ties [2, pp. 168–176]. For practically all tests, midranks violate some necessary assumptions (in A_s); they are likely to be nonintegral, and a given midrank occurs more than once. Therefore, using midranks alters the distribution of the test statistic to a degree which depends upon the number and upon the sizes of the group of equal observations. Thus the distribution of the test statistic is conditional upon an idiosyncrasy of the sample. In rare cases tables of probabilities have been prepared giving the exact sampling distribution of the test statistic conditional upon so many sets of two tied ranks, so many sets of three, etc., when midranks have been used. In all other cases one must either refer the test statistic to conventional tables which do not allow for the ties and are therefore inexact in this context, or, when sample sizes are large, one may sometimes use "asymptotic" formulas corrected for ties. Frequently the use of midranks reduces the variance, but does not affect the mean, of the null distribution of the test statistic, and when sample sizes are infinite that distribution often becomes normal. In such cases "asymptotic" probabilities, conditional upon the obtained ties in the same sense as above, may be found by referring to the unit normal distribution a critical ratio based upon a variance corrected for the number and extent of ties that actually occurred. However, while formulas and tables corrected for ties take account of the distortion in the test statistic's distribution, they do not resolve the ambiguities that caused it.

3.3.3 Critique of Preceding Methods

All of the methods suffer from the same disadvantage—the true situation underlying the ambiguous data is indeterminable and the price of this indeterminateness is an effective loss of power. Method A loses power because, when the true probability level lies between α_M and α ($< \alpha_L$), the test does not reject although rejection is appropriate. Method B loses power because of the reduction in sample size. Neither method attempts to squeeze information out of the unknowable, so while their power is reduced, their validity is unimpaired. Methods C through F resolve ambiguities in a way which (in effect) assumes that H_0 is true, thus tending to force the test statistic to have its null distribution. This reduces power without taking that reduction into account, as do Methods A and B. The consequences of these various procedures can be seen by considering the case where all difference-scores are zero or all observations have the same value. This can happen when H_0 is false, as well as when it is true, if the measuring instrument is sufficiently imprecise. It will be assumed that sample size is large enough to permit rejection if no ambiguities had occurred and that α is one of the standard tail probabilities. Under Method A, α_M and α_L would fall on opposite sides of α, and under Method B sample size would be zero. In both cases it would be impossible either to accept or reject, which is a consequence entirely appropriate to the situation. Under Method C the actual probability of rejection would be α, irrespective of whether H_0 were true or false. Thus the power of the test would be no greater for a false hypothesis than for a true one. Under Methods D, E, and F, for most tests, the test would always accept H_0, even if it were quite false. Thus the power of the test would be zero. Furthermore, these latter three methods tend to distort even the null distribution of the test statistic.

Method A makes a probability statement that is exactly true of the entire sample. Method B yields a probability statement that is exactly true of the unambiguous part of the sample. Methods B through F yield probability levels that are only *estimates* of the exact probability level for the entire sample, without, however, accompanying them with any indication of the *precision* of estimate. (The reporting of a single probability level implies an accuracy that these methods do not possess.) For all these reasons, the order of excellence of the various methods corresponds roughly to their alphabetical order, and it is suggested that only Methods A, B, and C be used, C being used only in the discrete case (see the next section).

3.4 THE DISCRETE CASE

Up to this point, we have assumed that the sampled populations are continuous and that any equal observations or zero differences are due to imprecision of measurement. Suppose, however, that they are due instead to intrinsic discreteness in the sampled population. So far as the sample is concerned, the situation is the same. This point can be best appreciated by considering three situations: (a) the variate of interest is a continuously distributed variate V, but zeros or ties occur because of imprecision of measurement; (b) the variate of interest V_m is the measured value of V, so the distribution of interest is the discretely distributed population of all possible measurements upon the varying values of V; (c) the variate of interest is not a measurement but rather is an underlying variate whose distribution is intrinsically discrete and just happens to be identical to that of V_m, and measurement is always precise enough to identify the discrete values correctly. (Notice that in proceeding from (a) to (b) we have moved from the continuous to the discrete case merely by transferring the locus of the experimenter's interest.) The distributions of measurements under the three situations are identical, consequently there is nothing in a set of sample observations that distinguishes one situation from another. Furthermore, if one of the methods outlined in Section 3.3 is used upon the zero differences or equal observations, the numerical test outcome will be exactly the same for a given set of sample observations, irrespective of which one of the three situations obtained. And irrespective of whether H_0 is true or false, and of whether assumptions are met or failed, the *distribution* of the test statistic will be the same under all three situations. (In order for this statement to apply, when Method A is used, the "test statistic" must be a single value rather than a set of values, so the statement holds if we regard the test statistic as the value corresponding to either α_L or α_M.) All of these considerations point to the conclusion that the discrete case can be treated in the same way as the continuous case.

The important difference between the two cases is that if the variate of interest is continuous, zeros or ties are sample-located artifacts which merely obscure a true situation which corresponds to one of the possible resolutions of zeros or ties. But if the variate of interest is discrete (and measurement is precise enough to identify the discrete values correctly), the true situation is not a resolution at all. It is the obtained set of sample data including zeros or ties, i.e., the zeros or ties are "real." Furthermore, the pattern, i.e., "existence" and relative frequencies, of tied or zero observations in the sample need not correspond even roughly with that of their counterparts in the population, especially when sample size is moderate. Thus, without the continuity assumption the bounds of Method A cannot be considered as absolute upper and lower bounds for the "true value" of the test statistic. They can, however, be regarded simply as defining an interval containing an *estimate* of the true value of the test statistic, in which case the probability statement becomes $\alpha_M \leqq \widehat{P(T)} \leqq \alpha_L$. But whereas Method A becomes more vague, and therefore less appropriate, in the discrete case, Method B, i.e., dropping the zeros, makes more sense in the discrete case where the zeros are intrinsically unsigned than in the continuous case where they represent truly signed, nonzero values. The remaining methods assume equiprobability in the continuous case, which becomes meaningless in the discrete case. However, while we cannot directly reconcile equiprobability with discreteness, we can simply omit continuity from the sufficient assumptions that guarantee equiprobability, thereby, in a sense, removing a "pure continuity" factor from the conditions which justify the use of the methods.

It would appear, therefore, that if all sufficient assumptions other than continuity are met (including the sufficient "method assumptions" of unbiased measurement, symmetry of the difference-score population about zero, under H_0, for tests based on signs, and identical and unchanging populations, under H_0, for tests based on ranks), distribution-free tests should be just as *valid* when zeros or ties, treated by any of the Methods B through F, are due to intrinsic discreteness as when they are due to imprecision in measuring a continuous variate. (The loss of power, however, *may* be greater.)

3.5 EFFICIENCY AND RELATED CHARACTERISTICS

Certain mathematical properties of a test are important in evaluating its usefulness. The **power** of a test is the probability of its rejecting a specified false hypothesis. (It is equal to $1 - \beta$, where β is the probability of committing a Type II error—failing to reject a false null hypothesis.) Power, then, depends upon at least four variables: (a) the amount by which the hypothesis is in error, i.e., the size of the discrepancy δ between the hypothesized and true condition; (b) the size α of the significance level chosen; (c) the location of the rejection region, e.g., whether the test is one-tailed or two-tailed; (d) the sizes of the samples used in the test. A **power function** is a curve in which all but one of these variables are held constant and power $1 - \beta$ is plotted as ordinate against that one variable, usually δ, as abscissa.

A test is **unbiased**, for a given alternative, if the probability of rejecting the null hypothesis is greater when the alternative hypothesis is true than when the null hypothesis is true.

A test is **consistent** for a given alternative to the null hypothesis if, when that alternative hypothesis is true, the probability of rejecting the false null hypothesis, i.e., the power of the test, approaches 1 as the sample size n on which the test is based approaches infinity. The test is consistent with respect to a class of alternatives if it is consistent for each of the alternatives of which the class is composed.

Efficiency is a relative term comparing the power of one test with that of a second test which acts as a standard of comparison and is often the most powerful test for the conditions under which the comparison is made. The **relative efficiency** of test A with respect to test B is generally defined as b/a, where a is the number of observations required by test A to equal the power of test B based on b observations, when both statistics test the same null hypothesis H_0 against the same alternative hypothesis H_a at the same significance level α (both either one-tailed or two-tailed). Relative efficiency therefore depends upon, and is implicitly qualified by, the comparison statistic used, the size of sample upon which it is based, the alternative hypothesis, the significance level, and the location of the rejection region. It is an index which is both highly peculiar to experimental test conditions and highly realistic to them. (It is also highly peculiar to the mathematical procedures used to obtain it; other perfectly realistic definitions of relative efficiency, based on slightly different procedures, may lead to quite contrasting results. See [1] and Sign test for the median.)

By far the most common index of efficiency for distribution-free tests is **asymptotic relative efficiency,** abbreviated A.R.E., and sometimes called Pitman efficiency. It is defined roughly as follows. Let A and B be two tests based on a and b observations respectively, each test statistic being asymptotically normally distributed (i.e., having a distribution which becomes normal when sample sizes are infinitely large), and each testing the same null hypothesis H_0 against the same class of one-sided alternatives $H_a > H_0$, against which both tests are consistent. The A.R.E. of A with respect to B is the limiting value of the ratio b/a as a is allowed to vary in such a way as to give A the same power as B while simultaneously b approaches infinity and H_a approaches H_0. The purpose of the "approach" of H_a to H_0 is to prevent the powers of each test from assuming a limiting value of 1 which they would otherwise do since at extremely large sample sizes the power of a consistent test against a *fixed* alternative is virtually 1. The approach is achieved by letting $H_a = H_0 + (K/b^r)$ where K and r are constants and b, the sample size of the comparison statistic, approaches infinity. Corresponding to H_a each test statistic will have an expected value E and a variance V, and as H_a approaches H_0 that expected value will change at a certain rate E'. In uncomplicated cases, the A.R.E. of test A relative to test B equals the limiting value, as b approaches infinity, of $(E_A')^2/V_A$ divided by $(E_B')^2/V_B$. For a formal mathematical definition of A.R.E., see [4, 10, 11].

The A.R.E. of test A relative to test B is related in an interesting way to the slopes of the power functions of the two tests or to their rate of change. Let P_A and P_B be points corresponding to an abscissa of H_0 on the respective power functions of the two asymptotically normal and consistent tests when the tests are based upon a common infinite sample size. Then not only do P_A and P_B have a common abscissa H_0, but also a common ordinate $1 - \beta = \alpha$, the common significance level of the tests. If the tests differ in sensitivity, however, then either the slopes, i.e., rates of change, of their power functions at the common point $P_A = P_B$, or the rate of change of the slopes, should differ. Let P'_A and P'_B be the respective slopes (i.e., the first derivatives of the power functions) and let P''_A and P''_B be the respective rates of change of P'_A and P'_B(i.e., the second derivatives) at the points P_A and P_B. If the tests are one-tailed, then the ratio P'_A/P'_B usually is the square root of the A.R.E. and the ratio of sample sizes b/a (where b is infinite) necessary to make $P'_A = P'_B$ is the A.R.E. itself. If the tests are two-tailed, then each power function has a minimum at H_0 and consequently the slopes P'_A and P'_B are zero. However, the ratio P''_A/P''_B is a meaningful index of limiting efficiency and is usually equal to the A.R.E. The A.R.E. may also be quite directly related to estimation efficiency; see Stuart [11].

A.R.E.'s are independent of α, the size of the significance level. Otherwise they are qualified by test conditions just as are the relative efficiencies for finite samples. They have the advantage that the qualifying conditions do not vary from one A.R.E. to another and are therefore, in a sense, standardized. But they suffer from the serious disadvantage that two of their qualifying conditions are totally unrealistic to experimental practice. A.R.E.'s represent the relative efficiency of the test under the conditions (implicit in their derivation) that sample size is infinite, that the alternative hypothesis is essentially identical to the null hypothesis, and that the test is one-tailed. They are further qualified by the comparison statistic and by whatever compromises or adjustments were made in the usual forms of the null hypotheses tested by the two statistics in order to equate them and thereby render the tests comparable in this respect. (For example, a test for identical populations may be compared with a test for equal means by holding shapes and variances constant under both H_a and H_0, so that the only form that nonidentity can take is a difference between means. Under these conditions a test for nonidentity is artificially forced to become a test for unequal means for purposes of comparison. The A.R.E. ignores the fact that one test statistic, or one form of the null hypothesis, may be more relevant to the experimenter's objectives than the other, i.e., may be better suited to the inferences which the experimenter wishes to extend to the original physical problem.)

Clearly, it is precisely under the conditions that qualify it that the A.R.E. has the least practical utility. No experimenter takes infinitely large samples, and virtually no one is interested in power to reject hypotheses that differ only infinitesimally from the null hypothesis. Furthermore, since the power of any consistent test approaches 1.00 as sample size approaches infinity and since A.R.E. applies only to consistent tests in the case of infinite sample sizes, it follows that a test with a very low but nonzero A.R.E. has as much power at infinite n as could be desired to reject *any fixed* $H_a \neq H_0$. Thus, while relative efficiency is realistic but not sufficiently general, A.R.E. is general (at least in the sense of being "standardized") but not sufficiently realistic. Fortunately, however, when relative efficiencies are obtained for a variety of tests using a common comparison statistic and comparable conditions, the order of excellence into which the tests are placed by their small-sample relative efficiencies tends to be about the same as that into which they are placed by their corresponding A.R.E.'s when the same comparison statistic is used. Thus, while the practical meaningfulness of the A.R.E. *magnitudes* is rather dubious, there does appear to be a good rank-order correlation between A.R.E.'s and small-sample relative efficiencies representing comparable tests and test conditions. In short, there is nothing sacred about the A.R.E.; however, viewed in its proper perspective, it can serve as a useful index.

Many of the A.R.E.'s for distribution-free tests use a classical parametric test as the comparison statistic. In evaluating such A.R.E.'s, several important qualifications should be borne in mind: (1) Often the comparison was made under common conditions which met all of the assumptions of the parametric test, thus making conditions "optimal" for the comparison test while giving no comparable advantage to the distribution-free test. The results simply reflect the fact that a test which is expressly designed for a certain highly specific situation is likely to be more effective in that situation than is a more general and more versatile test. (2) The two tests practically never were designed to test exactly the same null hypothesis (see next to last paragraph). They are often conceived as doing so only by the semantic device of designating the null hypothesis and its alternatives in very general terms. Two tests for different aspects of dispersion are, of course, both tests for "dispersion," but one may test for identical variances and the other for an aspect of dispersion implicit in the squares of ranks or in the interquartile range. Sometimes meeting all of the assumptions of the parametric comparison statistic makes the two hypotheses equivalent. For example, if distributions are normal, then a distribution-free test for the median becomes equivalently a test for the mean (because of the symmetry of the population) and can therefore be legitimately compared with Student's t test. However, in other cases, the equivalence is somewhat strained. (3) The A.R.E. is generally a lower bound for efficiency in the neighborhood of H_0. The relative efficiency of a distribution-free relative to a classical test is generally largest (often very close to 1.00) when samples are smallest, diminishing with increasing sample size toward a limiting figure represented by the A.R.E.

Table 3.5.1 gives the A.R.E.'s of many distribution-free tests relative to the most powerful parametric test or to another distribution-free test, often the most powerful one available. Many A.R.E.'s other than those listed can be easily calculated from those presented in the table. If the A.R.E. of test A relative to test C is A_C and the A.R.E. of test B relative to test C is B_C, then the A.R.E. of test A relative to test B is A_C/B_C (provided, of course, that all three tests are being used to test the "same" null hypothesis). Tests having zero

Table 3.5.1: *Asymptotic Relative Efficiencies of Some Distribution-Free Tests Relative To Tests That Are Optimal under Normal-Theory Conditions*

Distribution-Free Test	Comparison Test	Population Distribution: Normal	Uniform	Double Exponential	Logistic	Cauchy	Parabolic	Exponential	Bounds for A.R.E.: Lower	Upper
Tests of Location against One-Sided[a] Alternatives of Change in Location Only										
One-Sample or Two-Matched-Sample Tests										
Fisher's Randomization	t	1.000								
Fraser's Normal Scores (*NS*)	t	1.000								
Wilcoxon Signed-Rank	t	.955							.864	
Sign	t	.637	.333							
Wilcoxon Signed-Rank	*NS*	.955		1.178	1.047					
Sign	*NS*	.637		1.571	.785					
Two-Sample Tests										
Fisher's Randomization	t	1.000								
Terry-Hoeffding Normal Scores (*NS*)	t	1.000							1.000	
van der Waerden Inverse Normal Scores	t	1.000							1.000	
Wilcoxon Rank-Sum	t	.955	1.000				.864		.864	∞
Grouped-Data Wilcoxon	t	.637–.955								
Westenberg-Mood Median	t	.637								
Exceedances (r = Median)	t	.637								
Wald-Wolfowitz Number-of-Runs	t	0							0	
Wilcoxon Rank-Sum	*NS*	.955	0	1.178	1.047	1.413		0	0	1.910
Multi-Sample Tests										
Pitman's Randomization	F	1.000								
Kruskal-Wallis	F	.955	1.000						.864	
Friedman	F	.637–.955								
Brown-Mood Median	F	.637	.333							

		Two-Sample Tests of Dispersion against Alternatives of Change of Scale Only								
Klotz Normal Scores for Scale (*NS*)	*F*	1.000							0	∞
Mood	*F*	.760	1.000	1.08					0	∞
Siegel-Tukey Rank-Sum	*F*	.608	.60	.94					0	∞
Freund-Ansari Rank-Sum	*F*	.608	.60	.94					0	∞
Number-of-Runs	*F*	0							0	
Mood	*NS*	.760	0	.900	.896	1.670		0	0	
Siegel-Tukey	*NS*	.608	0	.774	.750	1.783		0	0	∞
		Two-Sample Tests of Independence against Alternatives of Common Elements								
Kendall	*r*	.912	1.000	1.266			.857			
Hotelling and Pabst	*r*	.912	1.000	1.266			.857			
Blomqvist	*r*	.405	.250	1.000			.316			
		One-Sample Tests of Randomness against Normal Regression Alternatives								
Mann-Kendall	*b*	.985								
Daniels	*b*	.985								
Cox and Stuart S_3	*b*	.827								
Cox and Stuart S_2	*b*	.782								
Median	*b*	.782								
Foster and Stuart Records	*b*	0							0	
Number-of-Runs-Up-and-Down	*b*	0							0	

[a]Except, of course, for multi-sample tests.

A.R.E. with respect to the same comparison test do not necessarily have zero A.R.E. with respect to one another. Also, as has already been implied, a test with zero A.R.E. may have good relative efficiency at small sample sizes and will have power close to 1.00 at very large sample sizes. (From the definition of A.R.E., it follows that all test statistics listed in Table 3.5.1 are consistent against the indicated alternative hypotheses and are asymptotically normally distributed.)

It is important to avoid a parochial view of power or efficiency which considers them to be an exclusive property of the statistical *test*, when actually they are, practically speaking, a property of the *experiment*. The power of an experiment depends on many things of which the selection of a statistical test is only one. The connection between the original problem and the probability level of the test statistic is often a loose one, involving a succession of relationships in the nature of substitutions, approximations, correlations, and implications. The inductive logic of the experimental process depends upon the closeness of approximation between the original problem and the aspect of it chosen for experimentation, between that aspect and its simulation in the laboratory, upon the relevance of the laboratory variables and hypotheses to their counterparts in the original problem, and upon the appropriateness of the statistical test to those laboratory variables and hypotheses. There is almost always some degradation or attenuation (other than the unreliability due to chance) in the process of proceeding through the various steps and substitutions of "relevant" variables or phenomena for their archetypes and finally back to the original problem by way of inductive inference; and any reduction in degradation at any point increases the power of the experiment. Often a deeper analysis of the general problem or a closer look at a proposed experimental plan will reveal opportunities for enormous increases in power (sometimes accompanied by spectacular reductions in cost) simply by increasing relevance or tightening the closeness of an approximation. It is not at all unusual to find that experimental power can be greatly increased by investigating a more relevant aspect of the problem for which a *less* statistically efficient test than the one originally proposed is now appropriate. For example, the objective of providing useful information to infantrymen may be far more satisfactorily met by testing hypotheses about the first percentile, than by testing hypotheses about the mean, of the population of time intervals between activation and explosion of grenades, even though the tests about means may be considerably more statistically efficient.

3.6 BASIC FORMULAS

Many of the tests that follow are based upon the elementary formulas associated with permutations, combinations, and the relative frequency definition of a probability. These terms are defined and their probability formulas are developed below.

3.6.1 Permutations

A **permutation** is a spatial arrangement or temporal sequence of objects or events with respect to one another. It is a relative order which is usually conceptualized as a linear arrangement, but which, spatially at least, can be an arrangement in more than one dimension. Specifically, if there are n different possible fixed locations or positions, each of which can contain only one object at a time, then each of the different possible allocations of n objects to the n positions is a different permutation of the n objects within the set of n positions.

In allocating the n objects to the n fixed positions, one could pick any one of the n objects to occupy the first position, so there are n ways of assigning an object to the first position. For each of these n ways, there are $n - 1$ ways of choosing an object from the remaining $n - 1$ objects and assigning it to the second position. So there are $n(n - 1)$ ways of assigning objects to the first two positions. For each of these ways there are $n - 2$ ways of choosing an object from among the remaining $n - 2$ objects for assignment to the third position, etc. So in all there are $n(n - 1)(n - 2) \ldots (3)(2)(1)$ or $n!$ different ways of assigning all n objects to the n different positions in such a way that each position is occupied by exactly one object. Therefore *the number of different possible permutations of n things is n!.*

3.6.2 Distinguishable Permutations

Suppose we had six billiard balls, numbered 1 to 6, occupying six different fixed positions. Then there would be 6! or 720 different possible permutations of the six balls within the set of six fixed positions. Now if we were to paint two of the balls orange and three of them green, thereby obscuring their numbers, there would *still* be 6! different possible permutations, but not all of them would be visually distinguishable from each other. Specifically, those permutations differing from each other only because same-colored balls had exchanged positions with each other would be indistinguishable. Consider any visually distinguishable permutation. No matter where they are located, the two orange balls can be switched in position without creating a new distinguishable permutation, so the distinguishable permutation of the six balls remains the same for each of the 2! permutations of the two orange balls within the same two positions. And for each of these 2! permutations of the orange balls, the three green balls can be permuted within their three positions in 3! ways without changing the distinguishable permutation of the superordinate set of six balls. So for every *distinguishable* permutation there are (2!)(3!) permutations of the six balls that are *in*distinguishable from each other. Therefore if the number of distinguishable permutations is d, then the number of indistinguishable permutations is

$$(2!)(3!)d = 6! \qquad \text{and} \qquad d = (6!)/[(2!)(3!)]$$

In general, then, let there be n things of k different distinguishable kinds, the members of each single kind being all alike and therefore indistinguishable from each other, and let n_i be the total number of things of the ith kind. Then for each distinguishable arrangement or permutation of the n things, the n_1 things of the first kind can be permuted among themselves in $n_1!$ ways without changing the appearance of the overall arrangement. For each of these ways the n_2 things of the second type can be permuted among themselves in $n_2!$ ways without altering the appearance of the overall arrangement of the n things, etc. So for every distinguishable arrangement or permutation of the n things there are

$$(n_1!)(n_2!) \ldots (n_i!) \ldots (n_k!) \quad \text{or} \quad \Pi_{i=1}^{k} n_i!$$

*in*distinguishable permutations for which the overall arangement of the n things has the same appearance. Therefore, the total possible number of permutations $n!$ must equal the number of distinguishable permutations times $\Pi_{i=1}^{k} n_i!$. And the total number of *distinguishable* permutations of the n things within their n fixed positions must be

$$\frac{n!}{(n_1!)(n_2!) \ldots (n_i!) \ldots (n_k!)} = \frac{n!}{\Pi_{i=1}^{k} n_i!}$$

3.6.3 Combinations

A **combination** is a subset of objects or events whose order of arrangement within the subset is immaterial. Thus a particular combination is defined by the aggregation of the individual identities of the objects or events of which it is composed, but not by their relative positions within the subset.

The number of different combinations, or subsets, of r objects that can be drawn from a set of n objects can be obtained as follows. Imagine the n objects arranged in a horizontal line, and suppose that the leftmost r of them are a sample of size r drawn from the n. There are $n!$ possible different permutations of the n objects within their n fixed positions, and during the course of these $n!$ permutations every object will occupy every position and will do so with the other $n - 1$ objects taking every one of the $(n - 1)!$ possible relationships to it. Therefore the leftmost r positions will surely receive every possible different combination of r objects. Consider now a single overall permutation of the n objects. The contents of the leftmost r positions will be one of the combinations whose total number we seek. However, the $n - r$

objects in the remaining $n - r$ positions can be permuted in $(n - r)!$ ways without altering the r objects in the r leftmost cells, and each of these ways results in a different one of the $n!$ permutations of the overall set of n objects. Furthermore, for each of these $(n - r)!$ ways there are $r!$ ways in which the spatial order of the r leftmost objects can be changed without changing their identities, i.e., there are $r!$ ways of permuting the r leftmost objects within the r leftmost positions without changing the combination of objects involved. Each of these ways of rearranging the leftmost r objects while holding constant the arrangement of the $n - r$ rightmost objects results in a different overall permutation of the n objects. Therefore, for every unique combination of r objects in the r leftmost positions there are $r!\,(n - r)!$ different possible permutations of the overall set of n objects. So, since there are $n!$ permutations in all, the number of different combinations of r objects in the r leftmost positions must be

$$\frac{n!}{r!(n-r)!}$$

which is denoted by the symbol $\binom{n}{r}$. The qualification about the r leftmost positions merely identifies our method of sampling. By permuting the objects and always investigating the r leftmost positions, we have let the sample come to us. Had we held the locations of the n objects constant and then sampled from all combinations of location, we would have achieved the same result. Therefore, the number of different combinations of r things that can be drawn from a set of n things, or as it is usually expressed, *the number of combinations of n things taken r at a time*, is

$$\binom{n}{r} = \frac{n!}{r!\,(n-r)!}$$

Since the contents of the complementary subset of $n - r$ things is completely dependent upon the contents of the subset of r things, this is also the number of different possible ways of dividing a set of n things into nonoverlapping subsets of r and $n - r$ things.

The method can be extended to the case of more than two subsets. Suppose that we wish the number of different possible ways of dividing a set of n objects into nonoverlapping subsets of $n_1, n_2, \ldots, n_i, \ldots, n_k$ objects, where $\sum_{i=1}^{k} n_i = n$. Again, imagine the n objects arranged in a horizontal line. Let the first n_1 positions identify the first subset, let whatever objects occupy the immediately following n_2 positions be defined as the second subset, etc., and let the last n_k objects be the kth subset. For any given permutation of the n objects, the first n_1 can be permuted within their n_1 positions in $n_1!$ ways without changing the contents of the first subset. And for each of these ways, the n_2 objects in the next n_2 positions can be permuted within those positions in $n_2!$ ways without altering the constitution of the second subset, etc. So for every unique division of the n objects into subsets of the specified sizes, there are

$$(n_1!)(n_2!)\ldots(n_i!)\ldots(n_k!) = \Pi_{i=1}^{k} n_i!$$

different permutations of the overall set of n objects that can be generated by varying *only* the order of arrangement of the objects within their subsets. And since the total number of different possible permutations is $n!$, the total number of different possible divisions of the n objects into subsets of sizes $n_1, n_2, \ldots, n_i, \ldots, n_k$, where order of arrangement of objects within their subset is immaterial, must be

$$\frac{n!}{(n_1!)(n_2!)\ldots(n_i!)\ldots(n_k!)} = \frac{n!}{\Pi_{i=1}^{k} n_i!}$$

Notice that this is the same expression obtained earlier for the number of distinguishable permutations of n objects of k different distinguishable kinds, the n_i objects of the ith kind being indistinguishable from each other and this being the case for every value of i. In the earlier situation, it was certain objects that were indistinguishable. In the present case, it is, in effect, certain positions that are indistinguishable. But since the total number of objects equals the total number of positions, and since the subset sizes for objects are the same as the subset sizes for positions, we arrive at the same mathematical expression.

The expression

$$\frac{n!}{\Pi_{i=1}^{k} n_i!} = \frac{n!}{(n_1!)(n_2!)\ldots(n_i!)\ldots(n_k!)}$$

giving the number of different subdivisions of n things into subsets of sizes $n_1, n_2, \ldots, n_k$ is known as the **multinomial coefficient,** while the expression

$$\binom{n}{r} = \frac{n!}{r!\,(n-r)!}$$

giving the number of combinations of n things taken r at a time is known as the **binomial coefficient,** the latter being a special case of the former resulting when $k = 2$.

3.6.4 Probability of an Outcome

If an event can occur in N different equally likely ways and if exactly n of those ways cause it to have the outcome A, then the a priori probability of the outcome A is $P(A) = n/N$. Usually the qualification that ways be equally likely can be met by defining a way so narrowly that it represents only a single possibility, i.e., only a single one of the configurations of circumstances that produce A, or by making certain that each way represents the same number of different possibilities.

REFERENCES

1. Blyth, C. R., "Note on Relative Efficiency of Tests," *Annals of Mathematical Statistics*, **29** (1958), 898–903.
2. Bradley, J. V., *Distribution-Free Statistical Tests*, Wright Air Development Division Technical Report 60-661, Aerospace Medical Research Laboratories, Wright-Patterson Air Force Base, Ohio, August 1960.
3. Hemelrijk, J., "A Theorem on the Sign Test when Ties are Present," *Proceedings Koninklijke Nederlandse Akademie van Wetenschappen, Series A*, **55** (1952), 322–326.
4. Noether, G. E., "On a Theorem of Pitman," *Annals of Mathematical Statistics*, **26** (1955), 64–68.
5. Noether, G. E., "The Efficiency of Some Distribution-Free Tests," *Statistica Neerlandica*, **12** (1958), 63–73.
6. Owen, D. B., *Handbook of Statistical Tables*. Reading, Mass.: Addison-Wesley Publishing Company, Inc., 1962.
7. Pratt, J. W., "Remarks on Zeros and Ties in the Wilcoxon Signed Rank Procedures," *Journal of the American Statistical Association*, **54** (1959), 655–667.
8. Putter, J., "The Treatment of Ties in Some Nonparametric Tests," *Annals of Mathematical Statistics*, **26** (1955), 368–386.
9. Savage, I. R., *Bibliography of Nonparametric Statistics*. Cambridge, Mass.: Harvard University Press, 1962.
10. Stuart, A., "The Asymptotic Relative Efficiencies of Tests and Derivatives of their Power Functions," *Skandinavisk Aktuarietidskrift*, **37** (1954), 163–169.
11. Stuart, A., "The Measurement of Estimation and Test Efficiency," *Bulletin of the International Statistical Institute*, **36** (1956), Part III, 79–86.
12. Thornton, G. R., "The Significance of Rank Difference Coefficients of Correlation," *Psychometrika*, **8** (1943), 211–222.

4

FISHER'S METHOD OF RANDOMIZATION

Consider the following data table containing two X observations and two Y observations, the observations in each column being recorded in the sequence in which they were obtained:

X	Y
3	5
9	1

By simply permuting the four observation-values within the four cells of the table, we can generate the following set of twenty-four tables containing the same four values but in different arrangements:

1	5
3	9

1	9
3	5

3	5
1	9

3	9
1	5

1	3	1	9	5	3	5	9
5	9	5	3	1	9	1	3
1	3	1	5	9	3	9	5
9	5	9	3	1	5	1	3
3	1	3	9	5	1	5	9
5	9	5	1	3	9	3	1
3	1	3	5	9	1	9	5
9	5	9	1	3	5	3	1
5	1	5	3	9	1	9	3
9	3	9	1	5	3	5	1

Under the conditions described by certain null hypotheses and assumptions, all or some of the twenty-four tables were just as likely a priori to have been obtained as was the actually obtained table. For example, if H_0 is that the X- and Y-variates are identically distributed, then if H_0 is true and sampling is random and independent, *all* of the twenty-four tables have the same a priori probability. For in that case neither the sample to which an observation belongs nor the sequence in which it was drawn helps to identify which of the four values it will have. As a second example, suppose that the X- and Y-observations in the same row of the original table are measurements upon a single unit drawn from a bivariate XY-population. Then if H_0 is that X- and Y-measurements are independent and if sampling is random and H_0 is true, each of the four tables in the fifth row of the above matrix of twenty-four tables had equal a priori probability of becoming the actually obtained data. This is the case because the H_0 of independent variates does *not* imply identically distributed variates, which restricts us to tables having the same sets of values in the X-column and in the Y-column as originally recorded. The H_0 *does* imply that any of the possible reassignments of the original Y's to original X's has equal a priori probability, and the four tables in the fifth row of the matrix represent just such reassignments.

Next consider the following three observations:

$$4 \quad -7 \quad 9$$

By simply varying the algebraic signs of the three numbers, we can generate the following eight sets of three observations each:

−4	−7	−9	4	−7	−9	−4	7	−9	−4	−7	9
4	7	9	−4	7	9	4	−7	9	4	7	−9

Now if the three observations are a random sample from a single population and H_0 is that the sampled population is symmetrically distributed about a mean of zero, then if H_0 is true all eight of the above sets had the same a priori probability of becoming the actually obtained sample of data. For in that case the algebraic sign of any observation's absolute value is as likely to be positive as negative, and any of the distinguishable assignments of algebraic signs to absolute magnitudes was as likely as any other to represent the data.

Once having determined the sets of data that are equally likely under H_0 and the assumptions, one can choose a test statistic T and calculate its value for each such set. If a proportion p of these calculated T values are as extreme or more so as the T value for the actually obtained set of data, then an event having probability $\leq p$ under H_0 has occurred. If T tends to take these extreme values under an alternative condition against which one wishes to guard, and if p is less than or equal to a preselected significance level α, one can appropriately reject H_0. For example, in the first case considered if one wished to guard against alternatives to the H_0 of identically distributed populations in which the X-population has a larger mean than the Y-population, one may choose $\bar{X} - \bar{Y}$ as the test statistic and reject when it takes values that are "too large." The test statistic $\bar{X} - \bar{Y}$ has the same value for all four tables in each row of the matrix of twenty-four tables, having the values −5, −3, 1, −1, 3, and 5, for all tables in the first, second, third, fourth, fifth, and sixth rows respectively. And its value is 3 for the actually obtained table. So 8 of the 24 tables or a proportion of $\frac{1}{3}$ of them yield values of $\bar{X} - \bar{Y}$ as great as or greater than that for the obtained table, and H_0 would be rejected only if one had chosen an α greater than .333. In the second case considered, if one wished to guard against alternatives in which the X- and Y-variates are negatively correlated, one might choose the correlation coefficient r as the test statistic and reject the H_0 of independent measurements if the obtained table was among a proportion α of tables yielding the most extremely negative r's. For the four tables in the fifth row of the matrix, i.e., for the tables that are equally likely under H_0, r has the values 1, −1, −1, and 1, and is −1 for the actually obtained table. So half of the four tables yield an r as small as or smaller than the one for the obtained table, and the H_0 of independent measurements would be rejected only if one had chosen an $\alpha \geq .50$. Finally, in the third case considered, if one wished rejection to be especially likely when the sampled population's mean exceeds zero, one might take the sample mean as the test statistic and reject when it lies among the proportion α of means having the largest values. For the eight equally likely sets of observations under H_0, the means of the sets, in the same order in which the sets were listed, are −6.67, −4.00, −2.00, −.67, 6.67, 4.00, 2.00, and .67; and

2.00 is the mean of the actually obtained set. So three of the eight sets, or 37.5% of them, yield a mean as great as or greater than that for the obtained set. Therefore, H_0 would be rejected only if the previously chosen value of α had been $\geq .375$.

At the risk of unpleasant redundancy, we have briefly sketched some applications of the Method of Randomization to be taken up in more elaborate detail in a later section of this chapter. Our purpose in doing so is to provide some concrete material with which to associate the abstract principle to be enunciated next and therefore to facilitate its comprehension.

4.1 THE PRINCIPLE

The logical basis for a large class of distribution-free tests is rooted in a principle first conceived by R. A. Fisher, the application of which is known as the Method of Randomization. The essence of Fisher's principle is as follows. For any set of obtained data we may distinguish between (a) the absolute magnitudes of the observations; (b) their algebraic signs; and (c) their "locations" in the data table identifying the combination of sampling conditions under which they were obtained (such as treatment under which recorded, population from which the measured unit was drawn prior to treatment, i.e., "matching" condition, etc.). Thus we may regard any set of obtained data consisting of N observations (perhaps distributed among several samples) as but one member of a whole family of distinguishably different sets of N observations *having the same absolute magnitudes* but differing with regard to algebraic signs, "locations," or both. Within this family of distinguishably different sets of data will be a family of sets all of which, if the null hypothesis and all assumptions are true, had the same a priori probability of being drawn as had the obtained set. By holding the N obtained absolute magnitudes constant and systematically varying algebraic sign or "location" in ways that are conscionable under the null hypothesis, assumptions, and other restrictive conditions (such as fixed sample sizes), we can "build" this family of sets of data, each member of which was just as likely as the *actually* obtained set to have been obtained if H_0 and the assumptions are true. We may then choose a test statistic and calculate its value for every member of this family, thereby obtaining its null distribution, at least for that family of events. We can now refer the value of the test statistic calculated from the actually obtained set of data to that null distribution and accept or reject in the usual fashion.

Actually, it only becomes necessary to distinguish between an observation's absolute and algebraically signed values when the null hypothesis and assumptions together imply that a sampled population is symmetric about zero. In that event, the algebraic sign for any given observation magnitude was just as likely, a priori, to be positive as negative. Therefore the family of equally likely sets of data, in the one-sample case where there are n observations, consists of the 2^n different sets corresponding to the 2^n distinguishable ways of assigning algebraic signs to the n obtained observation magnitudes. (There are two ways of assigning a sign to the first observation, for each of these ways there are two ways of assigning a sign to the second, making $(2)(2) = 4$ ways, etc., so that for the entire set one must take the product of n 2's.) This type of application, involving the variation of algebraic signs, accounts for a single important test and is unique in that one must consider not merely the redistributions of the *obtained* number of minus signs among the n observations, but rather of *all* numbers of minuses from zero to n.

In all other cases, it is necessary only to distinguish between the values (including algebraic sign) and the "locations" of the observations. And for these (rather than all) cases it becomes much easier to state the general principle. For any set of obtained data, consisting of a total of N observations, we may imagine the N observations to be recorded into N cells of a table whose rows, columns, blocks, etc., identify the relevant experimental conditions under which the recorded data were obtained (columns might correspond to treatments, rows to replications or to common sampling conditions). Thus the location of a cell identifies the conditions under which its contents were obtained. However, we wish to distinguish sharply between the cell and its contents, i.e., between the conditions and the observation measurement obtained under them. There are $N!$ different ways in which the N observations can be permuted within the N different cell locations, and if there are no tied observations, each such "way" or permutation results in a distinguishably different pattern of association between cells and observations and therefore in a distinguishably different "table." Let W be the number of these "ways," i.e., the number of patterns of association, that had equal a priori probability of becoming the actually obtained data under the null hypothesis that one wishes to test, assumptions one is able to make, and test conditions one knows to exist (such as collection of observations in "matched pairs"). It should be emphasized that the W ways that are equally likely under H_0 constitute a *sub*set of the $N!$ distinguishable ways, i.e., that W is not necessarily equal to $N!$. Now let T be a test statistic that is sensitive to the alternative hypothesis against which one wishes to guard, and let T be calculated for each of the W ways of reshuffling the data. (Although the W patterns are all distinguishably different, this will not necessarily be the case for the corresponding T's, i.e., there may be ties among the T's.) Finally, let w be the number of these W ways that result in a T whose value is as extreme, or more so, in the direction implied by the alternative hypothesis, as the T value for the data as originally recorded. Then, *given that* one's data consist of the N measurement values originally recorded, an event has occurred (the event being the originally obtained association between the N measurements and the N measured units) which belongs to a set of events having probability

w/W under the null hypothesis, but which should be considerably more likely under the alternative hypothesis. Thus one has fulfilled all of the requirements for a statistical test and the probability level for the T test, calculated from the data as originally recorded, is w/W.

Given that the null hypothesis and assumptions, together, do *not* imply that any observation was just as likely, a priori, to have values with the same absolute magnitudes but opposite algebraic signs, the above paragraph expresses the principle in a form sufficiently general to include all special cases. In specific types of application one can usually obtain the probability fraction w/W by investigating a far smaller set and subset of events than W and w (for example, in certain multi-sample tests, by excluding within-sample permutations and considering only those assignments that result in the various samples having distinguishably different *combinations* of the N observation-values).

Although not confined to this situation, the principle is easiest to grasp when the null hypothesis, assumptions, and test conditions together imply that all of several samples have been drawn from identical populations. In that case, the identical populations may be regarded as a single common population yielding a pooled sample whose "component samples" are merely arbitrary labels for groupings of observations. That is, the several samples may be regarded as one large sample whose observations have been randomly assigned to subsamples or component samples of the sizes actually drawn. Each of the different random assignments was, prior to sampling, equally likely to be the actually obtained set of data, if the null hypothesis of identical populations is true, but unequally likely (except, perhaps, for some highly special cases) to be if the null hypothesis is false.

The null hypothesis actually tested is that each of the ways of reshuffling the data, whose number is represented by the denominator of the probability fraction, had equal probability, prior to sampling, of identifying the data as actually obtained. However, "ways" can be unequally probable not because populations render them so but rather because of bias in sampling. And the cumulative probability fraction is generally calculated from formulas that regard all observations as distinguishable. Therefore random sampling must be assumed and the nonexistence of ties in the obtained data usually is also. Additional assumptions are sometimes required in specific types of application

4.2 APPLICATIONS

4.2.1 Pitman's Test for Correlation

Suppose that an X-observation and a Y-observation have been made on each of n randomly selected units and that it is desired to test the null hypothesis that the X- and Y-variates are independent. We may imagine the data recorded in a table with n columns, representing the n different units upon which paired measurements were taken, and two rows representing the X- and Y-measurements. There are, of course, $(2n)!$ possible permutations of the tabled observations. But since the null hypothesis of independence does not imply identically distributed variates, W must exclude permutations involving a change in an observation's row. Therefore, $W = (n!)^2$, since under H_0 each of the $n!$ within-row permutations of the X's is equally likely and for each of them, each of the $n!$ within-row permutations of the Y's is equally likely. But one of these two sets of permutations is "redundant," amounting, in effect, to permutations of entire columns of the table, so we may restrict our considerations to the $n!$ within-row permutations of the Y's under a single fixed permutation of the X's. Under the null hypothesis of independence, each of the $n!$ different patterns of association, corresponding to the $n!$ permutations of the Y's, was equally likely, a priori, to have become the obtained data. Suppose that for each of these $n!$ ways we had calculated the Pearson product-moment correlation coefficient r and that the frequencies of occurrence of the various values of r had been tabulated. We would then have the null distribution of r, given that the data consisted of the obtained set of n observations on X and n observations on Y. And if the r-value for the data actually recorded fell among the $n!\,\alpha$ most "extreme" of these $n!$ values of r, we could reject H_0 at the α level of significance (assuming, of course, that the value of α and the location of the rejection region, i.e., the definition of "most extreme," had been decided upon prior to sampling). More specifically, let L, S, and A be the number of the $n!$ "ways" that result in a correlation coefficient as large or larger, as small or smaller, and as great or greater in absolute value, respectively, as the correlation coefficient for the data as recorded. Then the probability levels for the obtained r as a test of the null hypothesis of independence against alternatives of positive correlation, negative correlation, and either correlation are $L/n!$, $S/n!$, and $A/n!$, respectively.

The formula for the Pearson product-moment correlation coefficient is

$$r = \frac{\sum (X - \bar{X})(Y - \bar{Y})}{\sqrt{\sum (X - \bar{X})^2 \sum (Y - \bar{Y})^2}}$$

and, since our $n!$ permutations of the obtained data include only the within-row permutations of the Y's "among themselves," the denominator of r is the same in all $n!$ cases. Therefore by using only the numerator as the test statistic we save labor and accomplish the same result as if we had used r; the two tests are mathematically equivalent. Furthermore, although r is the classical index of amount of correlation, the classical *test* for correlation is based upon the test statistic

$$t = \frac{r\sqrt{n - 2}}{\sqrt{1 - r^2}}$$

which is a monotonic increasing function of r. Therefore, the randomization tests based upon the numerator of r, the entire r, or

$$t = \frac{r\sqrt{n-2}}{\sqrt{1-r^2}}$$

as test statistics are all equivalent tests yielding identical probability levels.

As an example, suppose that an X-measurement and a Y-measurement have been made on each of five randomly drawn units and that the obtained data are as follows:

	Unit				
	I	II	III	IV	V
X-Measurement	29	32	39	40	60
Y-Measurement	5	6	11	8	20

The mean of the X-measurements is 40, and that of the Y's is 10, so replacing the original observations by their deviations from their sample means we obtain

	Unit				
	I	II	III	IV	V
$X - \bar{X}$	−11	−8	−1	0	20
$Y - \bar{Y}$	− 5	−4	1	−2	10
$(X - \bar{X})(Y - \bar{Y})$	55	32	−1	0	200

for which $\sum (X - \bar{X})(Y - \bar{Y}) = 286$. Now there are $n! = 5! = 120$ ways of permuting the entries in the second row of the table, all of which had equal a priori probability of being the "obtained" permutation under the null hypothesis of independence. But, aside from the actually obtained permutation, only one of them, the one obtained by interchanging the third and fourth cell entries in the second row, results in a positive value of $\sum (X - \bar{X})(Y - \bar{Y})$ as large as, or larger than, that for the data as originally obtained, and none of them produces a negative value as large or larger in absolute magnitude as that for the data as obtained. (The negative value with greatest absolute magnitude is -216, which is obtained when the $Y - \bar{Y}$'s are, from left to right, 10, 1, −2, −4, −5.) So since 2 of the 120 permutations yield a test statistic as extreme or more so in the positive direction, and none yield a test statistic as extreme or more so in absolute magnitude but opposite in sign, the right-tailed, two-tailed, and left-tailed cumulative probability levels for the actually obtained value of the test statistic are 2/120, 2/120, and 119/120, respectively.

4.2.2 Fisher's Test of Difference-Score Population Symmetry about Zero

An $X - Y$ difference-score may be regarded simply as an arithmetic operation involving an observation drawn from an X-population and an observation drawn from a Y-population, or it may be regarded as a single observation drawn from a population of $X - Y$ difference-scores. A test of whether or not an $X - Y$ difference-score population is symmetric about zero is equivalent to a test of the H_0 that the X- and Y-populations either have symmetric shapes and a common axis of symmetry or are identical, since either of these conditions results in the population of difference-scores being symmetrically distributed about zero if sampling of the X- and Y-observations is random and independent. It is mainly the legitimacy of this inference about the X- and Y-populations that gives the test of the symmetry of the $X - Y$ difference-score population its practical utility. If it is known that the X- and Y-populations have symmetric or identical shapes, the test becomes a test of whether or not the X- and Y-populations have identical locations, e.g., means.

Suppose that we randomly draw a pair of units from each of n different populations and subject a randomly selected one of the members of each pair to treatment X and the other to treatment Y. (Alternatively, if treatments produce no carryover effects, we may draw one unit from each population and subject it to both treatments.) Let X_i and Y_i be measurements on the appropriate unit from the ith population subsequent to treatment X and to treatment Y, respectively. We wish to test the null hypothesis that, for every i, the populations to which X_i and Y_i *now* belong are either identical or both symmetric about a common axis (implying, perhaps among other things, that the X- and Y-treatments had either identical effects or, at least, identical mean effects) and we wish the test to be especially likely to reject when the populations to which X_i and Y_i now belong have different means (implying that mean treatment effects differ).

If the above H_0 is true then each of the n $X_i - Y_i$ difference-scores is, in effect, a randomly drawn observation from one of n different difference-score populations, each of which is symmetrically distributed about zero. And because of this symmetry about zero, for each obtained $X_i - Y_i$ difference-score there is, in the ith difference-score population, a "mirror-image" difference-score of equal absolute magnitude but opposite algebraic sign, which was just as likely a priori to have become the difference-score actually obtained. Under H_0, therefore, each of the absolute magnitudes of the n obtained difference-scores was just as likely to be accompanied by a plus sign as by a minus. Therefore each of the 2^n distinguishably different sets of n difference-scores that can be obtained by assigning one or the other of the 2 algebraic signs to each of the n absolute difference-score magnitudes had equal a priori probability under H_0. Now suppose that the mean difference-score $\overline{X - Y}$ had been calculated for each of the 2^n sets and that the frequencies of occurrence of the various mean differences had been tabulated. We would then have the null distribution of $\overline{X - Y}$ given that the data consisted of the obtained set of n absolute difference-score magnitudes. And if the $\overline{X - Y}$ for the data actually recorded fell among the $\alpha\, 2^n$ most extreme of these 2^n mean differences, we could reject H_0 at the α level of significance (provided, of course, that the value of α and the definition of "extreme," i.e., the location of the rejection region, had been decided upon prior to sampling).

Instead of using $\overline{X - Y}$ as the test statistic, we could have used Student's t statistic for matched pairs of observations,

$$t = \frac{\overline{X - Y}}{\sqrt{\dfrac{\sum_{i=1}^{n} [(X_i - Y_i) - (\overline{X - Y})]^2}{n(n-1)}}} = \frac{\overline{X - Y}}{\sqrt{\dfrac{-n(\overline{X - Y})^2 + \sum_{i=1}^{n} (X_i - Y_i)^2}{n(n-1)}}}$$

In the latter expression, over the 2^n sets of data, $\sum_{i=1}^{n}(X_i - Y_i)^2$ is a constant and t is a monotone function of the only variable $\overline{X - Y}$. Therefore $\overline{X - Y}$ and t are entirely equivalent test statistics yielding identical probability levels, the latter test statistic, however, being more cumbersome because of the additional unnecessary computations.

As an example, suppose that each of seven individuals (representing different "matching conditions") have been subjected to each of two treatments, X and Y, that there are no sequential or interaction effects between treatments, and that one wishes to test the null hypothesis that, for each individual, the populations associated with the effects of the two treatments are either identical or coaxially symmetric (in both of which cases the two populations have equal means) and wishes rejection to be especially likely when there is a difference between the mean effects produced by the two treatments. The data are presented in the accompanying table. There are 2^7

	Scores							Σ	*Mean*
Treatment X	23	16	11	12	9	5	1	77	11
Treatment Y	8	5	2	7	6	4	3	35	5
Difference: X − Y	15	11	9	5	3	1	−2	42	6

or 128 different ways of distributing plus and minus signs among the seven difference-scores, and under H_0 they are equally probable. Three of these ways result in a positive mean difference, and six result in an absolute mean difference, as great as or greater than that actually obtained. They are shown in the second table.

X − *Y* *Difference-Scores*							Σ	*Mean*
15	11	9	5	3	1	2	46	6.57
15	11	9	5	3	−1	2	44	6.29
15	11	9	5	3	1	−2	42	6.00
−15	−11	−9	−5	−3	−1	−2	−46	−6.57
−15	−11	−9	−5	−3	+1	−2	−44	−6.29
−15	−11	−9	−5	−3	−1	+2	−42	−6.00

As indicated, only a small number of the 2^n mean differences need actually be calculated, specifically those equal to or more extreme than that actually obtained or those constituting the rejection region, whichever is less. Therefore, assuming that sampling was random, the null hypothesis can be rejected at the 6/128 or .047 level of significance in favor of the ("two-tailed") alternative hypothesis that the mean effects of the two treatments probably differ, or at the 3/128 or .023 level of significance in favor of the ("one-tailed") alternative hypothesis that treatment X probably has more mean effect than treatment Y.

4.2.3 Fisher's Two-Sample Test for Identical Populations, Sensitive to Unequal Locations

Suppose that, randomly and independently, n observations have been drawn from a population of X's and m observations from a population of Y's and that it is desired to test the null hypothesis that the two populations are identical with a test that is especially likely to reject when the two populations are nonidentical in ways that include inequality of population means. We may imagine the data recorded in a table having n cells in an "X"-column and m cells in a "Y"-column. If the null hypothesis of identical populations is true and sampling is random and independent, the $n + m$ observations are, in effect, a single homogeneous sample from a common parent population; and each of the $(n + m)!$ possible permutations of the data within the $n + m$ cells of the table was equally likely a priori to become the actually obtained set of data.

However, since we intend to use $\bar{X} - \bar{Y}$ as the test statistic, we do not need to investigate all $(n + m)!$ permutations. For *any* value of $\bar{X} - \bar{Y}$ the n observations in the X-column may be permuted in $n!$ ways, each of which results in the same value of $\bar{X}$; and for each of these ways, the m observations in the Y-column may be permuted in $m!$ ways for each of which $\bar{Y}$ is the same. So there are $n!\,m!$ permutations, involving only within-column exchanges of observations, for *any* and therefore for *every* value of $\bar{X} - \bar{Y}$ that leave it unchanged, and we may therefore confine our attention to the $(n + m)!/(n!\, m!)$ permutations involving transference of observations across columns. By so doing we are ignoring the irrelevant factor of the order in which the observations within a sample were recorded and concentrating upon the sample to which an observation belongs. The $(n + m)!/(n!\, m!)$ "permutations" to which we have restricted ourselves are simply the $\binom{n+m}{n}$ combinations of $n + m$ things taken n at a time, i.e., the number of ways of dividing the $n + m$ observations into samples of sizes n and m containing different combinations of observations.

For each of these $\binom{n+m}{n}$ arrangements of the data there corresponds a value of $\bar{X} - \bar{Y}$ and the relative-frequency distribution of these values is the null distribution of the difference between sample means, given that sampling is random and independent, that the two samples are of the specified sizes, and that the sample data consist of the $n + m$ obtained observations. The $\binom{n+m}{n}\alpha$ most extreme negative, positive, or absolute values in the null distribution of $\bar{X} - \bar{Y}$ may be taken as the lower-tailed, upper-tailed, or two-tailed rejection regions, respectively, for a test at the α level of significance. If the value of $\bar{X} - \bar{Y}$ for the data as originally recorded falls in such a region and it had been decided prior to sampling to use that rejection region for the test, H_0 may be rejected at the α level of significance.

The alternative hypothesis for the test as described above is that the populations are not identical, but the test statistic and the rejection region are such that the form of nonidentity to which the test is most sensitive is that in which the population means differ. Only if it is known (i.e., only if one can justifiably "assume") that the two populations have identical shapes does the test become exclusively a test for means. In that case the null hypothesis is that the population means are equal, and, when rejection is not due to chance, the test rejects only because of inequality of population means (provided, of course, that the assumptions of random and independent sampling are true).

Over the $\binom{n+m}{n}$ combinations of the obtained data, Student's t statistic for unmatched data

$$t = \frac{\bar{X} - \bar{Y}}{\sqrt{\dfrac{\sum_1^{n_X}(X - \bar{X})^2 + \sum_1^{n_Y}(Y - \bar{Y})^2}{n_X + n_Y - 2}\left(\dfrac{1}{n_X} + \dfrac{1}{n_Y}\right)}}$$

$$= \frac{1}{\sqrt{\dfrac{(n_X + n_Y)[\sum_1^{n_X} X^2 + \sum_1^{n_Y} Y^2] - [n_X\bar{X} + n_Y\bar{Y}]^2 - n_X n_Y(\bar{X} - \bar{Y})^2}{n_X n_Y(n_X + n_Y - 2)(\bar{X} - \bar{Y})^2}}}$$

is a monotone function of $\bar{X} - \bar{Y}$. (The first expression inside square brackets is simply the sum of all squared observations and the second is the sum of all observations, both of which are constant, leaving $\bar{X} - \bar{Y}$ as the only variable contributing to t. Thus the entire expression under the radical, which must be positive, reduces to the form $C(\bar{X} - \bar{Y})^{-2} - K$, where C and K are constants, and its square root takes the algebraic sign of $\bar{X} - \bar{Y}$.) So t and $\bar{X} - \bar{Y}$ are entirely equivalent test statistics for this application of the Method of Randomization yielding identical probability levels.

As an example, suppose that one has randomly and independently obtained five X-observations and three Y-observations and wishes to test the null hypothesis that the X- and Y-populations are identical against the alternative that they are nonidentical in a way which (it is hoped) includes inequality of means. The data, as obtained, are given below.

	Observations					*Sum*	*Mean*	$\bar{X} - \bar{Y}$
X-Sample	29	25	22	16	8	100	20	14
Y-Sample	10	5	3			18	6	

The eight obtained observations can be assigned five to the X-sample and three to the Y-sample in $\binom{8}{5} = 56$ ways. Of these 56 ways, three (including the "obtained" assignment) result in an $\bar{X} - \bar{Y}$ difference as great or greater in absolute magnitude as that calculated from the data as originally obtained. These three ways result in the following hypothetical pairs of samples:

	Observations					*Sum*	*Mean*	$\bar{X} - \bar{Y}$
X-Sample	29	25	22	16	8	100	20.0	14.0
Y-Sample	10	5	3			18	6.0	
X-Sample	29	25	22	16	10	102	20.4	15.1
Y-Sample	8	5	3			16	5.3	
X-Sample	16	10	8	5	3	42	8.4	−16.9
Y-Sample	29	25	22			76	25.3	

Therefore, the probability level for a two-tailed test (sensitive to the alternative that $\mu_X \neq \mu_Y$) is $\frac{3}{56}$ while that for a right-tailed test (sensitive to $\mu_X > \mu_Y$) is $\frac{2}{56}$ and that for a left-tailed test (sensitive to $\mu_X < \mu_Y$) is, excluding from the numerator the single case where $\bar{X} - \bar{Y} = 15.1$, $\frac{55}{56}$.

4.2.4 Multi-Sample Tests for Identical Populations, Sensitive to Unequal Locations

The two preceding tests by Fisher (if regarded as two-sample tests for identical populations) may each be generalized to become multi-sample tests for identical populations, appropriate to the cases of matched and uncorrelated observations, respectively.

Following Pitman [9], suppose that we randomly draw C units from each of R different populations and randomly assign the C units in each set to C different treatments. (Alternatively, if treatments produce no carryover effects, we may draw one unit from each population and subject it to all C treatments.) We wish to test the null hypothesis that the C treatments have identical effects upon the units drawn from the same "matching" population and that this is the case for each of the R matching populations. We wish rejection to be especially likely when the mean effects (i.e., the average effects across all R original matching conditions) produced by the C treatments are not all equal. We may imagine the data recorded in a table having C columns and R rows. Under the null hypothesis all rearrangements of the obtained data which do not remove any observation from its original row were equally likely, prior to sampling, to become the actually obtained data. The C observations within a single row can be permuted in $C!$ ways. And, since there are R rows, there are $(C!)^R$ rearrangements of the table observations that had equal a priori probability under H_0. But for every value of our intended test statistic there will be $C!$ permutations of entire columns of observations which leave that value unchanged, so we may avoid unnecessary labor by confining our attention to the $(C!)^R/(C!) = (C!)^{R-1}$ rearrangements obtainable by permuting observations within rows other than the first, which we shall hold fixed. We need a test statistic that is sensitive to differences between mean treatment effects. We cannot use the sum of the column totals because they simply add up to the sum of all observations, which is a constant. However we can avoid this difficulty by first squaring each column total and taking as our test statistic T the sum of the squared column totals. We may imagine that T has been calculated for each of the $(C!)^{R-1}$ "tables" that are equally likely under H_0 and the assumption of random sampling. If the T for our actually obtained table lies in a predesignated rejection region associated with the $\alpha\,(C!)^{R-1}$ most extreme of these T's, we reject H_0 at the α level of significance. Over the $(C!)^{R-1}$ rearrangements of the obtained data, T is a monotone function of the appropriate analysis-of-variance F statistic, so T and F are equivalent randomization test statistics yielding identical probability levels.

Now consider the multi-sample generalization of Fisher's two-sample test for identical populations when observations are not matched across populations. Suppose that n_1 observations have been randomly and independently drawn from the first of C populations, n_2 from the second, and, in general, n_i from the ith, where $\sum_{i=1}^{C} n_i = N$. (Alternatively, we may suppose that N units have been randomly drawn from a single population and randomly assigned to C treatments, n_1 being assigned to the first treatment, n_2 to the second, etc., so that subsequent to treatment the n_i units assigned to the ith treatment are now, in effect, n_i randomly and independently drawn observations from the ith treatment-population.) We wish (in either case) to test the null hypothesis that the C populations are all identical. The data could have been recorded in a table with C columns and n_i cells in the ith column. Under H_0 all of the $N!$ permutations of the N observations among the N cells in the table were equally likely a priori to have become the obtained set of data. But since within-column permutations of observations leave our intended test statistic unaffected, we may legitimately confine our attention to the $N!/(n_1!\,n_2!\ldots n_C!)$ tables having different *combinations* of observations in some columns.

For our test statistic T, we could, when all n_i are equal, take the sum of the squared column sums $\sum_{i=1}^C S_i^2$, as before. However, this is undesirable in the general case since it amounts to permitting each column's squared mean to influence the test in proportion to the square of the number of observations contained in the column, i.e., since

$$\sum_{i=1}^{C} S_i^2 = \sum_{i=1}^{C} (n_i \bar{X}_i)^2$$

Alternatively, we could use the sum of squared column means $\sum_{i=1}^C \bar{X}_i^2$, and this would give each column equal influence, but would give greater weight to observations in the more sparsely occupied columns. Finally, in closest analogy with the analysis-of-variance F statistic, we could let $T = \sum_{i=1}^C n_i \bar{X}_i^2$, which would give every observation in the table equal weight in the test. For each of the $N!/(n_1!\,n_2!\ldots n_C!)$ tables there corresponds a value of T. If the value of T for the data as originally recorded lies among a predesignated $\alpha N!/(n_1!\,n_2!\ldots n_C!)$ of these T's, and sampling was random and independent, H_0 may be rejected at the α level of significance.

Instead of using $T = \sum_{i=1}^C n_i \bar{X}_i^2$ as our test statistic, we could have used the analysis-of-variance F statistic,

$$F = \frac{\dfrac{\sum_{i=1}^{C} n_i(\bar{X}_i - \bar{\bar{X}})^2}{(C-1)}}{\dfrac{\sum_{i=1}^{C}\sum_{j=1}^{n_i} (X_{ij} - \bar{X}_i)^2}{\sum_{i=1}^{C} (n_i - 1)}} = \frac{\sum_{i=1}^{C} (n_i - 1)}{C-1}\left(\frac{\sum_{i=1}^{C} n_i\bar{X}_i^2 - \bar{\bar{X}}^2 \sum_{i=1}^{C} n_i}{\sum_{i=1}^{C}\sum_{j=1}^{n_i} X_{ij}^2 - \sum_{i=1}^{C} n_i\bar{X}_i^2}\right)$$

$$= K\left(\frac{T-K'}{K''-T}\right)$$

where the various K's are constants over the "permutations" of the obtained set of X_{ij}, and where $K'' \geqq T \geqq K'$ ("sums of squares" are progressively diminished by using means of larger and larger blocks of original observations) so that increasing T always increases $T - K'$ and decreases $K'' - T$

while leaving both of them positive, thus necessarily increasing F. Thus T and F are equivalent randomization test statistics, yielding identical probability levels.

4.2.5 Relationships with Other Tests

The above examples of types of application illustrate the Method of Randomization but do not begin to exhaust its possibilities. They were chosen because each involves a unique randomization test which has been described in the statistical literature and because for each of the five tests there is a highly analogous, more practical test based on the ranks, rather than the magnitudes, of the observations. Hotelling and Pabst's test of correlation based on Spearman's rank difference correlation coefficient is the rank analogue of Pitman's test for correlation; the one-sample and two-sample Wilcoxon tests are the rank analogues of Fisher's two tests; and the Friedman and Kruskal-Wallis tests are the rank analogues of the two tests described in Section 4.2.4. These rank tests are among the best and most powerful distribution-free tests, and it is largely its role in the evolution of these rank tests that gives the Method of Randomization its importance. In fact the primary purpose of this chapter is to provide the conceptual background for the far more important chapter to follow.

4.3 CRITIQUE

So far as mathematical properties are concerned, tests based on the Method of Randomization and using a maximally appropriate test statistic are among the best tests to be found. A randomization test using a conventional, parametric test statistic, or a monotone function of it, appears to "capture" most of the good mathematical-statistical properties (such as "most powerfulness," consistency, unbiasedness, etc.) of the classical test when the corresponding normal-theory assumptions happen to be true, without however being forced to "make" these assumptions as the price of obtaining those benefits. Thus randomization tests appear to be superior or equal to their parametric counterparts in the generality of cases, superiority giving way to equality in those cases where the parametric assumptions are met. Of the tests discussed under "Applications," it has been formally proved that (a) Fisher's two tests have the same limiting power (i.e., at infinite sample sizes) as the corresponding parametric t tests, and Pitman's matched-observations test of identical treatment effects has the same limiting power as the corresponding analysis-of-variance F test, under certain common conditions meeting the necessary assumptions of the comparison parametric test [4]; (b) if all populations are normally distributed with the same variance, Fisher's two tests and Pitman's test for correlation are the most powerful one-tailed distribution-free tests of the hypothesis of "equally likely permutations"against the respective alternatives implicit in their test statistics [6]. The one important qualification upon which the optimality of randomization tests' efficiency seems to depend is that one must choose a test statistic that is maximally sensitive to the condition under which one wishes the test to reject. Generally, this will involve using a test statistic based on maximally efficient estimates of the population parameters concerned.

But despite their excellent mathematical properties, tests employing the original Method of Randomization are little more than statistical curiosities. The fact that the distribution of the test statistic is conditional upon the obtained set of observation values makes each separate application a "unique" test and prevents the compilation of tables that can be used repeatedly. Thus for every instance of application the experimenter must laboriously investigate every assignment of the observations that gives the test statistic either a value placing it in its rejection region or a value as extreme, or more so, as its value for the "obtained" assignment, whichever involves the fewer computations. As the number of observations increases, the amount of labor involved rapidly becomes prohibitive except for cases involving the tiniest of rejection regions or the most extreme "obtained" values of the test statistic. Thus the fact is that though statistically efficient, tests applying the Method of Randomization to observation values are almost never quick, are seldom practical, and often are not even feasible. These comments apply to exact tests based upon the original, Fisherian Method of Randomization. Pure feasibility can be extended by use of a clever, but somewhat unnerving, modification [1] which does nothing to correct the other impractical features of the tests. And by sacrificing exactitude, one can obtain tests for which generally applicable tables exist; Pitman [7, 8, 9] has devised test statistics which are monotone functions of those described in sections 4.2.2, 4.2.3, and 4.2.4, and which have approximately a Beta distribution.

There are a number of misconceptions about tests employing the original Method of Randomization. One is that because the null distribution of the test statistic is conditional upon the obtained data, statistical inference is therefore confined to the "subpopulation" represented by the W possible "permutations" of those data that are equally likely under H_0. This criticism appears to be based upon the false premise that in order for a test statistic to afford inferences applicable to an entire population (or group of populations), its null distribution must be the same as the null sampling distribution of the test statistic based upon an infinite number of samples (or groups of samples) from that population (or group of populations). However, while the above relationship holds for most parametric statistics, the situation described is by no means essential to the extension of statistical inference from an obtained sample back to the entire parent population. All that is necessary is the fulfillment of the simple logical relationship: "If *this* (population) condition occurs, then *that* (sample) condition results." And that relationship is, in fact, satisfied by the Method of Randomization.

Another misconception is the belief that the test statistic has the same distribution when an alternative hypothesis is true as it does under the null hypothesis, since in either case it is constructed from the same W ways of arranging the obtained data. This criticism overlooks the fact that while the W arrangements are the same in both cases, their probabilities of occurrence are not, being equal under H_0 but tending to be excessive in a properly constructed rejection region when the appropriate alternative hypothesis is true. The fact that the sample space for the test statistic is the same under the two conflicting hypotheses—the points in the space differing, however, in probability—is no defect of the test. It is just as true of Student's t, which can assume any value between minus and plus infinity under either hypothesis, as it is of randomization test statistics.

A final misconception concerns the relationship between a randomization test and the parametric test using the same (or the equivalent) test statistic. Certain eminent statisticians have stated that the randomization test is the truly correct one and that the corresponding parametric test is valid only to the extent that it results in the same statistical decision. However, they have said so in such a way that, if removed from context and hastily read, certain of their remarks can be (and have been) construed to imply that the randomization test somehow proves that its parametric counterpart is distribution-free! The fact that the classical F statistic affords a distribution-free test when referred to the *randomization* null distribution of F does not mean that it continues to do so when referred to the classical F distribution, i.e., when used as a parametric test. By reviewing the figures of Chapter 2, the reader can verify that the parametric assumptions of normality and homogeneity of variance are neither superfluous nor academic but are, in fact, often crucial to the validity of the commonest parametric tests. The critical difference between the two types of test lies in what they actually test and in what they claim to test. Multi-sample parametric tests for location using unpaired observations, for example, actually test the "hypothesis" that all samples come from identical, normal populations, whereas their randomization counterparts actually test the hypothesis that all samples come from identical populations. The superiority of the randomization test lies in its freedom from the qualification of normality and in its greater honesty about what it is really testing, e.g., identity against alternatives sensitive to location, rather than equal locations under the unproven assumptions of homogeneity and normality.

REFERENCES

1. Chung, J. H. and D. A. S. Fraser, "Randomization Tests for a Multivariate Two-Sample Problem," *Journal of the American Statistical Association*, **53** (1958), 729–735.
2. Fisher, R. A., " 'The Coefficient of Racial Likeness' and the Future of Craniometry," *Journal of the Royal Anthropological Institute of Great Britain and Ireland*, **66** (1936), 57–63.
3. Fisher, R. A., *The Design of Experiments*, 6th ed. New York: Hafner Publishing Company, 1951, pp. 43–47.
4. Hoeffding, W., "The Large-Sample Power of Tests Based on Permutations of Observations," *Annals of Mathematical Statistics*, **23** (1952), 169–192.
5. Kempthorne, O., "The Randomization Theory of Experimental Inference," *Journal of the American Statistical Association*, **50** (1955), 946–967.
6. Lehmann, E. L., and C. Stein, "On the Theory of Some Non-Parametric Hypotheses," *Annals of Mathematical Statistics*, **20** (1949), 28–45.
7. Pitman, E. J. G., "Significance Tests Which May be Applied to Samples from Any Populations," *Journal of the Royal Statistical Society* (*Series B*), **4** (1937), 119–130.
8. Pitman, E. J. G., "Significance Tests Which May be Applied to Samples from Any Populations. II. The Correlation Coefficient Test," *Journal of the Royal Statistical Society* (*Series B*), **4** (1937), 225–232.
9. Pitman, E. J. G., "Significance Tests Which May be Applied to Samples from Any Populations. III. The Analysis of Variance Test," *Biometrika*, **29** (1938), 322–335.
10. Wilks, S. S., *Mathematical Statistics*. New York: John Wiley & Sons, Inc., 1962, pp. 462–468.
11. Wilks, S. S., "Non-Parametric Statistical Inference," in *Probability and Statistics*, ed. Ulf Grenader, New York: John Wiley & Sons, Inc., 1959, pp. 331–354.

5

TESTS BASED UPON THE METHOD OF RANDOMIZATION APPLIED TO RANKS

The Method of Randomization applied to the original observations produced stunningly efficient tests which were dismally impractical. This was the case because the sample space for the test statistic varied from one application of the test to the next, thereby making it impossible to provide generally applicable tables for the null distribution of the test statistic. However, by the simple expedient of replacing the original observations by their size ranks, the sample space for the test statistic can be standardized, thereby standardizing the null distribution of the test statistic, which can then be profitably tabled. The resulting "rank-randomization" tests maintain much of the high efficiency of their "observation-randomization" counterparts while becoming vastly superior to the latter in practicality and ease of application. Many of the most successful distribution-free tests belong to this class.

5.1 NATURAL RANKS

Although proper and highly efficient under normal-theory conditions, or when applied to (most) continuously distributed variates, rank tests are especially appropriate when the original observations are in the natural form of ranks. These natural ranks may be entirely objective, such as a "hardness" ranking of materials in which each material will scratch all those having lower rank and be scratched by all those with higher rank. Or they may be subjectively determined ratings of order assigned to the sampled units by a judge, as when a professional wine-taster rates vintages as to excellence. There are many legitimate reasons for making the original (objective or subjective) measurements on an ordinal scale. The measured attribute may be so complex that it is impossible to quantify it any more precisely. A more sophisticated scale of measurement may be theoretically possible but may simply never have been devised. The experimenter may be more interested in order relationships than in relative magnitudes, his "voluntary" null and alternative hypotheses concerning the former rather than the latter, in which case rank-order measurements are the truly appropriate ones and measurements on a more sophisticated scale might simply confuse the issue and reduce the efficiency of the test. Finally, ordinal measurements may be used simply for convenience, i.e., because they are easier, quicker, or less expensive than more sophisticated measurements which could have been made but were not. When the rankings are subjective, they may be highly reliable in the sense that almost any judge would have assigned the ranks in the same way. This would be the case, for example, if objects differing from one another by easily perceptible increments of size were ranked according to size. On the other hand, subjective ratings for the same set of objects may vary considerably from one judge to another, as when paintings are ranked in order of artistic excellence. In such cases, it is generally desirable to have the ranks assigned by an established "expert," or by the net consensus of a large panel of judges.

5.2 SOME CONSIDERATIONS OF POWER, EFFICIENCY, AND VALIDITY

Let $P(\theta)$ be a classical parametric test based upon the test statistic θ. Let $F(\theta)$ be a test applying Fisher's Method of Randomization to the original observation values and based upon the same test statistic θ. Finally, let $R(\theta')$ be a test applying the Method of Randomization to the ranks of the original observations, based upon a test statistic θ' which involves exactly the same mathematical operations as θ, except that they are performed upon the ranks of the observations rather than upon the observations themselves. (Actually the two generic randomization tests are generally based upon a test statistic that is a monotone function of θ or θ' over the set of "equally likely permutations" of the obtained data under H_0. However, this is a mere convenience, and they could have been based upon θ and θ' with equal validity.)

Now the assumptions of $F(\theta)$ and $R(\theta')$ are subsumed under those of $P(\theta)$, so if all assumptions of the parametric test are met, the three tests are equally valid, i.e., all three are exact tests. Furthermore, since they are based upon identical test statistics, the only difference between $F(\theta)$ and $P(\theta)$ is that the former obtains its probability level by referring θ to the randomization distribution of θ over the obtained data, while the latter *should* obtain it by referring θ to the sampling distribution of θ over an infinite number of replications of the experiment when H_0 is true. It *does* obtain it by referring θ to the classical "θ distribution" which, when all assumptions are met, is identical to that sampling distribution. Therefore, when all parametric assumptions are met, the $P(\theta)$ and $F(\theta)$ tests refer their common test statistic to equally valid null probability distributions.

If H_0 is false, the only way its falsity can influence the test is through its effect upon the test statistic. The two tests have the *same* test statistic, and in both cases it is referred to a *valid* null distribution whose rejection region consists of the "most extreme" $100\alpha\%$ of its θ values. This tends to give the two tests equal power, but does not insure it because of differences between the null randomization and null sampling distributions of θ. For example, suppose that θ is Student's t statistic. Over the randomization distribution of t, the rank-order correlation between t and its numerator $\bar{X} - \bar{Y}$ is perfect at all sample sizes. So the $100\alpha\%$ most extreme t's, in an α-sized rejection region, contain the $100\alpha\%$ most extreme $(\bar{X} - \bar{Y})$'s in their numerators. On the other hand, over the classical distribution of t, the rank-order correlation between t and its numerator is poor at small sample sizes although it approaches 1.00 as sample sizes approach infinity. Thus at practical sample sizes, a t value whose numerator is among the $100\alpha\%$ most extreme $(\bar{X} - \bar{Y})$'s may be prevented from falling in its classical α-sized rejection region due to the entirely irrelevant influence of a counteractingly large sample variance in its denominator. The power of the test against alternatives of unequal locations would be expected to suffer. Another difference between the two tests is that for Student's t test, α can be any value we choose; but for the randomization t test, α can be no smaller than $1/\binom{n+m}{n}$ and its exact value must be an integral multiple of that lower bound. It seems clear, however, that so long as $F(\theta)$ and $P(\theta)$ are compared at a common α value that they can both assume, $F(\theta)$ should be *at least* as powerful on the average (although not necessarily in every specific instance of application to a common set of data) as $P(\theta)$ even though all of the latter's assumptions are completely met. These ambiguities about relative power are dispelled when sample sizes become infinite. In that case the "sample" takes on the essential characteristics of the population, the relative frequency distributions of observations in the sample becoming essentially the same as the relative frequency distribution of observations in the population. Consequently, the randomization distribution of θ becomes essentially the same as the sampling distribution of θ, α can take on an infinite gradation of values in both cases, and in both cases t has a perfect rank-order correlation with its numerator. Clearly, then, $F(\theta)$ and $P(\theta)$ are equally powerful when the latter's assumptions are met and sample sizes are infinite.

The favorable conclusions about $F(\theta)$ in comparison with $P(\theta)$ do not extend to $R(\theta')$, primarily because its test statistic θ' is not the same as θ but also partly because its actually tested H_0, its randomization distribution, and its rejection region all have been somewhat modified by the rank transformation. If $P(\theta)$ has optimum power properties under the assumed parametric conditions, as is generally the case (at least for large samples), it follows that so long as those conditions hold a test, $R(\theta')$, using a different test statistic (and, in fact, being appropriate for testing a different H_0) cannot be more powerful than $P(\theta)$ for testing $P(\theta)$'s hypothesis and would be expected to be less powerful, which it is for all tests treated in this chapter. The loss of power, however, is often quite slight and is seldom disturbing.

Next consider the case where the parametric assumptions are not met but the assumptions of the two randomization tests hold and where the observations are not in the natural form of ranks. In that case only $F(\theta)$ and $R(\theta')$ are valid (i.e., yield exact null probabilities when H_0 is true). $P(\theta)$ *would* be just as valid, and have the same, average, large-sample power, as $F(\theta)$, if only its probability levels could be obtained by referring θ to its true, but unknown, sampling distribution under H_0 [in which case, however, it would not be the $P(\theta)$ test]. Instead θ is referred to the classical, tabled "θ distribution," which is not its null distribution when the parametric assumptions are false. Thus any difference in average power between $P(\theta)$ and $F(\theta)$ in the large-sample case is a spurious power increment or decrement (i.e., one which affects the probability of *both* Type I and Type II errors) which is entirely attributable to the fact that $P(\theta)$ refers its test statistic to the wrong distribution. Clearly $P(\theta)$ is inferior to $F(\theta)$ with regard to validity, and also with regard to power if we discount, as we should, power increments that are attributable to α increments (i.e., that are due to the fact that the true rejection region occupies a greater proportion of the true null distribution than was intended and specified); $P(\theta)$ is also inferior to $R(\theta')$ with regard to validity. Whether or not its power will be inferior to that of $R(\theta')$ will depend upon the specific situation. Aside from the spurious changes in power of the type mentioned above, there are real power differences between the two tests which can be expected to vary with the types of populations sampled. These legitimate power differences are attributable to differences between the sensitivities of θ and θ' to the alternative hypothesis, i.e., the true condition.

Finally, consider the case where the original observations are in the natural form of ranks—the same ranks they would have received from the rank transformation—and the assumptions of the two randomization tests are met. [The assumptions of $P(\theta)$ are not met because natural ranks are not normally distributed.] In that case θ becomes identical to θ', so that all three tests have the same test statistic. Furthermore, $F(\theta)$ has the same null, randomization distribution as $R(\theta')$, so $F(\theta)$ and $R(\theta')$ are identical tests having identical powers in every individual application to a common set of data. Only $F(\theta)$ and $R(\theta')$ are valid, but $P(\theta)$ *would* be valid if only it could be referred to its true null sampling distribution. Therefore, since all three tests are based upon the same test statistic, any difference in average, large-sample power between $R(\theta')$ and $P(\theta)$ are due to the fact that $P(\theta)$ refers its test statistic to the wrong distribution. Thus when observations are in the natural form of ranks, $R(\theta')$ is identical to $F(\theta)$, and $P(\theta)$ is inferior to both of them in regard to validity and legitimate power. So while $P(\theta)$ may have optimal properties under the "parametric" conditions assumed under normal-theory, $F(\theta)$ is at least as good a test under "parametric" conditions and a better test otherwise, and $R(\theta')$ is a better test when observations are in the natural form of ranks.

5.3 HOTELLING AND PABST'S TEST FOR RANK-ORDER CORRELATION

This test is essentially the same as Pitman's test for correlation (as described in Section 4.2.1) except that, prior to applying the test, the observations in each sample are replaced by their within-sample size ranks. (The present test, however, was published first and therefore did not develop out of Pitman's test.) Pitman's test used the Pearson product-moment correlation coefficient r, or some monotone function of it, as the test statistic. However, when observations are replaced by their within-sample size ranks (the smallest observation being replaced by a 1, the next smallest by a 2, etc.) and no within-sample observations are tied, Pearson's r based on the ranks, rather than the original observations, reduces to Spearman's rank-difference correlation coefficient r_s, given below. The present test uses a monotone function of r_s as the test statistic, so the effect is the same as if the present test had used the same test statistic employed by Pitman. Thus (since the two tests are entirely analogous in all other ways) the present test is simply (i.e., is mathematically equivalent to) Pitman's test applied to ranks.

5.3.1 RATIONALE

Suppose that an X-measurement and a Y-measurement have been made on each of n randomly drawn units and that the experimenter wishes to test the null hypothesis that X and Y are independent and hence uncorrelated in the bivariate sampled population. Suppose further that there are no ties among the n X-measurements nor among the n Y-measurements and that each of the X's has been replaced by its rank in order of increasing size (the smallest X receiving a rank of 1, the largest a rank of n) and likewise for each of the Y's. We may arbitrarily arrange our n units in order of increasing X-ranks. Then our obtained data will be uniquely identified by the corresponding linear arrangement of the n associated Y-ranks. Now if H_0 is true, each of the $n!$ distinguishable permutations of the n Y-ranks had equal a priori probability of becoming the pattern of Y-ranks associated with increasing X-ranks. For each of these $n!$ distinguishable patterns of association between X-ranks and Y-ranks, there will be a value of Spearman's rank-difference correlation coefficient

$$r_s = 1 - \frac{6 \sum_{i=1}^{n} d_i^2}{n(n^2 - 1)} = 1 - \frac{6D}{n(n^2 - 1)}$$

where d_i is the difference between the ranks of measurements taken on the same unit, i.e., the difference between the X-rank and Y-rank for the ith unit, and D is the sum of the squared d_i's. The frequency distribution of these $n!$ values is the null distribution of r_s, to which the value of r_s for the actually obtained data may therefore be referred as a test of H_0. However, since at a given n value r_s is a monotone function of D, one may save labor by using D as the test statistic and referring the obtained value of D to the frequency distribution of D values corresponding to the $n!$ associations between X- and Y-ranks that are equally likely under H_0. If H_0 is false, i.e., if the X- and Y-variates are correlated, then the obtained values of r_s or D should be excessively likely to fall at one of the tails of their null distributions.

5.3.2 Null Hypothesis

Each of the $n!$ distinguishable patterns of association between X-ranks and Y-ranks for the n units was equally likely a priori. This will be the case if the X- and Y-variates are statistically independent and all assumptions are met.

5.3.3 Assumptions

In order for each of the $n!$ patterns of association between X-ranks and Y-ranks to be equally likely, not only must there be independence between the X- and Y-variates in the bivariate sampled population, but also there must have been no bias (conducive to pseudocorrelation) in the actual selection of sample units. Therefore *sampling* must be *random*.

In order for each of the $n!$ patterns to be distinguishable, and for the obtained value of the test statistic to be unequivocal, there must be *no tied observations* among the n X's or among the n Y's. This will be the case if the

X- and Y-variates are continuously distributed and measurement is perfectly precise. If intravariate ties occur, they should be dealt with as described in Section 3.3.1.

5.3.4 Efficiency

This test has an A.R.E. of $9/(\pi^2)$ or .912 relative to the parametric t test based on Pearson's r when both tests are applied under common conditions meeting all of the assumptions of the latter. (Spearman's r_s also has an efficiency of .912 relative to Pearson's r as an *estimator* of the population product-moment correlation coefficient ρ when sample size is infinite and the population is bivariate normal with a ρ of zero. In this case the A.R.E. merely reflects the estimating properties of the two test statistics, or the critical variables upon which they are based.) Although the A.R.E. of the present test relative to the parametric test based upon Pearson's r is .912 when the sampled population has a bivariate normal distribution, it has A.R.E. of 1.000 if the population has a uniform distribution, .857 if it has a parabolic distribution, and 1.266 if it has a Laplace distribution. Since ranks are uniformly distributed and since Pearson's r applied to ranks is the same as Spearman's r_s, it is perhaps not surprising that the A.R.E. is 1 for a uniform distribution.

5.3.5 Tables

Table I of the Appendix gives lower-tail critical values of D, i.e., of $\sum_{i=1}^{n} d_i^2$. More precisely, it gives the largest value of D' for which $P(D \leqq D') \leqq \alpha$, and does so for α's of .001, .005, .01, .025, .05, and .10, and for $n = 4(1)30$, i.e., for n's progressing from 4, in steps of 1, to 30. The smaller the value of D the larger is the value of r_s, so a lower-tail test of D corresponds to an upper-tail test of r_s, and, if significant, indicates positive correlation. The distribution of D is symmetric, ranging (with D assuming only even-integer values) from 0 to $\frac{1}{3}n(n^2 - 1)$, so one may perform an upper-tail test by referring $\frac{1}{3}n(n^2 - 1) - D$ to the above table. That is, if D has an upper-tail cumulative probability $\leqq \alpha$, then $\frac{1}{3}n(n^2 - 1) - D$ will be less than or equal to the critical value, in the above table, associated with a lower-tail cumulative probability $\leqq \alpha$. The table is exact for $n \leqq 10$ and is based upon what appears to be the best available approximation at higher n's.

Kendall [30] has provided tables giving the exact cumulative probability distribution of D for $n = 4(1)10$. His tables give to at least three decimal places and to at least two significant figures the upper-tail cumulative probabilities of all values of D in the upper half of its distribution. Exact probabilities of D are obtained only by laborious methods which become increasingly difficult as n increases; therefore, exact tables are limited to values of $n \leqq 10$. For $n = 4(1)30$ and α's of .001, .005, .01, .025, .05, and .10, Glasser and Winter [21] have tabled the largest value of D^* for which $P(D < D^*) \leqq \alpha$ and the smallest value of r_s^* for which $P(r_s > r_s^*) \leqq \alpha$, the table entries being exact at $n \leqq 10$ and approximate at n's above 10. Since the above probability statements do not include the events $D = D^*$ or $r_s = r_s^*$, the tabled values of D^* and r_s^* lie just *outside* of the rejection region corresponding to α, and this must be borne in mind when using their tables.

If there is independence between the X-variate and the Y-variate in their bivariate population, then as n increases the sampling distribution of r_s approaches a normal distribution whose mean is zero and whose variance is $1/(n - 1)$. The corresponding sampling distribution of D approaches a normal distribution whose mean is $(n^3 - n)/6$ and whose variance is $[(n^3 - n)/6]^2/(n - 1)$. See Kendall [30]. Therefore for large samples

$$r_s\sqrt{n-1} \qquad \text{or} \qquad \frac{\sum_{i=1}^{n} d_i^2 - [(n^3 - n)/6]}{(n^3 - n)/6\sqrt{n-1}}$$

may be treated as normal deviates with zero mean and unit variance, and probabilities may be obtained by referring these critical ratios to normal tables. Likewise if the population correlation is zero, then as n increases, the distribution of

$$t = r_s\sqrt{\frac{n-2}{1-r_s^2}}$$

approaches closer and closer to the distribution of Student's t with $n - 2$ degrees of freedom [21], and this approximation appears to be better than the above normal approximation. For greater accuracy, therefore, take as the approximate cumulative probability of r_s the cumulative probability obtained for the above value of t when referred to tables of probabilities for Student's t based on $n - 2$ degrees of freedom.

5.3.6 Example

A speculator wishes to test whether the price X of a certain stock is correlated with the number Y of investment advisory services recommending its purchase. More precisely, he wishes to test at the .05 level the null hypothesis that X and Y are independent against the one-tailed alternative hypothesis that they are correlated positively. He randomly selects five points in time and determines the values of X and Y at each of these points, obtaining the following data:

X	39	29	60	40	32
Y	11	5	20	8	6

Now replacing the X's by their size ranks, and likewise for the Y's, he obtains the following:

X-rank	3	1	5	4	2
Y-rank	4	1	5	3	2
Rank difference, d_i	−1	0	0	1	0
d_i^2	1	0	0	1	0

$$D = \sum_{i=1}^{5} d_i^2 = 2 \qquad r_s = 1 - \frac{6D}{n(n^2 - 1)} = 1 - \frac{12}{120} = .90$$

Entering Table I of the Appendix with $n = 5$, he finds that $D = 2$ is the critical value of D for a lower-tail test at $\alpha = .05$. He therefore rejects the null hypothesis of independence in favor of the alternative hypothesis that X and Y are positively correlated.

5.3.7 Discussion

An alternative to the present test, which can be used under the same conditions to test the same H_0, is Kendall's test for correlation (taken up in a different chapter). The two tests are not mathematically equivalent, but their respective test statistics are highly correlated when applied to samples from a bivariate normal population whose product moment correlation coefficient ρ is not too extreme. When $\rho = 0$, i.e., when the population correlation between the two variates is zero, the product-moment correlation between the two test statistics, i.e., Kendall's and r_s, increases from .980 when $n = 5$, to .990 when $n = 20$, to 1.00 when $n = \infty$ [30, p. 80]. And when $n = \infty$ and the population is bivariate normal, the correlation decreases from 1.00 when $\rho = 0$, to .9996 when $\rho = .2$, to .9981 when $\rho = .4$, to .9843 when $\rho = .8$, "though it tends to zero as ρ approaches unity" [30, p. 131]. Since the correlation is 1.00 when $\rho = 0$ and $n = \infty$, the two tests have the same A.R.E. relative to the parametric test based on Pearson's r, and an A.R.E. of 1.00 relative to each other, when the data come from a bivariate normal population.

The present test has the advantage that, when n is moderately large, it can be performed more quickly and with perhaps less likelihood of error than Kendall's test, which involves a relatively complex counting task. Also, because rank differences are squared only in the present test, it is particularly desirable when one wishes to weight large discrepancies between ranked X's and Y's more heavily than small ones. In most other respects, however, the test appears to be inferior to Kendall's. Kendall's test statistic is much more mathematically tractible than r_s. Its probabilities can be obtained from a simple recursion formula and exact tables for it extend to a much higher n than is the case for r_s. And at a given n the distribution of Kendall's test statistic is generally smoother and better fitted by a normal distribution than is the distribution of r_s, which tends to be very jagged ordinatewise, presenting a sawtoothed appearance. (For a dramatic pictorial contrast between the two distributions at $n = 8$, see [30, pp. 50–51 and 59–60].) Thus, not only do exact tables extend to much higher n's for Kendall's test, but even if both tests resorted to approximate probabilities at the same n value the normal approximation would be better for Kendall's test, and one would also expect the normal approximation for Kendall's test to be better than the t approximation for the present test.

5.4 DANIELS' TEST FOR TREND

In the preceding test, if the X-observation is the *time* at which the unit was drawn from its population and the Y-observation is some measurement upon the unit itself, we have a test of whether or not Y is correlated with time, i.e., a test for trend. Thus we are, in effect, testing the hypothesis that the Y-population (or potential population, if the Y's are being sequentially generated by some "process") remains identical over the time period covered by the X-observations. This test has an A.R.E. of $(3/\pi)^{1/3}$ or .98 relative to the parametric test based on the regression coefficient b, when both tests are applied as tests of randomness against normal regression alternatives.

5.5 WILCOXON'S SIGNED-RANK TEST OF DIFFERENCE-SCORE POPULATION SYMMETRY ABOUT ZERO (And Therefore for Treatment Effects Upon Matched Units)

This test is essentially the same as Fisher's test of difference-score population symmetry about zero (as described in Section 4.2.2) except that prior to applying the test, the difference-score absolute magnitudes are replaced by their ranks in order of increasing absolute magnitude, the algebraic sign of the difference-score then being transferred to the rank with which it is replaced. Fisher's test used $\overline{X - Y}$ as the test statistic. However, it could just as legitimately have used the sum of the difference-scores having a given algebraic sign, since

$$\overline{X - Y} = \frac{\sum_{1}^{n} (X - Y)}{n}$$

$$= \frac{\sum \textit{positive values of } (X - Y) - |\sum \textit{negative values of } (X - Y)|}{n}$$

$$= \frac{\sum pos\,(X - Y) - \left[\sum_{1}^{n} |X - Y| - \sum pos\,(X - Y)\right]}{n}$$

$$= \frac{2 \sum pos\,(X - Y) - \sum_{1}^{n} |X - Y|}{n}$$

$$= \frac{c \sum pos\,(X - Y) - c'}{c''}$$

where the various c's are constants over the randomization distribution of $\overline{X - Y}$ obtained by varying the algebraic signs of difference-scores in the set of obtained data. [And by substituting for $\sum pos(X - Y)$ its equivalent, $\sum_1^n |X - Y| - |\sum neg(X - Y)|$, it can be shown that $\overline{X - Y} = [c \sum neg (X - Y) + c']/c''$.] So $\overline{X - Y}$, $\sum pos (X - Y)$, and $\sum neg (X - Y)$ are all equivalent test statistics for Fisher's test. Not only do the latter two involve fewer computations, but, because of the complementary relationship between them, their null randomization distributions are mirror-images of each other about an axis through the common point of zero [so that the lower-tail cumulative probability of $\sum pos (X - Y)$ is always identical to the upper-tail cumulative probability of the corresponding $\sum neg (X - Y)$]. Therefore, computations can be reduced still further by using as test statistic whichever one of $\sum pos (X - Y)$ and $\sum neg (X - Y)$ has the smaller absolute value and referring it to the (lower-tail cumulative) randomization distribution of $\sum pos (X - Y)$, only the lower half of which need be calculated. The present test takes this approach except, of course, that the absolute magnitudes of the algebrically signed difference-scores are replaced by their ranks in order of increasing absolute size.

5.5.1 Rationale

Suppose that an experimenter draws two units from each of n different populations (so that the units in each pair are matched), randomly assigns one unit from each pair to treatment X, the other to treatment Y. And he wishes to test the null hypothesis that the two treatments have identical effects against the alternative hypothesis that the treatments produce X- and Y-populations which, if asymmetric, differ in *any* way and if symmetric differ in location. Let X_i and Y_i be measurements, subsequent to treatment, on the units from the ith matched pair assigned to treatments X and Y respectively. Let S_i be the algebraic sign of the difference-score $X_i - Y_i$. And let R_i be the rank in increasing order of size of $|X_i - Y_i|$ among the n absolute values of the difference-scores for the n obtained pairs of measurements. Then if the experimenter's H_0 is true (or if the treatments produce X- and Y-populations that are symmetric about a common axis), the true distribution of the difference-score variate $X_i - Y_i$ will be symmetric about zero for every value of i (see Section 4.2.2). Thus if H_0 is true, for every obtained $X_i - Y_i$ there is in the true distribution of the $X_i - Y_i$ variate an equally probable $X_i - Y_i$ value having the same absolute magnitude as the obtained value but opposite algebraic sign. So for every obtained $|X_i - Y_i|$, and therefore for every obtained value of R_i, S_i was exactly as likely a priori to be a plus as to be a minus. Therefore, each of the 2^n possible assignments of algebraic signs to the n ranks R_i was as likely a priori to have become the set of signed ranks, i.e., the n values of S_iR_i, as was the actually obtained set. For each of these 2^n sets of n signed ranks, we may calculate the test statistic

$$W_+ = \sum_{S_i=+} S_i R_i$$

(or we could have summed over the negative $S_i R_i$ values and gotten a complementary test statistic), thereby obtaining its null distribution. If treatments produce differences of consistent direction between the locations of the distributions of the X_i and Y_i variates, we would expect $\overline{X - Y}$ and therefore W_+ to tend to fall in appropriately extreme portions of their null randomization distributions. W_+ is therefore an appropriate test statistic for testing H_0 against the stated alternatives.

5.5.2 Null Hypothesis

Each of the 2^n sets of data obtainable by varying the algebraic signs associated with the n actually obtained values of $|X_i - Y_i|$ was equally likely, prior to sampling, to have become the obtained set of data. Phrased differently, for each of the n obtained values of $|X_i - Y_i|$ it was true a priori that P (S_i will be $+$) $= P$ (S_i will be $-$) $= .5$. This will be the case if all assumptions are met and if for every value of i, i.e., under every matching condition, the X- and Y-treatments either (a) have identical effects so that the potential population of X-treated units is identical to that of Y-treated units; or (b) have effects such that the potential populations of X-treated and Y-treated units are both symmetrical and have a common axis of symmetry.

5.5.3 Assumptions

In order for nonchance effects between members of a matched pair to be attributable to differences between treatments, they must not be attributable to bias in the assignment of paired units to treatments. More precisely, in order for X_i and Y_i to belong to populations that are identical after identical treatments, the units upon which the measurements were taken must have belonged to identical populations subsequent to assignment to treatments but prior to receiving the treatments. This means that there must be no bias associated with assignment to treatment. In order to meet this requirement, we have the assumption that for every matched pair of units *assignment* to treatments *is random*. If the *same* unit receives *both* treatments, it must be assumed that the unit's responsiveness to the second treatment is altered neither by the lingering effects of the first treatment nor by the passage of time between treatments. This assumption is seldom entirely justified. But the effects of its falsity can usually be somewhat mitigated by counterbalancing or randomizing the order in which the two treatments are administered.

If assignment of the two units in a matched pair to treatments is random, there is no need to assume that the two units were drawn independently from their common population. That is, there is no need to assume either that the common population is infinite or that sampling was with individual replacements. Indeed it is only considerations of power, not validity under H_0, that render a *common* population desirable for the two members of a "matched" pair.

The tabled null distribution of W_+ applies to the case where every S_i is either a plus or a minus and the R_i are the consecutive integers from 1 to n. Therefore, in a sense (i.e., if one ignores the possibility of modifying the procedure), the test assumes that there are *no zeros* and *no ties* among the n obtained values of $|X_i - Y_i|$. However, if difference-scores of zero are not due to imprecision of measurement but rather represent true zeros in the difference-score populations, they may be dropped from the data with n reduced accordingly, the test then being applied to the remaining nonzero difference-scores. This amounts to testing the hypothesis that the nonzero portions of the difference-score populations are symmetric about zero; if this is the case, the entire difference-score populations must be symmetric about zero, which is our original null hypothesis. (Objections have been raised against "dropping the zeros" [49], not on grounds of validity of the test but rather to preserve consistency of an unnecessarily restrictive interpretation of the implications of rejection.) Furthermore, in the case of ties, if all difference-scores having the same absolute magnitude also have the same algebraic sign, the consecutive ranks for which a set of equal $|X_i - Y_i|$ are collectively tied may be assigned to them arbitrarily, and the validity of the test will be unimpaired. Thus while the assumption of no zeros and no ties is logically *sufficient* to eliminate the problems raised by zeros and ties, it is not always logically *necessary* to do so. For methods of dealing with zeros and ties in other situations, see Sections 3.3 and 3.4.

5.5.4 Efficiency

This test has an A.R.E. of $3/\pi$ or .955 relative to the matched-pair Student's t test when both tests are applied under conditions known to meet all the assumptions of the latter test and are used to test the latter test's null hypothesis. (If the t test's assumption of a normally distributed difference-score population is *known* to be true, the population must be symmetric about an axis through its mean as specified by the H_0 of the Wilcoxon test. The remainder of the Wilcoxon test's H_0 and the only part about which any question remains, is that the axis of symmetry—and therefore the mean—is zero, which is the H_0 of the t test.) Under the same conditions, as sample size diminishes from infinity toward zero, the efficiency of the Wilcoxon test relative to the t test increases from .955 toward 1.00. If the difference-score population is symmetric and continuous, its exact shape otherwise being left unspecified, the A.R.E. of the Wilcoxon test relative to Student's t test cannot drop below .864 [see reference 27], i.e., .864 is the lower bound for the A.R.E. over the class of continuous symmetric populations.

Klotz [31] has graphed the efficiency of the Wilcoxon test, relative to other powerful tests, as a function of the departure of the location parameter μ of the difference-score population, from its hypothesized value of zero. He has done this for populations of several different symmetric shapes in the case of infinite sample size. When the population is normal, with a variance of 1, the efficiency of the Wilcoxon relative to the t test rises from .955 at $\mu = 0$ to .981 at $\mu = .75$, which is about its highest point; it then declines toward a value of zero at $\mu = \infty$, but it is still as high as .598 at the relatively large departure of $\mu = 3.0$. When the population is logistic, the Wilcoxon test has certain optimum power properties at $\mu = 0$, and, in fact, is more efficient than either the normal scores test or the Sign test (the only tests with which it was compared) at all values of μ. When the population is double-exponential, the Sign test has certain optimum power properties at $\mu = 0$. However, the efficiency of the Wilcoxon test relative to the Sign test rises from .75 at $\mu = 0$, passing through 1.00 at about $\mu = .5$, to a maximum of 1.09, and thereafter descending to 1.00 at $\mu = \infty$. When samples are small the efficiency of the Wilcoxon relative to the t test behaves differently from the case of infinite-sized samples. When the population is normal, the efficiency of the Wilcoxon relative to the t test appears to be highest at $\mu = 0$, descending steadily with increasing values of μ to a nonzero value at $\mu = \infty$. For example, for $n = 5$ and $\alpha = .0625$ (for a one-tailed test), efficiency is .988 at $\mu = 0$, .979 at $\mu = 1.5$, and .881 at $\mu = \infty$; for $n = 10$ and $\alpha = .0527$, efficiency is .970 at $\mu = 0$, .964 at $\mu = 1.5$ and .756 at $\mu = \infty$. For powers and efficiencies at $n = 5\,(1)\,10$, and $\mu = 0\,(.25)\,1.5\,(.5)\,3\,(\infty)\,\infty$, for one-tailed tests of shift of a normal population, see Klotz [33]. Arnold [3] has tabulated the power of the Wilcoxon, Sign, and t tests for samples of sizes 5 to 10 from a Cauchy population with μ taking values from .25 to 3. In practically all cases (with exceptions explainable as inaccuracies) the Wilcoxon test was more powerful than the t test and the Sign test was the most powerful of all. The appropriately one-tailed Wilcoxon test is unbiased against a class of alternatives stated in [38].

5.5.5 TABLES

Table II of the Appendix gives lower-tail cumulative probabilities for $W_+ = \sum_{S_i=+} S_i R_i$, the sum of positive-signed ranks. This is done for values of W_+ that fall just within and just outside lower-tail rejection regions corresponding to significance levels α's of .05, .025, .01, and .005; and the table applies for $n = 5\,(1)\,50$. The randomization distribution of W_+ is symmetric, extending from zero to $n(n + 1)/2$. Therefore, the upper-tail cumulative probability of W_+ is the same as the lower-tail cumulative probability of $[n(n+1)/2] - W_+$, critical cases of which can be obtained from the table. However, one does not even need to perform the subtraction since $[n(n+1)/2] - W_+ = |\sum_{S_i=-} S_i R_i| = |W_-|$, the absolute value of the sum of the negative signed ranks. Thus for a one-tailed test that rejects if W_+ is too small (or $|W_-|$ is too large) refer W_+ to the table. For a one-tailed test that rejects if W_+ is too large (or $|W_-|$ is too small) refer $|W_-|$ to the table. And for a two-tailed test that rejects if W_+ (or $|W_-|$) is either too small or too large, refer the smaller of the two statistics W_+ and $|W_-|$ to the table and double the indicated probability level.

For $n = 4$ (1) 100, McCornack [41] tables critical values of W_+ for a lower-tail test at α's of .00005, .0005, .0025, and .005 (.005) .025 (.025) .15 (.05) .45; that is, he gives the largest integer I for which $P(W_+ \leqq I) \leqq \alpha$. The table is exact, and its inclusion of extreme α's is a particularly useful feature since it is when probability levels are extreme that approximation formulas tend to perform worst. The complete cumulative probability distribution of W_+ is given by Owen [47] for $n = 3$ (1) 20.

The null distribution of W_+ (or of $|W_-|$) has a mean of $[n(n+1)]/4$, a variance of $[n(n+1)(2n+1)]/24$, and a shape that becomes increasingly normal as n increases and perfectly normal when n becomes infinite. Therefore, when n is too large for the exact tables to apply, one can obtain the approximate probability level for W_+ by referring the critical ratio

$$\frac{W_+ - [n(n+1)/4]}{\sqrt{n(n+1)(2n+1)/24}}$$

to probability tables for the unit normal deviate.

5.5.6 Example

An experimenter wishes to test at the .05 level whether or not a certain new drug Y has any effect whatever upon reaction time. But he wishes the test to be especially sensitive to changes in the location of the reaction-time distribution. He decides to measure the reaction times of each of seven randomly selected subjects first under a "control" drug X which is a placebo known to have no biochemical effect upon the measured variable, and subsequently under drug Y (so he is assuming that neither the passage of time nor the psychological experience of having been tested once before has any effect upon reaction to Y). He knows that the hypothesis tested by the Wilcoxon test (i.e., difference-score population symmetry about zero) will be true not only if the X- and Y-populations are identical, but also if they have different symmetrical shapes and a common mean. He can dismiss the latter possibility, however, because reaction times are known to be positively skewed and therefore asymmetric. His knowledge that the X-population, at least, must be asymmetric thus enables him to use the Wilcoxon test as a test of whether or not the X- and Y-populations are identical. His experimental data and test computations are as shown in the accompanying table. The smaller of W_+ and $|W_-|$ is the latter, which equals 2. Entering Table II of the Appendix with $n = 7$ it is seen that a value of 2 for the test statistic W_+ or $|W_-|$ has a one-tailed cumulative probability of .0234. In the experimenter's case this means that $W_+ = 26$ has an upper-tailed cumulative probability of .0234 in the randomization distribution of W_+ or that $|W_-|$ has a lower-tailed cumulative probability of .0234 in the randomization distribution of $|W_-|$. But since he wants a *two*-tailed rejection region in the

	Subject						
	I	II	III	IV	V	VI	VII
Reaction Time under Drug X	.223	.216	.211	.212	.209	.205	.201
Reaction Time under Drug Y	.208	.205	.202	.207	.206	.204	.203
Difference: $X_i - Y_i$	.015	.011	.009	.005	.003	.001	−.002
S_i: *Sign of* $(X_i - Y_i)$	+	+	+	+	+	+	−
R_i: *Rank of* $\lvert X_i - Y_i \rvert$	7	6	5	4	3	1	2
S_iR_i: *Signed Rank*	7	6	5	4	3	1	−2

$$W_+ = \sum_{S_i=+} S_iR_i = 26 \qquad W_- = \sum_{S_i=-} S_iR_i = -2$$

$$\sum_{i=1}^{n} R_i - W_+ = \frac{n(n+1)}{2} - W_+ = 28 - 26 = 2 = |W_-|$$

randomization distribution of a *single* test statistic, he must double the indicated one-tailed cumulative probability, obtaining .0468 as the probability level for a two-tailed test. He therefore rejects at the preselected .05 level of significance the hypothesis that drugs X and Y have identical effects upon reaction time in favor of the implied conclusion that drug Y changes it through biochemical action.

5.5.7 Discussion

The most important rivals of the present test are the matched-pair Sign test (see Section 7.3) and the classical matched-pair t test. The essential population conditions which must be met if the three test statistics are to have their null distributions, and which are therefore the population characteristics that are actually being tested, are as follows: for the t test—that the difference-score population is normally distributed about a median of zero; for the Wilcoxon test—that the difference-score population is symmetrically distributed about a median of zero; for the Sign test—that the difference-score population has a median of zero. A normal population meets all of these conditions except the one specifying that the population median is zero. Therefore, if the population is *known* to be normal all three tests become, in effect, tests of the median and, in that sense, comparable. In that case, the A.R.E.'s of the three tests relative to Student's matched-pair t test are 1.00 for the t test, $3/\pi$ for the Wilcoxon test, and $2/\pi$ for the Sign test, so the A.R.E. of the Wilcoxon test relative to the Sign test is $\frac{3}{2}$ or 1.50.

It is clear, therefore, that if one *knows* (not if he is reckless, or hopeful, enough to take for granted, but if he *knows*) that the population is normal, the t test is the sensible choice. Unfortunately, while we sometimes know that a population is quasi-normal, we never know that it is exactly normal. Again, if one *knows* that the population is symmetric, but does not know its exact shape, and wishes to test an hypothesis only about its median, the Wilcoxon test is the sensible choice (although if the population's shape is double-exponential the Sign test may be better [31]). And it *is* possible to know on a priori grounds that a population is exactly symmetric, at least when the experimenter's H_0 is true. But if one does not know that the population is normal or symmetric and wishes to test a simple hypothesis about the median only rather than a compound hypothesis about both the location and shape of the difference-score population, the Sign test is the most reasonable selection. This will often be the situation, so the Sign test will often be the best of the lot. However, it will also frequently happen that the Wilcoxon test's *compound* hypothesis about the $X - Y$ difference-score population implies just the inferential information one wishes to test about the X- and Y-populations. This will be the case, as in our example, when one of the two populations is known to be asymmetric and one wishes to test whether or not the entire X- and Y-populations are identical.

The Wilcoxon signed-rank test has been generalized to a more comprehensive class of cases by a number of authors. A particularly interesting generalization of this kind is that to paired comparisons in the multiple-treatment case [42].

5.6 THE SIGNED-RANK TEST FOR THE DIFFERENCE BETWEEN MEDIANS

If it is known that the X- and Y-populations (i.e., that for every value of i the distributions from which the observations X_i and Y_i were drawn) either (a) are both symmetrical or (b) are identical in shape and variance, then one can test the hypothesis that the median of the Y-population is Δ units greater than that of the X-population. Let $Z_i = Y_i - \Delta$. Then if the hypothesis is correct, Z_i and X_i come from populations that are either coaxially symmetric or identical, each $X_i - Z_i$ difference-score comes from a difference-score population that is symmetric about zero, and W_+, based upon the n modified difference-scores $X_i - Z_i$, i.e., $X_i - (Y_i - \Delta)$, will have its null distribution. One need only substitute $Y_i - \Delta$ for Y_i in the preceding test, therefore, to test the new hypothesis about the exact value of the difference in location parameters, provided that the new assumptions about population shapes are true. Unfortunately, however, there are few research situations in which an experimenter would know these assumptions to be met.

5.7 THE SIGNED-RANK TEST FOR THE MEDIAN OF A SYMMETRIC POPULATION

As originally presented, the signed-rank test tests whether or not for every value of i the difference-score population to which the obtained difference-score $X_i - Y_i$ belongs is symmetric about zero. Actually there is no need for the variate (whose distribution constitutes the population) to be a difference-score. Thus instead of applying the test to n difference-scores, $X_i - Y_i$, it could have been applied to n scores, X_i, to test whether or not every X_i was drawn from a population that is symmetric about zero. Suppose that we know that every X_i was drawn from the same population and that the common X-population is symmetric. Then if M is the median of the X-population, the variate $X - M$ will be symmetric about zero. So if we hypothesize that M_0 is the median of the X-population, subtract M_0 from each X_i, and apply the signed-rank test to the n obtained values of $X_i - M_0$, we are testing the H_0 that the median of the known-to-be-symmetric X-population has the value M_0. Occasionally we do know that a population is symmetric, but often we do not. When the latter is the case, the Sign test for the median is much to be preferred.

When applied to observations from a normally distributed population, the signed-rank test for the median has A.R.E. of $3/\pi$ or .955 relative to the one-sample t test for an hypothesized population mean and A.R.E. of $\frac{3}{2}$ or 1.50 relative to the Sign test for the median.

If the X-population is known to be symmetrical, rough confidence limits for its median M can be found by the following trial-and-error method. Find a value of M_0 such that when the signed-rank test is applied to the n obtained values of $X_i - M_0$, the test statistic $|W_-|$ has a lower-tail cumulative probability of exactly $\alpha' \cong \alpha/2$. There will be a range of values about this M_0 for all of which $|W_-|$ and α' remain constant. Let M_0' be the smallest value of M_0 in this range. Now find a value of M_0 for which the test statistic W_+ has a lower-tail cumulative probability of exactly $\alpha'' \cong \alpha/2$, and let M_0'' be the largest value to which this M_0 can be increased without changing α''. Then, letting $\alpha' + \alpha'' = \alpha$, we can state at confidence level of at least

$1 - \alpha$ that the X-population median M lies in the interval between M_0' and M_0'', i.e., that $M_0' \leq M \leq M_0''$.

Walsh [61] has devised a test for the same situation to which the present test applies. The two tests test essentially the same hypotheses and require the same assumptions; and certain cases of the Walsh test are equivalent to the present test. The primary advantage of the Walsh test lies in the fact that its tables yield confidence limits for the median almost directly.

5.8 WILCOXON'S RANK-SUM TEST FOR IDENTICAL POPULATIONS, SENSITIVE TO UNEQUAL LOCATIONS

This test is essentially the same as Fisher's two-sample test for identical populations, sensitive to unequal locations (see Section 4.2.3), except that prior to applying the test, the observations are replaced by their size ranks in the pooled sample. Fisher's test, as we described it, used $\bar{X} - \bar{Y}$ as the test statistic. However, it could just as legitimately have used the sum of the observations of a given type (X or Y) since, letting S represent the sum of all $n + m$ observations in the combined sample,

$$\bar{X} - \bar{Y} = \frac{\sum_{i=1}^{n} X_i}{n} - \frac{\sum_{j=1}^{m} Y_j}{m} = \frac{m \sum_{i=1}^{n} X_i - n \sum_{j=1}^{m} Y_j}{nm}$$

$$= \frac{m \sum_{i=1}^{n} X_i - n \left[S - \sum_{i=1}^{n} X_i \right]}{nm}$$

$$= \frac{(m + n) \sum_{i=1}^{n} X_i - nS}{nm} = \frac{C \sum_{i=1}^{n} X_i - C'}{C''}$$

where the various C's are constants over the randomization distribution of $\bar{X} - \bar{Y}$. (And, of course, a similar relationship could have been established between $\bar{X} - \bar{Y}$ and the sum of the Y's.) Therefore, $\bar{X} - \bar{Y}$, $\sum X$, and $\sum Y$ are all equivalent test statistics for Fisher's test. The latter two, however, are easier to compute. And computations are minimized by taking as the test statistic the sum of observations in the smaller-sized sample or, when $n = m$, the smaller of the two sums. This is the approach taken by the present Wilcoxon test, except, of course, that the observations are replaced by their ranks in increasing order of size in the combined sample.

5.8.1 RATIONALE

Suppose that an experimenter who has randomly and independently drawn n observations from an X-population and m observations (where $m \geqq n$) from a Y-population wishes to test the H_0 that the two populations are identical against the alternative hypothesis that they are nonidentical, and wishes the test to be especially likely to reject when the populations have unequal locations. If the experimenter's H_0 is true, "X-sample" and "Y-sample" are merely arbitrary labels for subsets of n and m observations, respectively, in the pooled sample of $n + m$ observations, all of which were, in effect, drawn from a common population. And therefore each of the $\binom{n+m}{n}$ possible "splits" of the $n + m$ observations in the combined sample into subsets of n and m observations was just as likely a priori to have become the "X-sample" and "Y-sample" as was the actually obtained "split." For each of these $\binom{n+m}{n}$ splits, there exists a value of the test statistic $W_n = \sum_{i=1}^{n} R_i$, where R_i is the rank of X_i, in size, among the $n + m$ observations in the combined sample (the smallest of the $n + m$ observations receiving a rank of 1, the next smallest a 2, etc.). And the frequency distribution of all $\binom{n+m}{n}$ values of W_n is the randomization distribution of W_n and the distribution W_n will have if the experimenter's H_0 is true. Therefore, we may reject the H_0 of identical populations if the value of W_n for the actually obtained data falls in a preselected rejection region of size α in the null randomization distribution of W_n. If the two populations differ in location only, then the larger (smaller) the location parameter of the X-population relative to that of the Y-population, the larger (smaller) we would expect W_n to be. And even if the two populations differ in shape and variance as well, we would expect the difference in location generally (i.e., with relatively infrequent exceptions in practice) to be well correlated with obtained values of W_n. Thus the test should (generally) be especially sensitive to locational differences between the two populations. This is not to say, however, that the test is necessarily insensitive to differences between populations having the same location. Such differences tend to cause the $\binom{n+m}{n}$ splits to be unequally likely. And if they are, the randomization distribution of W_n will generally differ from its null distribution, i.e., from the true sampling distribution of W_n obtained from an infinite number of pairs of X- and Y-samples of sizes n and m. For example if the X-population is uniformly distributed from $-.5$ to .5 and from 10,050 to 10,150, and if the Y population is uniformly distributed from 9,850 to 9,950 and from 19,999.5 to 20,000.5, then both populations have a mean of 10,000. But rejection is certain, at reasonable sample sizes and significance levels, if all observations are drawn from the portions of the two populations lying between 9,849 and 10,151. And the probability of this happening is $(100/101)^{n+m}$, which is overwhelmingly large when sample sizes are small. Even if location is defined by the median, the point can still be made. If the X-population is uniformly distributed consisting of the integers from 1 to 49 and from 100 to 149, and the Y-population is uniformly distributed consisting of the integers from 51 to 100 and from 151 to 199, then both populations have a median of 100 but W_n will tend to be smaller than if the populations were identical and a lower-tail test will tend to reject.

5.8.2 Null Hypothesis

Each of the $\binom{n+m}{n}$ different pairs of "samples" obtainable by dividing the total of $n + m$ observations, i.e., the pooled sample, into two sets, one containing n observations, the other m observations was equally likely, a priori, to have become the actually obtained data. This will be the case if the X- and Y-populations are identical and all assumptions are met.

5.8.3 Assumptions

In order for nonchance differences between samples to be attributable to differences between the sampled populations, they must not be attributable to differences in the manner in which observations were selected for drawing, e.g., to bias. It is therefore assumed that *sampling* is *random*.

The derivation implicitly assumed that if H_0 is true the X-sample, Y-sample, and combined sample can all be regarded as having come, in effect, from the "same" population. But the very act of sampling depletes a population, and therefore changes it, unless the population is infinite or observations are drawn one at a time and each observation is replaced before the next one is drawn. Failing to replace when sampling from identical finite populations may prevent each of the $\binom{n+m}{n}$ splits from being equally likely. For example, suppose the X- and Y-populations are finite and identical, each containing 10 units (say the integers from 1 to 10), and suppose that 10 X's and 5 Y's are drawn without replacements. The derivation requires that each of the different sets of 10 units that can be drawn from the $10 + 5$ units in the pooled sample have equal a priori probability of becoming the X-sample. But the fact is that the actually obtained X-sample had an a priori probability of 1.00 and all others an a priori probability of zero, e.g., the X-sample could not have contained two 7's, nor, in fact, could the Y-sample, although the pooled sample could have done so. Therefore in order for the $\binom{n+m}{n}$ splits to be "equally likely" when the X- and Y-populations are identical it is necessary to assume that each population contains an *infinite* number of units or that sampling is with individual *replacements*. If this assumption is true and if every unit in a given population has equal probability of being drawn (which is what we mean here by the assumption of random sampling), then when several observations are to be drawn from a population, the probability of drawing any given observation will be completely unaffected by the outcomes of the drawings that preceded it, i.e., observations will be *independent*. Thus under random sampling the assumption of infinite populations or sampling with replacements is sometimes roughly equivalent to the assumption that observations are independent and is sometimes expressed in the latter form.

The tabled null distribution of W_n applies to the case where the ranks of the observations are the consecutive integers from 1 to $n + m$. Therefore, in a sense (i.e., if one ignores the possibility of modifying the procedure) the test assumes that there are *no tied observations*. However, if all of the members of a group of equal observations belong to the same sample, they may be arbitrarily assigned ranks from the group of consecutive ranks for which the equal observations are collectively tied. All such assignments contribute exactly the same quantity to $\sum_{i=1}^{n} R_i$. So if H_0 and the assumptions of random and independent sampling are true, if there are no intersample ties, and if the above procedure is followed, W_n will have exactly its null randomization distribution and the test will be just as valid as if the ties had not happened. And if, instead, H_0 is false, the test will be just as powerful, at least in this instance, as if the ties had not occurred.

5.8.4 Efficiency

When both tests are applied to data from continuously distributed populations having identical shapes and variances to test whether or not they also have identical locations, the Wilcoxon test has, relative to the two-sample t test for unmatched data, an A.R.E. which can be as high as infinity but no lower than .864, depending upon the particular form of the populations' common shape [27]. If the common shape is normal (so that all of the assumptions of the t test are satisfied), the A.R.E. is $3/\pi$ or .955. If the common shape is rectangular, i.e., if the X- and Y-distributions are uniform, differing only in location, the A.R.E. is 1.00. And if the common shape is that of the density function

$$f(\chi) = \frac{\chi^2 e^{-x}}{\Gamma(3)}$$

the A.R.E. is 1.266. The lower bound of .864 is reached when the populations have parabolic shapes. For these A.R.E.'s and for A.R.E.'s in other cases see references [25, 27]. It is emphasized that the lower bound of .864 for the A.R.E. is contingent upon the two populations having identical shapes and variances. If this condition is not met it is possible for the lower bound to drop as low as zero.

For the problem of testing whether or not two populations with identical shapes and variances also have identical locations, a number of distribution-free tests are appropriate. Which of several such tests will have the greatest power (say, when sample sizes are infinite, the ratio n/m is fixed, and locations differ infinitesimally—the conditions under which A.R.E.'s are obtained) will depend upon the particular form of the common population shape. The Wilcoxon test appears to be the most powerful rank test (for slightly unequal locations of otherwise identical populations) when the common population shape is that of a logistic distribution [38]. Tests of the normal scores family and Fisher's randomization test are more powerful than the Wilcoxon test when the common shape is normal. Indeed once the shape is specified it should be theoretically possible to invent an optimum distribution-free test which would often differ from the Wilcoxon test. To say this, however, is only to reaffirm that a test tailor-made for a particular situation naturally enjoys an advantage. What is more important is that under a wide variety of situations the Wilcoxon test competes well and has a high A.R.E. relative to rival distribution-free and parametric tests that have already been devised. Seldom does the A.R.E. of the Wilcoxon test, relative to whatever test has optimum efficiency, fall far enough below 1.00 to give the comparison test much practical advantage. (For exceptions to this statement see normal scores tests.)

When sample size is finite, the Wilcoxon test continues to compete well in relative power. Small-sample efficiencies of the Wilcoxon test relative to the t test under conditions assumed by the latter are given in references [27] and [67]; see also [11]. These small-sample relative efficiencies remain fairly constant with increasing differences between population means and exceed the A.R.E. when the difference between means is zero. They also seem to increase with decreasing sample sizes and with decreasing values of α.

The power and/or validity (i.e., correspondence between true and alleged probabilities of a Type I error) of the Wilcoxon test has been compared with those of the t test under common conditions meeting all the assumptions of both tests [7, 23, 55], under common conditions meeting all the assumptions of the Wilcoxon test but not of the t test [7, 23], and under common conditions meeting all the assumptions of neither test when they are used as tests of identical locations [7, 23, 50, 59, 62]. It has also been compared with that of rival distribution-free tests under strictly "nonparametric" conditions, i.e., conditions quite different from those demanded by normal-theory [19, 39, 50]. Comparisons between the powers of the Wilcoxon and t tests under violation of the latter's assumptions suffer from the fact that the violation affects the probability of a Type I error for the t test so that the two tests are not being compared at a common value of α. Thus a valid test is being compared with an invalid one under noncomparable conditions. To judge superiority on the basis of power alone, in such cases, is to ignore (a) the advantage of knowing one's true α level, and (b) the correlation between power and the true value of α. But despite the logical fuzziness of such comparisons and the bias implicit in letting an invalid test compete on equal terms (i.e., without penalty or disqualification) with a valid one, the Wilcoxon test still makes a good showing (having greater power under some conditions, less under others) as the authors of the comparisons have freely admitted. In comparisons with other distribution-free statistics, the Wilcoxon test typically either ranks first or, when the set of tests being compared includes the optimum test for the conditions of the comparison (and the optimum test is not the Wilcoxon test itself), ranks a close second. Graphic power functions for the Wilcoxon test can be found in references [7, 55; cf. 39], tabular power functions in references [19, 27; cf. 11].

Consider the one-tailed Wilcoxon test that rejects if the sum of the ranks of the X-observations is "too large." That test is consistent and unbiased if, when the null hypothesis of identically distributed X- and Y-variates is false, the true situation is that (ordinatewise) the cumulative distribution of the X-variate lies entirely below that of the Y-variate (the two cumulatives perhaps touching at certain points but never crossing) [36, 40; see also 38]. That test is also consistent against the broader class of alternatives in which the probability is greater than $\frac{1}{2}$ that a randomly drawn observation from the X-population will exceed a randomly drawn observation from the Y-population, i.e., against all alternatives for which $P(X > Y) > .5$ [36, 5]. The two-tailed test is consistent against alternatives for which $P(X > Y) \neq .5$ [5], but is not necessarily unbiased in all cases, especially when sample sizes are unequal.

5.8.5 Tables

Exact lower-tail critical values of W_n, i.e., of $\sum_{i=1}^{n} R_i$, the sum of the ranks in the smaller or equal-sized sample, are given in Table III of the Appendix. For $n = 1$ (1) 25, $m = n$ (1) 25, and "one-tailed" α's of .10, .05, .025, .01, .005, and .001 the table lists the largest value of W_n whose lower-tail cumulative probability is $\leq \alpha$ in the null, randomization distribution of W_n. That distribution is symmetric about its mean $\bar{W}$, and the value of $2\bar{W}$ is given in the table. To do an upper-tail test one simply enters the table with $2\bar{W} - W_n$ and rejects if it is smaller than the critical value listed. Because of the symmetry of the W_n distribution about $\bar{W}$, if the value $W_n = C$ has an upper-tail cumulative probability $\leq \alpha$ (cumulating from right to left) and is therefore significant at the α level for an upper-tail test, then the symmetrically opposite value $W_n = C - 2(C - \bar{W}) = 2\bar{W} - C$ has a lower-tail cumulative probability $\leq \alpha$ (cumulating from left to right) and is significant at the α level for a lower-tail test. To do a two-tailed test, enter the table with W_n or $2\bar{W} - W_n$, whichever is smaller, and double the listed α values.

For $n = 3$ (1) 50, $m = n$ (1) 50, and one-tailed α's of .05, .025, .01, and .005, Wilcoxon, Katti, and Wilcox [66] give: (a) the critical value of W_n for a lower-tail test at significance level α; (b) the critical value of W_n for an upper-tail test at level α; (c) their common, exact, one-tailed cumulative probability, cumulating from tail toward center; (d) the value of W_n falling just outside the lower-tail, α-sized, critical region; (e) the value of W_n falling just outside the upper-tail, α-sized, critical region; (f) their common, exact, one-tailed cumulative probability.

Aside from the above, not many tables give any more information than the table in the appendix. For the Mann-Whitney U statistic, which is a linear transformation of W_n and therefore an equivalent statistic if referred to U-tables, the complete cumulative probability distribution of U is given in [47] for all cases in which both sample sizes are $\leqq 10$; exact probability levels corresponding to critical values of U can be obtained in [29] for some additional cases in which $n + m \leqq 30$; and critical values of U for cases where $n = 1$ (1) m, $m = n$ (1) 40, and one-tailed α's of .10, .05, .025, .01, .005, .0025, .001, and .0005 can be found in [43] as well as references to other tables not mentioned here.

If the reader consults any of these "outside" tables, he will have to readjust to symbols and notation that are different, or even the reverse (in the case of sample sizes), from those used here.

The distribution of the test statistic W_n has a mean of $[n(n + m + 1)]/2$, a variance of $[nm(n + m + 1)]/12$, and a symmetric shape that approaches normality as n and m increase (proportionately, i.e., with n/m a constant) and becomes normal when n and m become infinite. Therefore, at sample sizes too large for the exact tables to apply, approximate probability levels can be obtained by referring the critical ratio

$$\frac{W_n - [n(n + m + 1)/2]}{\sqrt{nm(n + m + 1)/12}}$$

to probability tables for the unit normal deviate. To correct for continuity, reduce the absolute value of the numerator by $\frac{1}{2}$. This approximation is a reasonably good one if $n = m \geqq 25$ and $\alpha \geqq .025$, and it is fair if either of these conditions holds, but it is poor if α is extreme and n is much smaller than m (e.g., at $n = 5$, $m = 25$, and true, exact $\alpha = .0008$, the approximation yields an α-value of .0019) [60]. For a better but more complicated approximation see [60].

For treatment of ties see Sections 3.3 and 3.4.

5.8.6 Example

An experimenter wishes to test, at the one-tailed .05 level of significance, the null hypothesis that the distribution of weight increments for a certain class of individuals under diet X would be identical to that under diet Y. He is willing to settle for the broad alternative hypothesis that the two distributions are nonidentical, but he wishes the test to be especially sensitive to forms of nonidentity which include differences in central tendency, in which the X-distribution's location is below that of the Y-distribution. Ten individuals are randomly drawn from the class in question (which might be restricted to people of the same sex, age, and initial weight, etc.). Five of them are randomly assigned to diet X, five to diet Y, but, through misadventure unrelated to the experiment, two of the X-group fail to complete the experiment. The data for the three subjects receiving diet X and the five receiving diet Y, and the necessary steps and computations in performing the Wilcoxon test are as shown in the accompanying table. The number

	Weight increments				
Diet X	10	5	3		
Diet Y	29	25	22	16	8

	Ranks of weight increments					*Sum*	
Diet X	4	2	1			7	$= \sum_{i=1}^{n} R_i = W_n$
Diet Y	8	7	6	5	3	29	

n of observations in the sample with the fewer observations is 3; the number m of observations in the sample with the greater number of observations is 5; and the sum of the size ranks for the observations in the sample containing the fewer observations is

$$W_n = \sum_{i=1}^{n} R_i = 7.$$

Entering Table III with $n = 3$, $m = 5$, and $\alpha = .05$, the experimenter finds a critical value of 7 for W_n, meaning that a priori $P(W_n \leqq 7) \leqq .05$, so that the obtained value of 7 is significant for a lower-tailed test at the .05 level. He therefore rejects the H_0 that the distributions of weight increments under the two diets would be identical, accepts the alternative hypothesis that the potential distributions are nonidentical, and concludes that there is strong evidence that (a) the median, or other location parameter, of the X-distribution lies below that of the Y-distribution; (b) on randomly selected individuals from the tested class, diet X would tend to produce less weight increment than diet Y.

5.8.7 DISCUSSION

Those who are naïve in practical experimentation may be disheartened at the vagueness of the alternative hypothesis, preferring a definite statement of unequal means to a statement of nonidentity which is likely to include inequality of means. However, while it is theoretically possible for treatment-populations to be nonidentical while having exactly equal means, and while it is absurdly easy to invent such populations out of one's head, it is fantastically difficult to create them in the laboratory. That is, in most areas of research, it is virtually impossible by changing the manipulated variable (i.e., treatment or condition) to induce a change in *any* one aspect (i.e., mean, median, mode, variance, interquartile range, tenth percentile, shape, etc.) or combination of aspects of the population distribution of the measured variable without producing *some* change in *every* aspect. Nor is the sought situation any easier to find in nature. Thus, as a practical matter in most areas of research, a statistical test's verdict of nonidentical populations is tantamount to a verdict of unequal means, unequal variances, etc. The probability that the verdict of nonidentical populations is a false one, i.e., the probability of a Type I error, will generally greatly exceed the probability of erring in proclaiming means unequal when populations are actually nonidentical. Thus the problem is rather academic. Furthermore, in relatively few experimental situations is it actually desirable to reject only for a certain specific aspect of nonidentity, such as unequal means. More often the experimenter either is or should be interested in several or all aspects of possible nonidentity. This is especially the case when experimentation is "exploratory," as much experimentation is.

Tests that are equivalent to the Wilcoxon test have been published by a number of authors, e.g., Mann and Whitney, Festinger, van der Reyden, and Haldane and Smith. The best known of these alternative forms of the Wilcoxon test is the Mann-Whitney test, whose test statistic U is obtained by arranging the $n + m$ observations in the pooled sample in increasing order of size and counting the number of times an X precedes a Y. Thus if μ_j is the number of X's that are smaller than Y_j,

$$U = \sum_{j=1}^{m} \mu_j$$

The probability level for the test is obtained by referring the obtained value of U to the randomization distribution of U. Therefore, the Mann-Whitney U test must be equivalent to the Wilcoxon W_n test if U is a linear transformation of W_n. That it is can be demonstrated as follows. Let Y_j be the Y-observation having a rank of j (i.e., being the jth smallest) among the m observations in the Y-sample and a rank of R_j in the pooled sample, and let μ_j be the number of X's smaller than (i.e., preceding) Y_j. Also, let X_i and R_i be similarly defined for observations from the X-sample. Then,

$$R_j = j + \mu_j \quad \text{and} \quad \sum_{j=1}^{m} R_j = \sum_{j=1}^{m} j + \sum_{j=1}^{m} \mu_j$$

But the average of the integers from 1 to m is $(m + 1)/2$, so their sum is $m(m + 1)/2$ and therefore

$$\sum_{j=1}^{m} R_j = \sum_{j=1}^{m} j + \sum_{j=1}^{m} \mu_j = \frac{m(m + 1)}{2} + U$$

But it must also be true that

$$\sum_{j=1}^{m} R_j = \sum_{k=1}^{n+m} k - \sum_{i=1}^{n} R_i = \frac{(n + m)(n + m + 1)}{2} - W_n$$

Therefore,

$$U = \frac{(n+m)(n+m+1)}{2} - \frac{m(m+1)}{2} - W_n = \frac{n(n+2m+1)}{2} - W_n$$

and the equivalence between the two tests is established as well as the conversion relationship between their two test statistics. The primary advantages of the rank-sum form of the test are that it is easier to relate conceptually to its Fisherian predecessor based on observation randomization, and it is easier to apply and less conducive to errors in determining the value of the test statistic.

An interesting and useful feature of the test is that its test statistic yields a highly meaningful estimator. Let ρ be the probability that a randomly drawn observation from the X-population will be smaller than a randomly drawn observation from the Y-population, both populations being continuous, i.e., let $\rho = P(X < Y)$. Then $\hat{\rho} = U/(nm)$ is an unbiased estimator of ρ, and the distribution of $\hat{\rho}$ has a mean of ρ, a variance less than or equal to $\rho(1-\rho)/n$ (where n is the size of the smaller sample), and a shape that approaches normality (if $0 < \rho < 1$) as both sample sizes approach infinity while the ratio m/n remains constant. The above information can be used to establish appropriate confidence limits for ρ about the obtained value of $\hat{\rho}$. If sample sizes are neither small nor extremely unequal and if ρ (as estimated by $\hat{\rho}$) does not lie close to 0 or 1, the approximation should be a good one [5].

If observations become available sequentially and in order of increasing size, as in life-testing, it will often be possible to reach a statistical decision, i.e., to accept or reject, on the basis of less than the originally intended number of observations. If n X-units and m Y-units have been placed on life test at the same moment, and n' of the X's and m' of the Y's have expired, one ranks the known life-durations of the expired observations from 1 to $n' + m'$. Then W_n is calculated twice, once assigning the $n - n'$ largest of the ranks from 1 to $n + m$ to the unexpired X-units, once assigning the $m - m'$ largest of these $n + m$ ranks to the unexpired Y-units. If W_n falls within the rejection region in both cases or falls outside the rejection region in both cases, the statistical issue is already resolved and the experiment can be terminated [1]. Another approach is to treat the obtained X's and Y's as samples censored at a common point [22].

5.9 THE RANK-SUM TEST FOR AN HYPOTHESIZED DIFFERENCE IN LOCATION

If one knows that the X- and Y-populations have identical shapes and variances so that the populations can differ only in location, he can test the hypothesis that a location parameter, e.g., the median, of the Y-population is Δ units greater than the corresponding location parameter of the X-population. Let $Z = Y - \Delta$. Then, if the hypothesis is correct, the X- and Z-variates have identical distributions, and whether or not this latter condition is true can be tested by applying the previous rank-sum test to the n X-observations and m Z-observations, i.e., the m "relocated" or transformed "Y"-observations obtained by subtracting Δ from each of the m original Y-observations.

Confidence limits for Δ can be found in the following manner. Let Δ' (determined by trial and error) be the smallest (i.e., the most extreme negative) value of Δ for which W_n (based on X's and Z's) has a lower-tail cumulative probability of exactly $\alpha' \cong \alpha/2$. (There will generally be a range of Δ values for which W_n, and therefore α', remain constant; Δ' must be the smallest value of Δ in that range.) Let Δ'' be the largest value of Δ for which W_n has an upper-tail cumulative probability of $\alpha'' \cong \alpha/2$. Finally let $\alpha' + \alpha'' = \alpha$. Then we can state at confidence level of at least $1 - \alpha$ that the difference Δ between medians of the X- and Y-populations, known to be identical in shape and variance, lies in the interval between Δ' and Δ'', i.e., that $\Delta' \leq \Delta \leq \Delta''$. See also [37].

5.10 THE RANK-SUM TEST WITH GROUPED DATA

If of $n + m$ units from a common population, n are randomly assigned to Treatment X and the remainder to Treatment Y, the rank-sum test can be used without modification to test whether or not the sets of n and m units still belong to a common population subsequent to treatment, i.e., whether or not treatment effects are identical. It may be, however, that the original unit-population can be subdivided into components or subpopulations whose variances are smaller and which therefore represent common or matching conditions which can be exploited to increase the sensitivity of the test. For example, the original, overall population might be "people." But if response to treatment has any relationship to sex, age, or race, it might be well to subdivide the "people" population into populations of female–middle-aged–Caucasions, male–young–Negroes, etc., and to capitalize upon the greater homogeneity of units within the subpopulations. But in order to do so effectively, the units from each subpopulation, i.e., matching condition, must be dealt with on an individual-group basis, requiring the rank-sum test procedure to be modified as follows.

From the N_i units available for treatment from the ith subpopulation or matching condition, let n_i units be randomly selected and assigned to treatment X, the remainder, $m_i = N_i - n_i$, being assigned to treatment Y. After treatments, measurements upon these units will yield n_i X-observations and m_i Y-observations. Let these $n_i + m_i$ observations be ranked, from 1 to $n_i + m_i$, in order of increasing size. Let W_{n_i} be the sum of the ranks of the X-observations (irrespective of whether or not $n_i < m_i$) and let T be the sum of the W_{n_i} over all values of i, i.e., over all b blocks of data corresponding to the b different subpopulations or matching conditions. Then knowing the values of b and the n_i and m_i, it is possible to determine the randomization distribution of T. Wilcoxon [64, 65] has provided tables for T for $b = 2$ (1) 7 and $n_i = m_i = 2$ (1) 7 for every i. Considering the number of parameters upon which the distribution of T depends, exact tables for the general case are too much to hope for. However, unless the total number of observations is very small, the normal approximation to the distribution of T should be serviceable, and it is easily obtained. We know that the distribution of each W_{n_i} has a mean of $[n_i(n_i + m_i + 1)]/2$ and a variance of $[n_i m_i(n_i + m_i + 1)]/12$, that $T = \sum_{i=1}^{b} W_{n_i}$, and that the W_{n_i} are independent. Therefore, since the expected value and variance of a sum of independent variates are the sums of the expected values and variances, respectively, of the component variates, the mean and variance of the distribution of T must be

$$\sum_{i=1}^{b} \frac{n_i(n_i + m_i + 1)}{2} \quad \text{and} \quad \sum_{i=1}^{b} \frac{n_i m_i(n_i + m_i + 1)}{12}$$

Letting $E(T)$ stand for the former and $\mathrm{Var}(T)$ for the latter, approximate probability levels for an obtained value of T may be obtained by referring the critical ratio $[T - E(T)]/\sqrt{\mathrm{Var}(T)}$ to tables of probabilities for the unit-normal deviate.

Using the grouped-data form of the rank-sum test affects the power of the test in two opposite ways. To the extent that there are real and important differences among the subpopulations used as a basis for matching units prior to treatment, matching tends to reduce the purely chance differences between the sets of units receiving the two treatments. This greater initial homogeneity permits smaller treatment effects to become statistically detectable, and this tends to increase the power of the test. On the other hand, in the original rank-sum test the largest of all observations would receive a rank of $\sum_{i=1}^{b} (n_i + m_i)$, whereas in the grouped data form it would receive a rank of $n_i + m_i$, if it occurred in the ith block of observations. And unless the ith block contained more observations than any other block, the rank of the largest of all observations would not even be the largest of all ranks. Thus going from a single scale of rank measurement, extending from 1 to $\sum_{i=1}^{b} (n_i + m_i)$ (on which every observation receives a different rank) to b different and rather unrelated scales from 1 to $n_i + m_i$ (on which a smaller observation in a different group may receive a larger rank) tends to reduce the sensitivity of the test and therefore its power. The greatest sacrifice of ordinal (or, roughly speaking, of magnitudinal) information occurs when for every value of i, i.e., for every matching condition, $n_i = m_i = 1$. In that event, each observation receives a rank of either 1 or 2, depending upon whether it is smaller or larger than the other observation in the matched pair. Thus despite the ostensible use of ranks it is really only dichotomous directional information that is being exploited, and in this case the grouped data rank-sum test is equivalent to the Sign test. It would be a mistake, however, to conclude that in this case the grouped data form of the rank-sum test is inferior to the original form of the test. Clearly, there is a tradeoff between the benefits of greater homogeneity of original treatment groups due to matching and the attendant detriments of reduced ordinal information. If the various subpopulations used for matching are all identical, then there is no real benefit from matching, but there is a disadvantage from using abbreviated rank scales, so the original rank-sum test is superior to the grouped data form of the test. On the other hand, if no subpopulation's range of variate values overlaps with that of any other, then even if $n_i = m_i = 1$ for every value of i, we would expect the grouped-data form of the test to be superior. Certainly a good practical rule would be to use only such matching conditions as one has good reason to believe associated with strongly differing distributions of units, combining matching conditions whose subpopulations are suspected of being similar.

Methods of optimizing the test are discussed in [14; see also 26], and its efficiency is treated in [26, 46]. The test described in the preceding paragraphs is optimal without modification if $n_i + m_i$ is a constant. The A.R.E. of the optimal rank-sum test for grouped data relative to the t test for grouped data, under conditions assumed by the latter, ranges from $2/\pi$—the A.R.E. of the Sign test relative to the t test for matched pairs—to $3/\pi$—the A.R.E. of the original rank-sum test for ungrouped data relative to the t test for ungrouped data [26]. The reason that the A.R.E. of the rank-sum test for grouped data never rises above that for ungrouped data is presumably that the comparison t test benefits more from matching than does the rank-sum test. It will be recalled that $N_i = n_i + m_i$. If all $N_i = N$, i.e., if every block contains the same *total* number of X's and Y's, then the A.R.E. of the ordinary rank-sum test for grouped data relative to the t test for grouped data under the conditions assumed by the latter is

$$\left(\frac{3}{\pi}\right)\left(\frac{N}{N+1}\right)$$

irrespective of the value, or the constancy, of the ratio n_i/m_i, so long as there is at least one X and one Y in every block.

5.11 RANK TESTS FOR SCALE

In order for the rank sum to have its randomization distribution when the null hypothesis of identical X- and Y-populations is true, all that is necessary is that the integers from 1 to $n + m$ be assigned to the $n + m$ observations in the pooled sample in a way that takes no account of whether an observation is an X or a Y. Many different methods of integer assignment to observations fulfill this criterion. In the preceding tests, the advantage of letting these integers be the size ranks of the observations was that this made the rank-sum test statistic the rank analogue of $\bar{X} - \bar{Y}$ in Fisher's test based on the Method of Randomization applied to original observations. The rank-sum test was thereby made especially sensitive to differences in population location, since the difference between sample means is an unbiased estimate of the difference between population means. One might suppose that by modifying the method of assigning integers to observations one could obtain an equally successful rank test for dispersion. However this turns out not to be the case. "Rank-randomization" tests for dispersion are flawed by an overly strong dependence upon location and sometimes by erratic and undesirable behavior which may completely thwart the experimenter's purpose. Such tests *are* adequate and effective tests for differences in "scale" of otherwise identical populations. But one is unlikely to require such a test, since if he knows two populations to be identical in all respects other than scale, he probably also knows whether or not they differ in scale. Thus there would be no point in conducting the test.

5.11.1 The Siegel-Tukey Test

A modification of the Wilcoxon rank-sum test can be used to test whether or not two otherwise identical populations have the same scale parameter. Consider an X-population with median M_X and a Y-population with median M_Y, and let $x = X - M_X$ and $y = Y - M_Y$ be the deviations of the X- and Y-variates from their respective population medians. Now suppose that M_X equals M_Y and that for every population value of x there is a value of y equal to θ times x and having exactly the same point probability as x. (So that a graph of the x-distribution plotted against an abscissa scale where one x-unit equals one centimeter would have exactly the same appearance as a graph of the y-distribution plotted against the same ordinate scale and against an abscissa scale where one y-unit equals $1/\theta$ centimeters. In such cases the distributions of x and $z = y/\theta$ are identical, and the distributions of x and y are said to differ only as to scale parameter θ.) The following modification of the Wilcoxon procedure tests the hypothesis that $\theta = 1$, i.e., tests whether or not populations known to be identical in every respect except scale are also identical with regard to scale parameter.

Let n and m observations be drawn randomly and independently from the X- and Y-populations respectively, with $n \leqq m$, and assume that no observations are tied. The $n + m$ observations in the pooled sample are arranged in ascending order of size. Ranks are then assigned proceeding from the extremes toward the middle. The lowest observation is given a rank of 1, the highest a rank of 2, the second-highest a 3, the second-lowest a 4, the third-lowest a 5, the third-highest a 6, etc. The test statistic is W_n, the sum of the ranks of the X-observations, and its randomization distribution is the same as that for the W_n statistic in the Wilcoxon rank-sum test for location, so the latter test's tables can be used. As an example, suppose that we wished to test, at the .05 level, the hypothesis of identical scale parameters against the alternative that $\theta < 1$, i.e., that the X-population is more spread out than the Y-population (which would tend to cause W_n to be "too small"). If the X-observations were -28.1, -16.7, and 7.1, and the Y-observations were -11.0, -4.2, 1.5, 2.1, 3.6, 5.0, and 9.8, the test computations would be as shown in the table below, where the X-observations and their ranks are in boldface to facilitate identification. The sum

Observations in pooled sample	**−28.1**	**−16.7**	−11.0	−4.2	1.5	2.1	3.6	5.0	**7.1**	9.8
Their "ranks"	**1**	**4**	5	8	9	10	7	6	**3**	2

of the ranks for the three X-observations is 8. Entering Table III of the Appendix with $n = 3$, $m = 7$, we find $W_n = 8$ to be the critical value of W_n for a lower-tailed test at the .05 level. So we can reject the H_0 that $\theta = 1$ in favor of the alternative that $\theta < 1$ if we know that the two populations are identical in all respects other than scale and if all assumptions are met.

Mathematically, this test and the Wilcoxon rank-sum test make the same assumptions and test the same null hypothesis. Practically, both tests test the experimentally phrased null hypothesis that the X- and Y-populations are identical. And both assume random and independent sampling and the absence of ties among the observations. However, because they employ different procedures in assigning ranks to observations, the two tests are rendered unequally sensitive to alternative hypotheses and behave quite differently when H_0 is false. The difference is not simply in the *particular* alternatives against which they are most sensitive, nor in the *degree* of that sensitivity, but rather in the very existence of *any* sensitivity against a false H_0 of identical populations in certain experimentally realistic and uncontrived classes of situations. Although we were justified in labeling the Wilcoxon rank-sum test a test for identical populations sensitive to unequal locations, we shall see that we are *not* justified in analogously labeling the present test a test for identical populations sensitive to unequal *dispersions*. The present test does not necessarily tend to reject simply because the populations are spectacularly nonidentical—not even if that spectacular nonidentity includes spectacular differences in dispersion. Nor does the test necessarily tend to accept and reject when it should when the populations differ in dispersion but have equal locations. Rather, the test can be counted upon to behave properly only when possible population differences are confined exclusively to the "scale" aspect of dispersion. Thus the test is properly a test for scale, not dispersion, and in order to make it so it requires the additional assumptions that $M_X = M_Y$ and that population shapes can differ only because of differences in "scale."

The above points are easily exemplified. If the test is to be an effective test for the H_0 of identical populations, then not only must the sampling distribution of W_n coincide with the randomization distribution of W_n when H_0 is true, but obtained values of W_n must tend to fall at the extremes of the randomization distribution of W_n when H_0 is badly false. Consider the case where the X- and Y-populations are so spectacularly nonidentical that they are completely separate, i.e., their ranges do not overlap. Then whatever their respective shapes or dispersions, if $n = m$, the Siegel-Tukey procedure of assigning ranks will cause W_n to fall at about the middle of its randomization distribution, so the test is *certain* to *accept* the H_0 of identical populations (at least if a standard α value is used). But if n is sufficiently smaller than m, W_n is certain to fall at the lower tail of its randomization distribution, so a lower-tail or two-tail test is *certain* to *reject*, and an upper-tail is certain to *accept*. Thus acceptance or rejection of this false H_0 is determined exclusively by factors [the sizes of the samples and (sometimes) the location of the rejection region] which are completely irrelevant to the population condition in question.

A proposed solution is to restrict the test to the case where the difference between population medians, $M_X - M_Y = \Delta$, is known and to modify it in such a way as to eliminate the influence of location. This consists of subtracting Δ from each X-observation and substituting the $X - \Delta$'s for the X's, i.e., applying the test to the $X - \Delta$'s and Y's. Unfortunately, there are cogent objections to this solution. If one does not know whether or not the two populations have equal dispersions, he is quite unlikely to know the exact value of $M_X - M_Y$. And if he substitutes the difference between *sample* medians for the required difference between *population* medians, then even if the $X - (M_X - M_Y)$ and Y-populations (and therefore the $X - M_X$ and $Y - M_Y$ populations) are identical, the $\binom{n+m}{n}$ distinguishably different "assignments" of n ranks to one sample and m to the other will not be equally likely a priori, and the test will not be distribution-free [15].

But even if the influence of unequal locations can be completely and satisfactorily eliminated, the test does not necessarily become sensitive to all aspects of dispersion. It is easy to invent populations having identical medians and having identical dispersion indices of a given type but having shapes whose differences would tend to cause the test to reject. And this continues to be true even if we specify that both populations must be symmetric. For example, if the X-population is U-shaped with range greater than $R/2$ and the Y-population is rectangular with the same median M and variance and with range R, then more of the Y-population will lie within the verticals of the X-population than lies outside them. So, $P(|X - M| > |Y - M|)$ will exceed $\frac{1}{2}$ with the result that the average of the X-ranks will tend to be smaller than the average of the Y-ranks and a two-tailed or lower-tailed test will "tend to reject." By resorting to situations less likely to be encountered in practice, one can contrive much more spectacular examples.

Thus one is reluctantly forced to the conclusion that the test is an excellent one only in the highly unlikely situation where the experimenter has highly sophisticated information about all aspects of his populations except their scales. In the more usual practical situations the test is overly prone to accept or reject for the wrong reasons (at least so far as the experimenter is concerned). In such cases it is a tricky test at best and should certainly be avoided by the novice. Unfortunately the same comments apply to most rank tests for dispersion. Therefore in most cases the reader will probably find dispersion tests based on frequencies (see Sections 7.7.2, 7.7.4, 8.5, and 9.6) to be preferable.

The A.R.E. of the Siegel-Tukey test relative to the classical, variance-ratio F test can take values anywhere from zero to infinity, depending upon the type of population involved. When the populations are normally distributed the A.R.E. is .608. For A.R.E.'s relative to other tests, and for small-sample power functions and power comparisons, see [32].

5.11.2 The Freund-Ansari Test

This test preceded the Siegel-Tukey test. The latter differs from it primarily in slightly modifying the method of assigning ranks so as to give the test statistic the same null distribution as the Wilcoxon rank-sum test statistic. The advantage of the Siegel-Tukey test is that Wilcoxon tables can be used. However, brief tables exist [2] for the Freund-Ansari test, and the Freund-Ansari method of assigning ranks is the more logically appealing. It consists of assigning a rank of 1 to both the lowest and highest observations in the pooled sample, a rank of 2 to both the second-lowest and second-highest observations, a rank of 3 to both the third-lowest and third-highest, etc. The test statistic is the sum of the ranks assigned to the n observations in the X-sample, but its randomization distribution is different from that of the Wilcoxon rank-sum since half as many rank-integers are used and each integer (except the highest one when $n + m$ is odd) is used twice. Tables for the test statistic can be found in [2] and when these tables do not apply, approximate probabilities can be obtained by treating the test statistic as normally distributed with mean and variance given in [2 or 15]. The A.R.E. of this test relative to the parametric, variance-ratio F test has been obtained for the case where the alternative to the H_0 of identical populations is that the populations differ in scale alone. The A.R.E. is .61 when the two populations have normal distributions, .60 when they have uniform distributions, and .94 when they have double-exponential distributions [2]. The test is consistent if when the two populations differ they differ only in scale. It is conjectured [2] also to be consistent against alternatives for which

$$P(|X - M| < |Y - M|) \neq \tfrac{1}{2}$$

where M is the common median of the two populations. Presumably the same A.R.E.'s and conclusions about consistency would hold for the Siegel-Tukey test. Because of the similarity between the two tests, all of the criticisms of the Siegel-Tukey test apply equally to the present test.

5.11.3 Mood's Test

In closer analogy with classical tests for scale, Mood [44] has proposed a test using as test statistic

$$M = \sum_{i=1}^{n} \left(R_i - \frac{n+m+1}{2}\right)^2$$

where R_i is the size rank of X_i in the pooled sample (the smallest of whose observations receives a rank of 1, the next-smallest a 2, etc., and the largest a rank of $n + m$) and $(n + m + 1)/2$ is the average rank of all observations comprising the pooled sample. The null distribution of M has a mean of $[n(n + m + 1)(n + m - 1)]/12$ and a variance of $[nm(n + m + 1) \cdot (n + m + 2)(n + m - 2)]/180$, and the test apparently consists of treating M as a normal deviate, since no exact tables seem to have been prepared. Thus the test cannot be expected to be very accurate when sample size is small. Relative to the variance-ratio F test, when alternatives to the H_0 of identical populations involve differences in scale alone, Mood's test has A.R.E.'s of .76 when the populations are normal, 1.00 when they are uniform, and 1.08 when they are double-exponential [2]. These A.R.E.'s are higher than those for the Freund-Ansari test in corresponding cases (A.R.E.'s for Mood's test relative to the Siegel-Tukey test in these and other cases are given explicitly in reference [32]). One must attribute this to the squaring of the term $R_i - [(n + m + 1)/2]$, since otherwise Mood's test and the Freund-Ansari test use essentially the same information (although in somewhat different form). Because of this similarity, all of the criticisms of the Siegel-Tukey test must also apply to Mood's test, e.g., the fact that the test is certain to accept if $n = m$ and the two populations do not overlap.

5.11.4 Other Tests

Several other tests have been proposed (see Siegel and Tukey [52]). However they have little practical appeal, tending to be overly cumbersome.

The faults of rank tests for dispersion, as a class, are discussed by Moses [45], where the author points out that: "No rank test . . . can hope to be a satisfactory test against dispersion alternatives without some sort of strong restrictions (e.g., equal or known medians) being placed on the class of admissible distribution pairs." Gastwirth [18] has proposed some interesting modifications of the Siegel-Tukey and Wilcoxon tests which often increase their A.R.E. relative to their classical counterparts. The increase is brought about by " 'throwing away' part of the data."

5.12 FRIEDMAN'S MULTI-SAMPLE TEST FOR IDENTICAL TREATMENT EFFECTS, SENSITIVE TO UNEQUAL MEAN EFFECTS AND USING MATCHED OBSERVATIONS

This test is essentially the same as Pitman's test for identical populations, sensitive to unequal means and using matched observations (as described in Section 4.2.4) except that, prior to applying the test, the observations are replaced by their within-row ranks. Thus it is the rank-randomization analogue of Pitman's observation-randomization test.

5.12.1 Rationale

Suppose that from each of R possibly different populations (representing "matching conditions") an experimenter has drawn C units, has randomly assigned them to C different treatments, and wishes to test the hypothesis that the treatments have identical effects against the alternative that they do not, using a test that is especially sensitive to unequal medians in the treatment-effects distributions. (Alternatively, if there are no carryover effects, a single unit can be drawn from each of the R populations and subjected to all C treatments.) Let the observations, i.e., measurements, on the units subsequent to treatment be recorded in a table with C columns, representing treatments, and R rows, representing matching conditions. Then let the observations in each row be replaced by their size ranks from 1 to C, the rank of 1 going to the smallest observation, 2 to the next-smallest, etc. If treatments have identical effects under each matching condition, then the observations in a given row, subsequent to treatment, belong to the same population just as did the units in each row, prior to treatment. And since the units in a given row were randomly arranged prior to treatment, so are the observations in that row randomly arranged subsequent to treatment. Therefore, if within each matching condition treatment effects are identical, then for every row the arrangement of ranks is a random permutation of the integers from 1 to C, each of the $C!$ distinguishable arrangements of ranks in a given row is equally likely, and each of the $(C!)^R$ distinguishable arrangements of ranks in the entire table is equally likely. To each of these $(C!)^R$ distinguishably different and equally likely tables there will correspond a value of

$$S = \sum_{i=1}^{C} \left[T_i - \frac{R(C+1)}{2} \right]^2$$

where T_i is the sum of the ranks in the ith treatment column. And the relative frequency distribution of S values for these $(C!)^R$ cases will be the randomization distribution of S, which will be the true distribution of S when the null hypothesis and all assumptions are true. Thus the exact null probability for an S equal to or greater than that obtained is simply the number of the $(C!)^R$ different possible tables which yield such values of S, divided by $(C!)^R$. Now $(C + 1)/2$ is the average rank in any given row and $[R(C + 1)]/2$ is its sum over R rows, and is therefore the average column total, $\bar{T}$. So

$$S = \sum_{i=1}^{C} (T_i - \bar{T})^2$$

is simply the sum of the squared deviations of individual column sums from their mean. It is therefore a test statistic that should be sensitive to differences in locations of the treatment-effect populations, tending strongly to assume large values when median treatment effects differ. Therefore, by referring the obtained value of S to the randomization distribution of S and rejecting the H_0 of identical treatment effects when the obtained value falls in the α-sized upper tail of the randomization distribution, we have the test originally desired.

5.12.2 Null Hypothesis

For each row, each of the $C!$ permutations of the ranks 1 to C was equally likely a priori to be the sequence of cell entries actually recorded. This will be the case if, under each matching condition, treatments have identical effects (so that variations among observations are random) and if all assumptions are true. Note that the null hypothesis does not imply that the observations in different rows belong to the same population.

5.12.3 Assumptions

Assignment of the C units from each matching population to the C treatments must be *random*. Alternatively, if a single unit is subjected to all C treatments, it is assumed that there is *no carryover effect*. (The latter assumption is unlikely to be completely met; however, the deleterious effects of its violation can be mitigated by proper counterbalancing of the sequences in which treatments are administered.) It is, in fact, assumed that the response of one unit to treatment is uninfluenced by that of any other unit, e.g., departing subjects do not relate their experiences to arriving subjects. In a somewhat broader sense it is assumed that there is independence between rows, i.e., that the units in different rows are separate and noninteracting. Finally, it is assumed that there are *no tied observations* within any one row.

5.12.4 Efficiency

This test has A.R.E. of

$$\left(\frac{3}{\pi}\right)\left(\frac{C}{C+1}\right)$$

relative to the appropriate analysis-of-variance F test under the normal-theory conditions assumed by the latter. This A.R.E. thus depends upon the number C of treatments, ranging from $2/\pi$ for $C = 2$ to $3/\pi$ for $C = \infty$, some representative values being .637 (for $C = 2$), .716 ($C = 3$), .764 ($C = 4$), .796 ($C = 5$), .868 ($C = 10$), .910 ($C = 20$), and .955 ($C = \infty$). That the A.R.E. should diminish with diminishing number of treatments is not surprising. As the number of things being ranked gets smaller, the ordinal measurement scale becomes grosser and grosser, until when only two things are left to rank, the information contained in the ranks 1 and 2 amounts only to the directional information of "below" or "above" the other thing. Thus when $C = 2$ the present test becomes equivalent to the Sign test, which uses only dichotomous directional information. (And the F test becomes equivalent to the t test for matched pairs. So the A.R.E. of the present test relative to the F test becomes $2/\pi$, which is the A.R.E. of the Sign test relative to the matched-pair t test.) Because of this equivalence, power functions for the Sign test for the median difference apply equally to the present test in the case where $C = 2$. And efficiencies of that Sign test relative to the matched-pair t test apply equally to the efficiency comparison of the present test with the appropriate analysis-of-variance F test in the two treatment case.

5.12.5 Tables

Table IV-A of the Appendix lists exact critical values of S and their exact cumulative probabilities for an upper-tail test at nominal α values of .10, .05, .01, and .001; this is done for $C = 3$ and $R = 2(1)\,15$ and for $C = 4$ and $R = 2(1)\,8$. Complete, exact, cumulative probability distributions of S are given in [30] for the cases $C = 3$, $R = 2(1)\,10$; $C = 4$, $R = 2(1)\,6$; and $C = 5$, $R = 3$. Complete, exact, cumulative probability distributions for the statistic

$$\chi_r^2 = \frac{12S}{RC(C+1)}$$

are given in [47] for the cases $C = 3$, $R = 2(1)15$ and $C = 4$, $R = 2(1)8$. Other useful tables for χ_r^2 are given in [16, 17].

Table IV-B of the Appendix gives approximate critical values of S for upper-tail tests at α's of .05 and .01 for most combinations of $C = 3(1)7$ and $R = 3(1)6(2)10(5)20$ plus some additional values. The table is based upon the fact that

$$z = \frac{1}{2}\log_e \frac{12(S-1)(R-1)}{R^2(C^3 - C) - 12(S-3)}$$

is, to a close approximation, distributed as Fisher's analysis-of-variance z statistic, with degrees of freedom $V_1 = (C-1) - (2/R)$ and $V_2 = (R-1)V_1$, respectively. (For tables of critical values for Fisher's z, see [30].)

A less accurate but simpler way of obtaining probabilities when the exact tables are too limited is to refer the statistic

$$\chi_r^2 = \frac{12S}{RC(C+1)}$$

to tables of chi-square with $C - 1$ degrees of freedom. The rationale for this approximation is as follows. The mean and variance of the rectangular distribution of ranks from 1 to C are $(C+1)/2$ and $(C^2-1)/12$, and these are therefore the expected value and variance of a randomly selected rank from the C ranks within a given row. The sum T_i of the R ranks in the ith column therefore has a distribution with mean $[R(C+1)]/2$ and variance $[R(C^2-1)]/12$, and this is true for every column. By the Central Limit Theorem, as R increases, the distribution of T_i approaches a normal distribution with the same mean and variance, and therefore

$$\frac{T_i - [R(C+1)/2]}{\sqrt{R(C^2-1)/12}}$$

becomes more and more nearly a normally distributed variate with zero mean and unit variance. Now the sum of the squares of C random and independent values of such standardized normal variates (i.e., variates whose distributions are normal with zero means and unit variances) is distributed as chi-square with C degrees of freedom. But our C column sums are not independent. Knowing the values of $C - 1$ of them, we can obtain the value of the last one by subtracting the sum of the others from $RC(C+1)/2$. We could consider a randomly chosen $C - 1$ of the column sums to be roughly independent and base chi-square upon them, but this is too arbitrary, since the test outcome would be influenced by which of the C possible sets of $C - 1$ columns we chose. Alternatively, we could use all C column sums in obtaining our sum of squares, multiply this sum of squares by $(C-1)/C$ to obtain the average sum of squares based on $C - 1$ columns, and then treat the result as approximately a chi-square variate with $C - 1$ degrees of freedom. Thus we would have

$$\chi^2 \cong \frac{C-1}{C} \sum_{i=1}^{C} \left[\frac{T_i - [R(C+1)/2]}{\sqrt{R(C^2-1)/12}}\right]^2$$
$$\cong \frac{12}{RC(C+1)} \sum_{i=1}^{C} \left[T_i - \frac{R(C+1)}{2}\right]^2$$
$$\cong \frac{12S}{RC(C+1)} = \chi_r^2$$

The accuracy of this approximation is discussed in [16]. Accuracy can generally be increased by using a correction for continuity [30]. If this correction is employed, the relationship between χ_r^2 and S is changed to

$$\chi_r^2 = \frac{12R(C-1)(S-1)}{R^2(C^3-C)+24}$$

5.12.6 Example

An experimenter wishes to test at the .05 level the H_0 that four drugs have identical effects upon a person's visual acuity against the alternative that they do not. And he wishes his test to be especially sensitive against

	Drug			
Subject	I	II	III	IV
A	.39 (2)	.55 (4)	.33 (1)	.41 (3)
B	.21 (3)	.28 (4)	.19 (2)	.16 (1)
C	.73 (4)	.69 (3)	.64 (2)	.62 (1)
D	.41 (3)	.57 (4)	.28 (1)	.35 (2)
E	.65 (4)	.57 (2)	.53 (1)	.60 (3)
T_i	16	17	7	10
$T_i - \frac{R(C+1)}{2}$	3.5	4.5	−5.5	−2.5
$\left(T_i - \frac{R(C+1)}{2}\right)^2$	12.25	20.25	30.25	6.25

$$S = \sum_{i=1}^{C} \left(T_i - \frac{R(C+1)}{2}\right)^2 = 69.00$$

alternatives in which the true distributions of a person's acuity under the four drugs do not all have the same median. The visual acuity of each of five randomly selected subjects is tested under each of the four drugs, sufficient time being allowed between tests for the drug's effect to wear off. (It is assumed that there is no carryover effect of any type.) The acuity scores, followed in parentheses by their within-row ranks, are given in the accompanying table. Entering Table IV-A of the Appendix with $C = 4$ and $R = 5$, the critical value of S for $\alpha = .05$ is found to be 65. Since the obtained value of S exceeds this, being 69, the H_0 of identical drug effects upon acuity is rejected.

5.12.7 Discussion

It is to be noted that the test does not assume homogeneity of rows. The sets of C observations in different rows may differ tremendously in location, dispersion, or both. The test is not designed to detect such effects. It essays merely to detect any systematic tendency for observations in one column to exceed or be smaller than same-row observations in another column. It will fail if such a tendency exists in some rows but is balanced by an opposite tendency in other rows. Therefore, while the test does not assume (for the sake of validity) that the rows be similar in their responsiveness to treatments, it *is* desirable (for the sake of power) that they be at least *qualitatively* similar when they are quantitatively different in their responses to treatments.

The test is particularly useful in the behavioral sciences where individual differences between subjects are often much larger than differences in a single subject's responses to treatments. By letting each subject be a row (so that each "row population" is the population of potential responses from a single subject), as in the example given, the test more or less ignores the large, irrelevant differences between subjects and concentrates upon the small but relevant, treatment-induced, within-subject differences.

Since, in each row, ranks are substituted for original observations, the test is particularly suitable when original data are in intrinsic rank form, each row containing the ranks from 1 to C. This is, in fact, the case when C different objects are ranked, in order of preference, from 1 to C by each of R different judges. In that case the test can be used to test for concordance among the judges and therefore, by implication, for differences among the objects in the ranked characteristic. See Kendall [30].

The Friedman test is not limited to the one-way analysis where R observations are taken under each of C column conditions. It can be extended to multi-way, or higher-order, analyses, such as the case where R observations are taken under every combination of one of C column conditions with one of B block conditions, and it can be used to test for interactions as well as for main effects. The test can also be modified so as to be sensitive to a directional alternative hypothesis, i.e., one that specifies in advance the ranks of the column means among themselves. These subjects are treated in a later section of this chapter.

As described, the test requires that there be exactly one observation in each of the C cells of every row. Durbin [13] has generalized the test to apply to balanced incomplete block designs where certain cells are empty, and

Benard and van Elteren [4] have generalized it still further to the case where any cell can be empty or contain any number of observations. For an interesting variation of the Friedman test, using only the most extreme column sum as the test statistic, see [58].

5.13 KRUSKAL AND WALLIS' MULTI-SAMPLE TEST FOR IDENTICAL POPULATIONS, SENSITIVE TO UNEQUAL LOCATIONS

This test is the rank-randomization analogue of the observation-randomization test discussed in Section 4.2.4. It is also a direct generalization of the two-sample Wilcoxon rank-sum test to the multi-sample case.

5.13.1 Rationale

Suppose that an experimenter wishes to test the hypothesis that C infinite populations (which may be populations of true effects resulting from C treatments) are identical against the alternative that they are not, and wishes his test to be especially sensitive to differences in location. He may test this hypothesis as follows. Let R_i observations be randomly and independently drawn from the ith population and recorded in sequence in R_i cells under the ith column of a data table having C columns, and let this be done for every value of i from 1 to C, i.e., for each of the C populations. The total number of observations in the table is $\sum_{i=1}^{C} R_i$, which we shall call N. Now let the N observations in the table be ranked from 1 to N in increasing order of size and let each observation be replaced by its size rank. Let T_i and $\bar{T}_i = T_i/R_i$ be, respectively, the sum and the mean of the R_i ranks in the ith column. The mean of the N ranks in the entire table will be $(N + 1)/2$, and this will therefore be the expected value of the column mean $\bar{T}_i$, i.e., it will be the mean of the sampling distribution of $\bar{T}_i$, if the N ranks are randomly distributed to the occupied cells of the table. Finally, calculate

$$H = \frac{12}{N(N+1)} \sum_{i=1}^{C} R_i \left(\bar{T}_i - \frac{N+1}{2}\right)^2$$

and refer the obtained value to the randomization distribution of H or, if tables for the latter are unavailable at the sample sizes used, to the distribution of chi-square with $C - 1$ degrees of freedom for an approximate test. Reject if H falls in an upper-tail rejection region of size α, where α is determined prior to sampling.

If the C sampled populations are identical as hypothesized, the N observations are, in effect, a single sample from a common population, with the result that the magnitudes of the observations, and therefore the ranks with which they are replaced, are randomly distributed among the N occupied cells of the table. In that case, H (or any other test statistic) will have its randomization distribution. (That is, the null distribution of H will be the same as the randomization distribution of H—the relative frequency distribution of the $N!$ values of H calculated one for each of the $N!$ tables that can be made distinguishably different by permuting the N ranks among the N occupied cells.) On the other hand, if the C populations are not all identical, it will not be true that each of the $N!$ distinguishable tables had the same a priori probability of $1/N!$, so H cannot be counted on to have its randomization distribution. And if the populations have unequal medians, other things being equal, the higher ranks will tend to fall in the columns associated with populations having the higher medians, thus tending to cause H to take on the large values for which the test will reject.

Unfortunately the randomization distribution of H has been tabled only for tiny sample sizes. Consequently one must usually resort to approximations in order to obtain probability levels for the test. The justification for the chi-square approximation (and the reason for using H as the test statistic) in the case where all $R_i = R$, i.e., where all samples are the same size R, is roughly as follows. (For a formal proof that H is asymptotically distributed as chi-square in the general case, see [34].)

We want a test statistic the variable part of which is $\sum_{i=1}^{C} R_i \bar{T}_i^2$, i.e., we want each column's contribution to be the square of the column mean weighted by the number of observations in the column. Let R_i be the same value R for every i. Then under the null hypothesis T_i is the sum of R observations drawn randomly and without replacements from a finite rectangular population (the consecutive integers from 1 to N) with mean $(N + 1)/2$ and variance $(N^2 - 1)/12$. Therefore, considered as a variate, T_i has, under H_0, a distribution with mean $[R(N + 1)]/2$ and variance $[R(N^2 - 1)]/12$ times $(N - R)/(N - 1)$, the last fraction being the "correction" for sampling without replacement from a finite population. Furthermore, by the Central Limit Theorem, as R increases the distribution of the variate T_i approaches a normal distribution with the same mean and variance, so that

$$\frac{T_i - [R(N + 1)/2]}{\sqrt{R(N^2 - 1)(N - R)/12(N - 1)}}$$

can, if R is large enough, be treated as normally distributed with zero mean and unit variance. The sum of the squares of C of these unit normal deviates would be distributed as chi-square with C degrees of freedom if they were completely independent. But since only $C - 1$ of them can be regarded as independent, $(C - 1)/C$ times the sum should be roughly distributed as chi-square with $C - 1$ degrees of freedom (see Friedman test). Thus

$$\chi^2 \cong \frac{C - 1}{C} \sum_{i=1}^{C} \frac{\{T_i - [R(N + 1)/2]\}^2}{R(N + 1)(N - R)/12}$$

$$\cong \frac{12(C - 1)}{C(N - R)(N + 1)} \sum_{i=1}^{C} \frac{\{R\bar{T}_i - [R(N + 1)/2]\}^2}{R}$$

Now substituting $RC - R$ for $N - R$ and N/R for C in the fraction preceding the summation, the product $C(N - R)$ in the denominator becomes $N(C - 1)$ and we finally have

$$\chi^2 \cong \frac{12}{N(N+1)} \sum_{i=1}^{C} R\left(\bar{T}_i - \frac{N+1}{2}\right)^2$$

The formula already given for H can be reduced to a form that yields the value of H with fewer arithmetical computations:

$$\begin{aligned}
H &= \frac{12}{N(N+1)} \sum_{i=1}^{C} R_i \left(\bar{T}_i - \frac{N+1}{2}\right)^2 \\
&= \frac{12}{N(N+1)} \left[\sum_{i=1}^{C} R_i \bar{T}_i^2 - (N+1) \sum_{i=1}^{C} R_i \bar{T}_i + \left(\frac{N+1}{2}\right)^2 \sum_{i=1}^{C} R_i\right] \\
&= \frac{12}{N(N+1)} \left[\sum_{i=1}^{C} \frac{T_i^2}{R_i} - (N+1) \sum_{i=1}^{C} T_i + N\left(\frac{N+1}{2}\right)^2\right] \\
&= \frac{12}{N(N+1)} \left[\sum_{i=1}^{C} \frac{T_i^2}{R_i} - (N+1)N\left(\frac{N+1}{2}\right) + N\left(\frac{N+1}{2}\right)^2\right] \\
&= \frac{12}{N(N+1)} \left[\sum_{i=1}^{C} \frac{T_i^2}{R_i} - \frac{N(N+1)^2}{4}\right] \\
&= -3(N+1) + \frac{12}{N(N+1)} \sum_{i=1}^{C} \frac{T_i^2}{R_i}
\end{aligned}$$

5.13.2 Null Hypothesis

Each of the $N!$ distinguishable permutations of the N ranks within the N cells of the table (one rank to a cell) had equal a priori probability $1/N!$ of becoming the actually obtained table of ranks. Alternatively (and perhaps more accurately since the test statistic is the same for each of the $\prod_{i=1}^{C} R_i!$ possible ways of rendering a given table different solely by within-column permutation of ranks), each of the $N!/(\prod_{i=1}^{C} R_i!)$ different possible divisions of the N ranks into a *combination* of R_1 ranks for the first column, R_2 for the second, . . . and R_C for the Cth, results in tables which had equal a priori probability of becoming the actually obtained table. The stated H_0 tends to imply that all possible ways of dividing the N observations in the pooled sample into samples of the sizes drawn were "equally likely." This will be the case if all C sampled populations are identical and all assumptions are met.

5.13.3 Assumptions

Sampling is *random*, the populations are *infinite* or sampling is with *individual replacements*, and there are *no tied observations*.

5.13.4 Efficiency

When both tests are applied to data from continuously distributed populations having identical shapes and variances to test whether or not they also have identical locations, the A.R.E. of the Kruskal-Wallis test relative to the analysis-of-variance F test can be as high as infinity but no lower than .864 [see 27]. The value of the A.R.E. in this range depends upon the particular form of the common shape of the tested populations. The A.R.E. is $3/\pi$ or .955 when the common shape is normal and is 1.000 when the common shape is rectangular. These are the same as the A.R.E.'s of the two-sample Wilcoxon rank-sum test relative to Student's t test under the conditions stated above. This is not surprising, since when $C = 2$ the Kruskal-Wallis and F tests become equivalent to the two-tailed Wilcoxon and t tests, respectively. Under the conditions specified above, the A.R.E. of the Kruskal-Wallis test relative to the Brown-Mood median test is 1.5 for normal populations and 3 for rectangular populations. The Kruskal-Wallis test is consistent in the situation described above, i.e., is consistent against alternatives involving inequality of location only. In the words of Kruskal [34], "Roughly speaking . . . the test is consistent if and only if the variables from at least one population tend in the limit to be either larger or smaller than the other variables." This is a generalization of the consistency criterion for the two-sample Wilcoxon rank-sum test. The test will be *in*consistent if the C populations are not all identical but all do have symmetric shapes and a common axis of symmetry [34].

5.13.5 Tables

Tables of exact upper-tail cumulative probabilities for H values just within and just outside of α-sized rejection regions can be found in [35, 47] for cases where $C = 3$, all R_i are ≤ 5, and α is .10, .05, or .01. In other cases one may refer H to the distribution of chi-square with $C - 1$ degrees of freedom and reject the null hypothesis if the value of chi-square equal to H has an upper-tail cumulative probability $\leq \alpha$, the preselected significance level.

5.13.6 Example

An experimenter wishes to test, at the .05 level, whether or not three type styles have identical effects upon reading speed. He draws 15 subjects from a common population and randomly divides them into 3 groups of 5 subjects each, one group being randomly assigned to each type style. Each subject's reading speed is measured under the type style to which he was assigned. Through some misadventure (unrelated to type style or difficulty of reading it), one subject fails to complete the experiment, so one group is left with only four subjects. The reading speed measurements are given in the accompanying table, followed by their ranks (in parentheses), and these data are followed by the necessary test calculations. Entering Kruskal and Wallis'

Type style

	I	II	III
	135 (8)	175 (11)	105 (3)
	91 (2)	130 (7)	147 (9)
	111 (5)	514 (14)	159 (10)
	87 (1)	283 (13)	107 (4)
	122 (6)		194 (12)
T_i	22	45	38
T_i^2	484	2025	1444
T_i^2/R_i	96.8	506.25	288.8

$$H = -3(N+1) + \frac{12}{N(N+1)} \sum_{i=1}^{C} \frac{T_i^2}{R_i}$$

$$= -3(15) + \frac{12}{14(15)}(891.85)$$

$$= -45 + 50.9629 = 5.9629$$

tables [35 or 47] with sample sizes of 5, 5, and 4, the experimenter finds that an H value of 5.6657 has an upper-tail cumulative probability of .049. His H is larger, so it also is significant at the .05 level, and he can reject the null hypothesis that the three type styles have identical effects upon reading speed. Had the experimenter chosen to use the chi-square approximation, he would have entered the chi-square tables (see Table XVI of the Appendix) with $C - 1 = 2$ degrees of freedom and found that the critical value of chi-square for an upper-tail cumulative probability of .05 is 5.991. The obtained value of H is smaller than this, so the experimenter would not be able to reject the null hypothesis.

In the example given, the real differences between groups may have been due entirely to real differences in reading ability rather than to differences in type style. However, the process of assigning subjects randomly to type styles makes the above possibility a part of "chance," and as such it is properly taken account of in the chosen probability of a Type I error. There are many ways in which the experiment could have been made more sensitive (e.g., by using difference-scores between performance under the assigned type style and performance in a common "pretest" under a fourth condition) but not more valid.

In the example given we are, of course, testing the hypothesis that the populations of reading speeds under the three different type styles are identical.

5.13.7 Discussion

Instead of the overall test for *any* difference among the C-treatment effects, one may use the multiple comparison technique which compares treatment effects in pairs at a significance level adjusted to allow for the plurality of the comparisons. Multiple comparison techniques alternative to the overall Kruskal-Wallis test are discussed in [12, 53]. These techniques involve a series of pairwise tests analogous to the two-sample Wilcoxon rank-sum test.

Extensions of the undirectional, one-way analysis, Kruskal-Wallis test to testing directional hypotheses and to testing main effects and interactions in multi-way analyses are outlined in the next two sections of this chapter.

The Kruskal-Wallis test is only one of many generalizations of the Wilcoxon test to the multi-sample case. While the Kruskal-Wallis test weights each squared deviation of column mean from grand mean by the number of observations in the column, an otherwise analogous test by Rijkoort [51] weights each column equally, using as test statistic the sum of squared deviations of column sums from their expected values. It is therefore equivalent to the Kruskal-Wallis test when all $R_i = R$. Other authors have generalized the Mann-Whitney form of the Wilcoxon test to the multi-sample case. The "directional" quality of the Mann-Whitney U statistic makes such tests particularly adaptable to the testing of H_0 against directional alternatives. A test by Terpstra [57], for example, requires that columns be arranged in an order implied by the (directional) alternative hypothesis; the test statistic is then taken to be the number of pairs of observations in the table whose higher member lies in the higher column.

5.14 TESTING DIRECTIONAL HYPOTHESES IN THE MULTI-SAMPLE CASE

The Friedman test can be easily modified to test directional hypotheses, and so can the Kruskal-Wallis test in the case where all $R_i = R$, which will be assumed to be the case in what follows [8]. Most multi-sample tests for column effects in an $R \times C$ table (such as the Friedman, Kruskal-Wallis, and analysis-of-variance F tests) use a test statistic based upon the sum of the squared deviations of column means (or sums) from their expected values. In so doing they ignore the direction of the deviation and, in fact, ignore all information about the order of arrangement of the set of C column means. Let S be such a test statistic. Let S' be a value of S corresponding to an obtained $R \times C$ table and let α' be the upper-tail cumulative probability of S' in the null distribution of S. Now consider the particular table of observations that yielded S'. Without altering the positions of the columns or their labels, we can permute the contents of entire columns of observations (e.g., the R observations in the seventh column can be switched with the R observations in the third column, etc.). There are $C!$ such permutations, to each of which there corresponds a distinguishably different table having the same set of column means (although over the whole set of $C!$ tables the column means will be arranged in $C!$ different *orders* if all means differ, but fewer if some means are equal) and therefore yielding the same value S' of the test statistic. Furthermore, under H_0 each of these $C!$ tables was equally likely a priori, and not only is this true of the table that yielded S', but it is also true of any table yielding any value of S. In particular, it is true for any table that yields an $S \geqq S'$, i.e., of any table yielding an S whose upper-tail cumulative probability is $\leqq \alpha'$. Therefore the probability of obtaining an $S \geqq S'$ with a table whose columns of observations correspond to a single specifiable one of these $C!$ permutations is $\alpha'/C!$. To the $C!$ permutations of columns there will correspond $C!$ arrangements (some of which may be the same) of the same set of C column means. Now suppose that prior to obtaining data the experimenter had predicted the order of the column means, e.g., if $C = 4$, he might have predicted that $\bar{T}_1 < \bar{T}_2 < \bar{T}_3 < \bar{T}_4$. Consider all tables that yield an $S \geqq S'$. For some of them (and, in fact, usually for most of them), there will be no tied column means, and therefore given that this is the case the null probability of the predicted order for a single such table is $1/C!$. But in some, usually few, cases some of the C column means may be equal, and therefore, given that this is the case, the null probability of a predicted order that makes all column means unequal is zero. Therefore, over all tables that yield an $S \geqq S'$, the null probability of the predicted order is $\leqq 1/C!$. Therefore, of the proportion α' of the null distribution of S in which $S \geqq S'$, the proportion of S's obtained from tables meeting the predicted order of column means is $\leqq 1/C!$, and the proportion of this set of S's in the entire null distribution of S is $\leqq \alpha'/C!$. The probability level for a test that takes this conditional set of S's as its rejection region is therefore $\leqq \alpha'/C!$, and one can reject at less than or equal to an overall (preselected) α level of significance if $\alpha' \leqq C!\,\alpha$. In so doing, one is rejecting the H_0 of identical column effects in favor of the alternative hypothesis that column effects differ in such a way as to tend to cause $\bar{T}_1 < \bar{T}_2 < \bar{T}_3 < \bar{T}_4$ or, roughly speaking, in favor of the presumption that $M_1 < M_2 < M_3 < M_4$, where M_i is the true median of the population corresponding to the ith column.

Practical research workers may well object that they seldom feel confident in predicting the order relationship for the entire set of C column means. Even when one has good a priori reasons for predicting a given *true* order, the differences between some of the "true" column means may be so small that by chance alone there is considerable likelihood that the corresponding *obtained* column means will differ in the direction opposite the predicted one. Therefore, a properly cautious experimenter would want to confine his prediction to that portion of the set of C column means where differences between them are large or at least highly reliable. In other cases there may be good a priori reasons for predicting a partial ordering of column means, but no reason for predicting the order of the remaining means. Both situations can be handled. Let P be the number of means whose exact rank in the set of C means one wishes to predict. Then instead of only one of the $C!$ possible orders of unequal means having the predicted relationship, $(C - P)!$ of them will have it, since in an order relationship in which the means of the P "predicted" columns have their respective predicted ranks, the remaining $C - P$ column means may be permuted in $(C - P)!$ ways without violating the partially predicted order. The null probability of the predicted order, given that the means are all unequal, is therefore $(C - P)!/C!$. And the null probability that the obtained column means will fall in the partially predicted order while yielding an $S \geqq S'$ is $[(C - P)!\,\alpha']/C!$. For example, one might wish to predict that $\bar{T}_1$ will be the smallest, $\bar{T}_2$ the next-smallest, and $\bar{T}_7$ the largest of $C = 7$ column means. This can happen in $4!$ ways, since each of the $4!$ possible permutations of order among $\bar{T}_3$, $\bar{T}_4$, $\bar{T}_5$, and $\bar{T}_6$ represents (i.e., accompanies) a different way in which the prediction can be fulfilled. Therefore, of the $7!$ permutations without restrictions, a proportion $4!/7!$ meet the restrictions, i.e., fulfill the prediction, and therefore the overall probability level for the test is $\alpha \leqq (4!\,\alpha')/7!$.

The procedure can be generalized still further. One might wish only to predict that a specified L of the means will have the L lowest values and a specified H of them will have the H highest values, without attempting to predict relative size within the set of L, or within the set of H, means. Of the C means, the L lowest can be placed in the L lowest-ranking positions in $L!$ ways. For each of these ways, the H highest can be placed in the H highest-ranking positions in $H!$ ways. And, for each of these ways, the $C - L - H$ remaining can be placed in the $C - L - H$ remaining positions in $(C - L - H)!$ ways. So there are $L!\,(C - L - H)!\,H!$ ways of meeting the prediction out of the $C!$ ways of allocating ranks to means without restriction. Consequently the null probability of meeting the prediction, given that the means all differ, is $[L!\,(C - L - H)!\,H!]/C!$, and therefore the overall null probability of the test is $\alpha \leqq [L!\,(C - L - H)!\,H!\,\alpha']/C!$.

Even the grossest kind of information can be used effectively. For example, the experimenter may feel confident only in predicting that $\bar{T}_8$ will be in the higher half of the C means. If C is even, the null probability of the predicted result is $\frac{1}{2}$ and if the experimenter wishes to use an overall α of .05, he can reject if the prediction is confirmed and the obtained value of S has an upper-tail cumulative probability $\leqq .10$ in the null distribution of S. Thus he may use twice as large a rejection region for the mere *value* of S as he could without the information that $\bar{T}_8$ should be higher than the median of the $\bar{T}_i$'s. He is in the same situation if he can predict that $\bar{T}_8 > \bar{T}_2$, or that

$\bar{T}_i > \bar{T}_j$ for any specified i and j. And if he can predict that $\bar{T}_8$ will be among the four largest and $\bar{T}_1$ among the four smallest of eight column means, he can use $3\frac{1}{2}$ times as large a rejection region for the value of S as he could without the information about order.

When $C > 3$, there will seldom be any need to predict the order for the entire set of C means, since the null probability, $1/C!$, of simply fulfilling the prediction is less than .05 so that one could reject at the $\alpha = .05$ level without even obtaining the supplementary information afforded by the S test. The technique, therefore, of using prediction of order in conjunction with a test sensitive only to the absolute magnitudes of deviations from expected value is, in fact, particularly appropriate in precisely those situations where only a partial order relationship can be predicted with confidence.

It is important, however, to stress that rejection presupposes exact and complete fulfillment of the predicted order. If one chooses an overall significance level $\alpha = .001$ and predicts that $\bar{T}_1 < \bar{T}_2 < \bar{T}_3 < \bar{T}_4 < \bar{T}_5$ and it turns out that $\bar{T}_1 < \bar{T}_2 < \bar{T}_4 < \bar{T}_3 < \bar{T}_5$, then one is not entitled to reject even if the probability level for S' is $\alpha' = .0001$. One has initially chosen a rejection region consisting only of S's obtained from the predicted order. No matter how extreme they may be, S-values obtained from some other order cannot fall in the rejection region. This emphasizes the necessity of confining one's predictions to those that follow almost inevitably from the soundest of logical considerations (in the context of an untrue H_0).

The frustration of the almost-but-not-quite-fulfilled prediction can be avoided by using a test originated by Page [48]. His test presupposes matched observations and is identical to the Friedman test up to the point of obtaining column sums, T_i's. Each T_i is then weighted by its predicted rank P_i in order of increasing size among the C column sums. The test statistic is $L = \sum_{i=1}^{C} P_i T_i$, and Page has provided tables of critical values of L for $C = 3(1)10$, $R = 2(1)50$, and α's of .05, .01, and .001. The tables are exact, apparently being based upon the randomization-distribution of L, for $C \leqq 8$ and $R \leqq 12$ (as well as for $R \leqq 20$ in the case where $C = 3$); in other cases, the tables are based upon a normal approximation. Power curves for the L test and analysis-of-variance F test under comparable conditions are given in [6]. Page's test rejects for large values of L. These rejectable L's are not likely to be obtained unless there is a close correspondence between predicted and actual size ranks of the column sums, but the correspondence does not have to be exact. Thus perfect prediction is not always required for rejection. But because of the heavy weights prediction must be quite good. Furthermore prediction must be complete, i.e., all C columns must be ranked. Finally, in common with most tests using weights, this test has the disadvantages that it is a bit difficult to state exactly what rejection implies. That is it is hard to visualize in *experimental* terms exactly what the rejection region consists of, and it is hard to say exactly what alternative hypothesis is associated with rejection. Rejection does not imply that the experimenter has predicted correctly; it does seem to imply that he has predicted well. But "well" is left largely undefined.

5.15 TESTING MAIN EFFECTS AND INTERACTIONS AMONG SEVERAL VARIABLES

The Friedman and Kruskal-Wallis tests are one-way analyses, i.e., if columns represent C different levels of a single variable, they test the H_0 that all C levels have the same effect. By small modifications of procedure, however, both tests can be used to test for main effects and interactions in the case where observations are taken under combinations of levels involving several different variables. First we shall consider the case where the number of replications is constant, i.e., where the same number of observations is taken under every investigated combination of conditions [65]. (This typically is the case when observations are matched as assumed by the Friedman test, but is only a special case for the Kruskal-Wallis test.) The method can best be explained in terms of an example. Suppose that four observations have been taken under each of three levels, I, II, and III, of a column variable and that this has been done under each of two levels A and B of a block variable. Let the data be as shown in the accompanying table. If the observations in each row are not matched, we require that the four observations under a single combination of block and column be randomly assigned to their rows, i.e., the row location of unmatched observations must not be biased. Often this requirement will be met if the observations are simply recorded sequentially in the order in which they were drawn.

Block	*Row*	*Column*		
		I	II	III
A	1	7.2	5.3	6.5
	2	1.1	3.4	8.6
	3	5.7	4.2	9.0
	4	0.3	8.5	4.8
B	1	4.9	3.3	7.5
	2	0.7	1.9	8.8
	3	5.1	2.4	7.0
	4	6.7	6.4	9.1

If we wish to test the main effects of one variable, we sum correspondingly located observations over all levels of all other variables (except, of course, the row variable, which merely represents replications). And we then apply the appropriate test (Friedman, if observations in each row are matched, Kruskal-Wallis otherwise) to the resulting table. If we wish to test the main effect of columns, we add to each observation in block *A* the observation in the same row and column of block *B*, obtaining the table shown on the left. Or, if we wish to test the main effect of blocks, we sum the three column entries under each combination of row and block, obtaining the table shown on the right. We then apply the Friedman test to the resulting tables if the resulting same-row observations are matched or the Kruskal-Wallis test if they are not. (In the case of the test on blocks, since there are only two conditions we would apply the Sign test if the observations in each row are matched and the Wilcoxon rank-sum test if they are not.)

Table of "A + B's" for Testing Main Effect of Columns

Row	*Column*		
	I	II	III
1	12.1	8.6	14.0
2	1.8	5.3	17.4
3	10.8	6.6	16.0
4	7.0	14.9	13.9

Table of "I + II + III's" for Testing Main Effect of Blocks

Row	*Block*	
	A	*B*
1	19.0	15.7
2	13.1	11.4
3	18.9	14.5
4	13.6	22.2

If we wish to test for interaction between columns and blocks, we subtract from each observation in block *A* the correspondingly located observation in block *B*, obtaining the following table of $A - B$ difference-scores. To that table we then apply the Friedman test, if the difference-scores in

Table of "A − B's" for Testing Column × Block Interaction

Row	*Column*		
	I	II	III
1	2.3	2.0	−1.0
2	0.4	1.5	−0.2
3	0.6	1.8	2.0
4	−6.4	2.1	−4.3

each row are matched, or the Kruskal-Wallis test if they are not. If there had been three blocks, *A*, *B*, and *C'*, two tables would have been constructed, one for the $A - B$ difference-scores and one for the difference-scores

$[(A + B)/2] - C'$, or, equivalently in the end since they will be replaced by ranks when the test is conducted, the difference-scores $(A + B) - 2C'$. If the original observations were matched, the Friedman test is conducted upon each table and the test statistic is taken to be $\chi_r^2 = (12S)/[RC(C + 1)]$. The sum of the two resulting χ_r^2's is distributed approximately as chi-square with $2(C - 1)$ degrees of freedom (where C is the number of columns in the table of difference-scores, not in the table of original observations). If the original observations were unmatched, the Kruskal-Wallis test is conducted upon each table. The sum of the two resulting values of the test statistic H is also distributed approximately as chi-square with $2(C - 1)$ degrees of freedom. With four blocks A, B, C', and D, three tables would be constructed, one each for the differences $A - B$, $A + B - 2C'$, and $A + B + C' - 3D$. Ranks would be substituted for difference-scores and χ_r^2 or H would be calculated for each table. Since each test statistic is distributed approximately as chi-square with $C - 1$ degrees of freedom, by the additive property of chi-square the sum of the three values of χ_r^2 or H is distributed as chi-square with $3(C - 1)$ degrees of freedom. In general, therefore, one sums k values of χ_r^2 or H and rejects if the sum has an upper-tail cumulative probability $\leqq \alpha$ in the distribution of chi-square with $k(C - 1)$ degrees of freedom. The hypothesis tested (in the present example) is that, except for chance fluctuations, as one moves from one column to another the difference between observations in the same column (and same row, if observations are matched), but lying in two different blocks, remains constant (i.e., roughly that the difference between pairs of correspondingly located observations in any two different blocks is independent of the common column in which they lie).

If the number of observations taken under each investigated combination of levels of different variables is not constant, it is necessary to resort to means in place of sums. Let X_{ijk} be the observation recorded in the ith column, jth block, and kth row of the table (i.e., let it be the kth observation recorded under the jth level of the block variable and the ith level of the column variable). And let the replacement of a subscript by a dot indicate that we are using the average of existing values of X_{ijk} over all values of the replaced subscript under the combination of conditions indicated by the remaining subscripts, e.g., if there are two blocks $X_{i \cdot k} = (X_{i1k} + X_{i2k})/E$, where $E = 2$ if there is an observation in the kth row and ith column of both blocks and $E = 1$ if such an observation exists in only one block. Then the entries in the table we form to test main effects of columns will be $X_{i \cdot k}$'s, and those in the table to test main effects of blocks will be $X_{\cdot jk}$'s. And entries in the table we form to test column $\times$ block interaction will be $X_{i2k} - X_{i1 \cdot}$ if there are only two blocks. If there are three blocks, in addition to that table we will also have a table whose entries are $X_{i3k} - [(X_{i1 \cdot} + X_{i2 \cdot})/2]$; and if there are four blocks, in addition to the two tables already mentioned we will also have a table whose entries are $X_{i4k} - [(X_{i1\cdot} + X_{i2\cdot} + X_{i3\cdot})/3]$. In testing for interaction, we apply the Kruskal-Wallis procedure and obtain H for each table after which we treat $\sum H$ as chi-square with degrees of freedom equal to $(B - 1)(C - 1)$, where B is the total number of blocks in the original data table.

REFERENCES

1. Alling, D. W., "Early Decision in the Wilcoxon Two-Sample Test," *Journal of the American Statistical Association*, **58** (1963), 713–720.
2. Ansari, A. R., and R. A. Bradley, "Rank-Sum Tests for Dispersions," *Annals of Mathematical Statistics*, **31** (1960), 1174–1189.
3. Arnold, H. J., "Small Sample Power of the One Sample Wilcoxon Test for Non-Normal Shift Alternatives," *Annals of Mathematical Statistics*, **36** (1965), 1767–1778.
4. Benard, A., and Ph. van Elteren, "A Generalization of the Method of m Rankings," *Proceedings Koninklijke Nederlandse Akademie van Wetenschappen, Series A*, **56** (1953), 358–369.
5. Birnbaum, Z. W., "On a Use of the Mann-Whitney Statistic," in *Proceedings of the Third Berkeley Symposium on Mathematical Statistics and Probability*, ed. Jerzy Neyman. Berkeley and Los Angeles: University of California Press, 1956, Vol. I, pp. 13–17.
6. Boersma, F. J., J. J. DeJonge, and W. R. Stellwagen, "A Power Comparison of the F and L Tests—I," *Psychological Review*, **71** (1964), 505–513.
7. Boneau, C. A., "A Comparison of the Power of the U and t Tests," *Psychological Review*, **69** (1962), 246–256.
8. Chassan, J. B., "On a Test for Order," *Biometrics*, **16** (1960), 119–120 [see also **18** (1962), 245–247].
9. Daniels, H. E., "Rank Correlation and Population Models," *Journal of the Royal Statistical Society, Series B*, **12** (1950), 171–181.
10. van Dantzig, D., "On the Consistency and Power of Wilcoxon's Two Sample Test," *Proceedings Koninklijke Nederlandse Akademie van Wetenschappen, Series A*, **54** (1951), 1–8.
11. Dixon, W. J., "Power under Normality of Several Nonparametric Tests," *Annals of Mathematical Statistics*, **25** (1954), 610–614.
12. Dunn, Olive J., "Multiple Comparisons Using Rank Sums," *Technometrics*, **6** (1964), 241–252.
13. Durbin, J., "Incomplete Blocks in Ranking Experiments," *British Journal of Psychology (Statistics Section)*, **4** (1951), 85–90.

14. van Elteren, Ph., "On the Combination of Independent Two Sample Tests of Wilcoxon," *Bulletin de l'Institut International de Statistique*, **37** (1960), 351–361.

15. Freund, J. E., and A. R. Ansari, *Two-Way Rank Sum Tests for Variances*, Technical Report No. 34, Virginia Polytechnic Institute. Blacksburg, Va., August 1957.

16. Friedman, M., "A Comparison of Alternative Tests of Significance for the Problem of *m* Rankings," *Annals of Mathematical Statistics*, **11** (1940), 86–92.

17. Friedman, M., "The Use of Ranks To Avoid the Assumption of Normality Implicit in the Analysis of Variance," *Journal of the American Statistical Association*, **32** (1937), 675–701.

18. Gastwirth, J. L., "Percentile Modifications of Two Sample Rank Tests," *Journal of the American Statistical Association*, **60** (1965), 1127–1141.

19. Gibbons, Jean D., "On the Power of Two-Sample Rank Tests on the Equality of Two Distribution Functions," *Journal of the Royal Statistical Society, Series B*, **26** (1964), 293–304.

20. Glasser, G. J., "A Distribution-Free Test of Independence with a Sample of Paired Observations," *Journal of the American Statistical Association*, **57** (1962), 116–133.

21. Glasser, G. J., and R. F. Winter, "Critical Values of the Coefficient of Rank Correlation for Testing the Hypothesis of Independence," *Biometrika*, **48** (1961), 444–448.

22. Halperin, M., "Extension of the Wilcoxon-Mann-Whitney Test To Samples Censored at the Same Fixed Point," *Journal of the American Statistical Association*, **55** (1960), 125–138.

23. Hemelrijk, J., "Experimental Comparison of Student's and Wilcoxon's Two-Sample Tests," in *Quantitative Methods in Pharmacology*. Amsterdam: North Holland Publishing Company, 1961, pp. 118–134.

24. Hemelrijk, J., "Note on Wilcoxon's Two-Sample Test when Ties Are Present," *Annals of Mathematical Statistics*, **23** (1952), 133–135.

25. Hodges, J. L., and E. L. Lehmann, "Comparison of the Normal Scores and Wilcoxon Tests," in *Proceedings of the Fourth Berkeley Symposium on Mathematical Statistics and Probability*, ed. Jerzy Neyman. Berkeley and Los Angeles: University of California Press, 1961, Vol. I, pp. 307–317.

26. Hodges, J. L., and E. L. Lehmann, "Rank Methods for Combination of Independent Experiments in Analysis of Variance," *Annals of Mathematical Statistics*, **33** (1962), 482–497.

27. Hodges, J. L., and E. L. Lehmann, "The Efficiency of Some Nonparametric Competitors of the *t*-Test," *Annals of Mathematical Statistics*, **27** (1956), 324–335.

28. Hotelling, H., and Margaret R. Pabst, "Rank Correlation and Tests of Significance Involving No Assumption of Normality," *Annals of Mathematical Statistics*, **7** (1936), 29–43.

29. Jacobson, J. E., "The Wilcoxon Two-Sample Statistic: Tables and Bibliography," *Journal of the American Statistical Association*, **58** (1963), 1086–1103.

30. Kendall, M. G., *Rank Correlation Methods*. New York: Hafner Publishing Company, 1955.

31. Klotz, J., "Alternative Efficiencies for Signed Rank Tests," *Annals of Mathematical Statistics*, **36** (1965), 1759–1766.

32. Klotz, J., "Nonparametric Tests for Scale," *Annals of Mathematical Statistics*, **33** (1962), 498–512.

33. Klotz, J., "Small Sample Power and Efficiency for the One Sample Wilcoxon and Normal Scores Tests," *Annals of Mathematical Statistics*, **34** (1963), 624–632.

34. Kruskal, W. H., "A Nonparametric Test for the Several Sample Problem," *Annals of Mathematical Statistics*, **23** (1952), 525–540.

35. Kruskal, W. H., and W. A. Wallis, "Use of Ranks in One-Criterion Variance Analysis," *Journal of the American Statistical Association*, **47** (1952), 583–621 [and **48** (1953), 907–911].

36. Lehmann, E. L., "Consistency and Unbiasedness of Certain Nonparametric Tests," *Annals of Mathematical Statistics*, **22** (1951), 165–179.

37. Lehmann, E. L., "Nonparametric Confidence Intervals for a Shift Parameter," *Annals of Mathematical Statistics*, **34** (1963), 1507–1512.

38. Lehmann, E. L., *Testing Statistical Hypotheses*. New York: John Wiley & Sons, Inc., 1959, pp. 236–243.

39. Lehmann, E. L., "The Power of Rank Tests," *Annals of Mathematical Statistics*, **24** (1953), 23–43.

40. Mann, H. B., and D. R. Whitney, "On a Test of Whether One of Two Random Variables is Stochastically Larger than the Other," *Annals of Mathematical Statistics*, **18** (1947), 50–60.

41. McCornack, R. L., "Extended Tables of the Wilcoxon Matched Pair Signed Rank Statistic," *Journal of the American Statistical Association*, **60** (1965), 864–871.

42. Mehra, K. L., "Rank Tests for Paired-Comparison Experiments Involving Several Treatments," *Annals of Mathematical Statistics*, **35** (1964), 122–137.

43. Milton, R. C., "An Extended Table of Critical Values for the Mann-Whitney (Wilcoxon) Two-Sample Statistic," *Journal of the American Statistical Association*, **59** (1964), 925–934.

44. Mood, A. M., "On the Asymptotic Efficiency of Certain Nonparametric Two-Sample Tests," *Annals of Mathematical Statistics*, **25** (1954), 514–522.

45. Moses, L. E., "Rank Tests of Dispersion," *Annals of Mathematical Statistics*, **34** (1963), 973–983.

46. Noether, G. E., "Efficiency of the Wilcoxon Two-Sample Statistic for Randomized Blocks," *Journal of the American Statistical Association*, **58** (1963), 894–898.

47. Owen, D. B., *Handbook of Statistical Tables*. Reading, Mass.: Addison-Wesley Publishing Company, Inc., 1962.

48. Page, E. B., "Ordered Hypotheses for Multiple Treatments: A Significance Test for Linear Ranks," *Journal of the American Statistical Association*, **58** (1963), 216–230.

49. Pratt, J. W., "Remarks on Zeros and Ties in the Wilcoxon Signed Rank Procedures," *Journal of the American Statistical Association*, **54** (1959), 655–667.

50. Pratt, J. W., "Robustness of Some Procedures for the Two-Sample Location Problem," *Journal of the American Statistical Association*, **59** (1964), 665–680.

51. Rijkoort, P. J., "A Generalization of Wilcoxon's Test," *Proceedings Koninklijke Nederlandse Akademie van Wetenschappen, Series A*, **55** (1952), 394–404.

52. Siegel, S., and J. W. Tukey, "A Nonparametric Sum of Ranks Procedure for Relative Spread in Unpaired Samples," *Journal of the American Statistical Association*, **55** (1960), 429–445.

53. Steel, R. G. D., "Some Rank Sum Multiple Comparison Tests," *Biometrics*, **17** (1961), 539–552 [see also **15** (1959), 560–572, and *Technometrics*, **2** (1960), 197–207].

54. Stuart, A., "The Correlation between Variate-Values and Ranks in Samples from a Continuous Distribution," *British Journal of Statistical Psychology*, **7** (1954), 37–44 [see also **8** (1955), 25–27].

55. Sundrum, R. M., "The Power of Wilcoxon's 2-Sample Test," *Journal of the Royal Statistical Society, Series B*, **15** (1953), 246–252.

56. Taylor, W. L., and C. Fong, "Some Contributions To Average Rank Correlation Methods and To the Distribution of the Average Rank Correlation Coefficient," *Journal of the American Statistical Association*, **58** (1963), 756–769 [see also **59** (1964), 872–876].

57. Terpstra, T. J., "The Exact Probability Distribution of the T Statistic for Testing against Trend and Its Normal Approximation," *Proceedings Koninklijke Nederlandse Akademie van Wetenschappen*, **56** (1953), 433–437 [see also **57** (1954), 505–512].

58. Thompson, W. A., and T. A. Willke, "On an Extreme Rank Sum Test for Outliers," *Biometrika*, **50** (1963), 375–383.

59. van der Vaart, H. R., "On the Robustness of Wilcoxon's Two Sample Test," *Quantitative Methods in Pharmacology* (Symposium, Leyden, 1960). Amsterdam: North Holland Publishing Company, 1960, pp. 140–158.

60. Verdooren, L. R., "Extended Tables of Critical Values for Wilcoxon's Test Statistic," *Biometrika*, **50** (1963), 177–186.

61. Walsh, J. E., "Some Significance Tests for the Median which Are Valid under Very General Conditions," *Annals of Mathematical Statistics*, **20** (1949), 64–81.

62. Wetherill, G. B., "The Wilcoxon Test and Non-Null Hypotheses," *Journal of the Royal Statistical Society, Series B*, **22** (1960), 402–418.

63. Wilcoxon, F., "Individual Comparisons by Ranking Methods," *Biometrics*, **1** (1945), 80–83.

64. Wilcoxon, F., "Probability Tables for Individual Comparisons by Ranking Methods," *Biometrics*, **3** (1947), 119–122.

65. Wilcoxon, F., *Some Rapid Approximate Statistical Procedures.* New York: American Cyanamid Company, 1949 (pamphlet).

66. Wilcoxon, F., S. K. Katti, and Roberta A. Wilcox, *Critical Values and Probability Levels for the Wilcoxon Rank Sum Test and the Wilcoxon Signed Rank Test.* Lederle Laboratories Division, American Cyanamid Company, Pearl River, New York and Department of Statistics, The Florida State University, Tallahassee, Florida, August 1963.

67. Witting, H., "A Generalized Pitman Efficiency for Nonparametric Tests," *Annals of Mathematical Statistics*, **31** (1960), 405–414.

6

NORMAL SCORES TESTS

Tests based on Fisher's Method of Randomization applied to original observations were impractical but powerful, having A.R.E. of 1 relative to counterpart parametric tests when used to test the latter's H_0 under conditions meeting the latter's assumptions. Transforming the observations to ranks before applying the test made the resulting "rank-randomization" tests practical but did so at the sacrifice of a (usually) small amount of A.R.E. (under "parametric" H_0 and assumptions). This A.R.E. gap can be closed, restoring the A.R.E. to 1, by introducing a second transformation which replaces the ranks by the expected values of the corresponding "normal order statistics" before applying the Method of Randomization. (Thus the second transformation might replace the Rth smallest observation in a pooled sample of N observations by the expected value of the Rth smallest observation in a sample of size N drawn from a standard normal distribution.) The resulting Expected Normal Scores tests require two sets of tables, one for the expected values of the normal order statistics, the other for the randomization distribution of the test statistic. There are several different types of normal scores tests of which the expected normal scores type is only one. However, they are similar in principle and all require a transformation that, unlike the rank transformation, cannot be done without reference to tables. The tests are therefore considerably more cumbersome than the quick, simple, easy tests of the preceding chapter. As measured by efficiency they compete superbly well with their rivals, both parametric and distribution-free. However, they are not always superior in efficiency, and even when they are the extra statistical efficiency is bought at the expense of practical ease and conceptual simplicity. There is no gainsaying, however, that from the strictly theoretical point of view they represent a brilliant *tour de force* combining the versatility of distribution-free tests with the efficiency of classical ones (as measured by A.R.E.)

6.1 THE EXPECTED NORMAL SCORES TRANSFORMATION

In their classic book of statistical tables [4], Fisher and Yates suggested that when original observations were in the form of several sets of ranks (for example if N objects had been ranked from 1 to N by each of several subjects), the ranks might be replaced by the appropriate expected "normal scores" and the resulting data might then be analyzed by performing a conventional, parametric analysis of variance (for matched data, in this case). In general, the appropriate expected normal score for an object (or value) that was eligible to receive one of the ranks from 1 to N and actually did receive a rank of R is the expected value of the observation having a rank of R in a random sample of N observations from a normal distribution with zero mean and unit variance. For example, the expected values for the smallest, second smallest, middle, next-to-largest, and largest observations (i.e., for the observations whose respective size ranks are 1, 2, 3, 4, and 5) in a random sample of size $N = 5$ from a normal distribution with zero mean and unit variance are -1.163, $-.495$, 0, .495, and 1.163, respectively. And these are the expected normal scores that would replace observations having corresponding size ranks in a set of observations ranked from 1 to 5, where 5 is the maximum rank any of the observations was eligible to receive before their values were known.

Let E_{NR} be the expected value of the Rth smallest observation (i.e., the Rth observation in order of increasing size and therefore the observation whose size rank is R) in a random sample of N observations drawn from a standard normal distribution, i.e., a normal distribution with mean of 0 and variance of 1. Table V of the Appendix gives the absolute value of E_{NR} (therefore omitting the minus sign) for $N = 2\,(1)\,20$ and R's of $1\,(1)\,N/2$ when N is even and $1\,(1)\,(N-1)/2$ when N is odd (when N is odd and $R = N/2$, $E_{NR} = 0$ and these cases are omitted from the table). For given N the distribution of E_{NR} is symmetric about zero, so the $|E_{NR}|$ listed for the Rth smallest observation is also the E_{NR} for the Rth largest [or $(N - R + 1)$th smallest] observation. Fisher and Yates themselves tabled $|E_{NR}|$, to two decimal places, for $N = 2\ (1)\ 50$ and, in effect, $R = 1\ (1)\ N/2$ [cf. Table XX in ref. 4]. Since it will be needed in the denominator of t or F, they also tabled $\sum_{R=1}^{N} (E_{NR})^2$, to four decimal places, for $N = 2\ (1)\ 50$ [cf. Table XXI in ref. 4]. The first of these tables has been greatly extended by Harter, who tabled $|E_{NR}|$, to five decimal places, for $N = 2\ (1)\ 100(25)\ 250\ (50)\ 400$ and $R = 1\ (1)\ N/2$ in [7] and for these as well as 39 additional N values between 100 and 400 in [8]. Tables accurate to ten decimal places are given by Teichroew [20] for, in effect, all expected normal scores and all expected values of products of two normal order statistics for $N = 1\ (1)\ 20$.

Fisher and Yates advocated performing the conventional, parametric analysis of variance directly upon the substituted, expected normal scores. Such a test would not be an exact test of the hypothesis of equal population means. Rather it would be an approximate test of the hypothesis of identical populations. It is not a test for equal means because the "normality" or "homogeneity of variance" of the *transformed* scores do not rule out nonnormality or heterogeneity of variance as possible forms of *population* nonidentity, leaving unequal means as the only form of nonidentity not eliminated, as do the assumptions of population normality and homogeneity. (Of course, if one can assume population normality and homogeneity, the test does become a test for equal means, but in that case one had no need to abandon the classic analysis of variance.) The Fisher-Yates test is not an exact test because in order for the analysis-of-variance F statistic to have its null, tabled distribution, the scores upon which it is based must be *random* normal deviates, not the *constant expected values* of certain sample order statistics when N observations are made on a standardized random normal deviate. Another consequence of the transformation is that unusually (i.e., improbably) extreme sample observations are literally made usually (or at least averagely) extreme "for normality," thereby greatly altering their influence upon the test. The highest and lowest ranking observations, especially, may have far different influence upon the test outcome than they would have had if not transformed.

The expected normal scores transformation is an intuitively appealing one. It obviates the population assumption of normality. And although the resulting F test is *really* a test for identical populations, if one wishes to "assume" identical population shapes and variances it is just as much a test for means as is the classical F test which makes even less realistic assumptions. The only serious flaw in the procedure advocated by Fisher and Yates is that the null distribution of F based on expected normal scores is not exactly the same as the classical, tabled F distribution. This single serious defect can be remedied by referring the test statistic to its true null distribution. At given sample sizes the set of expected normal scores is simply a single, unvarying set of constants, subsets of which are allocated to the various samples. Therefore, the true null distribution of a test statistic based on expected normal scores is simply the randomization distribution of the test statistic over all allocations that could possibly result under the limitations imposed by sample sizes, matching, and other mechanics of test procedure. In essence, the tests to follow are based upon this simple modification of the Fisher-Yates procedure. Over the randomization distribution of the test statistic the numerators of t and F are monotone increasing functions of the entire t and F, respectively (when testing equality of means); and the denominator of r is a constant. Clearly, therefore, the test statistics for the modified Fisher-Yates procedure can be greatly simplified versions of their parametric counterparts, using the simplest term that is a monotone function of the parametric test statistic and dropping all others. They do not, therefore, necessarily beer a close resemblance to their parametric equivalents.

6.2 THE TERRY-HOEFFDING EXPECTED NORMAL SCORES TEST FOR IDENTICAL POPULATIONS, SENSITIVE TO UNEQUAL LOCATIONS

The hypothesis that two populations are identical can be tested by a variety of tests that use unmatched observations and are especially sensitive to differences in population location: Student's t test, which requires an additional assumption of normality; Fisher's observation-randomization test (see Section 4.2.3.), for which no tables exist; Wilcoxon's rank-randomization test (Section 5.8), which has an A.R.E. of less than 1.00 relative to Student's test when the populations have normal shapes with the same variance. The test to follow is the expected normal scores analogue of all three of the above tests and suffers from none of the above listed criticisms.

6.2.1 RATIONALE

Suppose than an experimenter has randomly and independently obtained n observations upon a variable X and m observations on a variable Y, that he wishes to test the hypothesis that the X- and Y-populations are identical, and that he wishes the test to be especially likely to reject when the nonidentical populations have unequal locations, e.g., medians. Let the $n + m = N$ observations in the pooled sample be arranged from smallest to largest, i.e., in order of increasing size, irrespective of whether they are X's or Y's. Then let these observations be replaced by the constant expected values of the observations having corresponding size ranks in a sample of N observations drawn randomly and independently from a standard normal distribution (i.e., a normal distribution whose mean is zero and whose variance is 1). Finally, let S be the sum of the expected normal scores that replaced the n observations that were originally X's. Thus if X_i is the ith observation in the sample of n X's, if R_i is the size rank of X_i in the pooled sample of $n + m = N$ observations, and if E_{NR_i} is the expected value of the observation having a size rank of R_i in a random sample of N observations from a standard normal distribution, our test statistic is

$$S = \sum_{i=1}^{n} E_{NR_i}$$

Now if the hypothesis of identical X- and Y-populations is true, each of the $\binom{N}{n}$ different possible sets of n observations that can be drawn from the $N = n + m$ observations in the pooled sample had equal a priori probability of becoming the X-sample. Therefore, each of the $\binom{N}{n}$ different possible sets of n expected normal scores drawn from the N expected normal scores E_{N1}, $E_{N2}, \ldots, E_{NN}$ also had equal a priori probability of replacing the n X's and determining the value of the test statistic. So the null distribution of the test statistic is simply its randomization distribution obtained by tabulating the relative frequency distribution of S over all $\binom{N}{n}$ different possible selections of n of the N constant, expected values, $E_{N1}, E_{N2}, \ldots, E_{NN}$, for the N observations in a sample of size N drawn from a standard normal distribution. If the hypothesis of identical populations is false, the distribution of S will differ from the null randomization distribution, and the difference will almost certainly cause the probability of rejection to differ from α, the direction of the latter difference depending upon the characteristics of the populations and the location (and perhaps size) of the rejection region. However, if the populations have identical shapes and variances and the median of the X-population lies below (above) that of the Y-population, S will certainly fall in an α-sized lower- (upper-) tailed rejection region with relative frequency greater than α. Furthermore, although it is less than certain, one would generally expect the same tendency even if the populations also differed in shapes and variances, provided that their shapes were not radically dissimilar.

6.2.2 NULL HYPOTHESIS

Each of the $\binom{N}{n}$ different possible selections of n of the expected normal scores, $E_{N1}, E_{N2}, \ldots, E_{NN}$ was equally likely, a priori, to be required by the test procedure to replace the n X-observations. This will be the case if the X- and Y-populations are identical and all assumptions are true.

6.2.3 ASSUMPTIONS

Observations are drawn *randomly* and *independently* from their respective populations, and *no observations are tied.*

6.2.4 Efficiency

This test is the most powerful rank-order test of the null hypothesis that two continuous populations are identical against the alternative that they are normally distributed with equal variances but slightly unequal means [15, 22]. As a test of shift (i.e., as a test of whether or not two continuous populations with identical shapes and variances also have identical locations) the test has an A.R.E. $\geqq 1$ relative to Student's t test, the equality sign holding only if the populations are normal (in which case the t test's assumptions are completely satisfied) and the A.R.E. being > 1 for any other continuous population shape [3, 9]. Under the same circumstances, i.e., as a test of shift for continuous populations, the test's A.R.E., relative to Wilcoxon's rank-sum test, which we shall call e_{sw}, can go no lower than $\pi/6$ (or .524), but can rise as high as infinity and can assume any value between these bounds depending upon the shape of the population [9]. If the common population shape is rectangular or exponential, e_{sw} is infinite; if the shape is normal, the e_{sw} is $\pi/3$ or 1.047 (so the A.R.E., e_{ws}, of the Wilcoxon test relative to the expected normal scores test is $3/\pi$, or .955, the same as its A.R.E. relative to Student's t test under conditions assumed by the latter); if the shape is logistic (in which case the Wilcoxon test is the most powerful rank-order test for shift), e_{sw} is $3/\pi$, or .955; if the shape is double-exponential, e_{sw} is $8/(3\pi)$, or .849; and if the shape is Cauchy, e_{sw} is .708 [9]. Additional information on A.R.E.'s can be found in [9, 16].

Small-sample power functions for both the expected normal scores test and Student's t test, when the populations are normal and differ only in location, are compared in [13—see also 22], showing the t test to be slightly more powerful. Small-sample power functions for the expected normal scores test and for a variety of other distribution-free tests, including the Wilcoxon test, are compared in [6] under several types of alternatives (both conventional, e.g., normal shift, and otherwise) to the tested H_0. The robustness against heterogeneity of variance of the expected normal scores test for shift in location, when sample sizes are large, is studied in [18] and compared with that of the t, Wilcoxon, and median tests for shift.

The appropriately one-tailed test is unbiased against the class of alternatives in which the cumulative distribution of the X-variate lies entirely above or entirely below that of the Y-variate (the two cumulatives perhaps touching at certain points, but never crossing) [15].

6.2.5 Tables

Before conducting the expected normal scores *test* (or, at least, before calculating the value of the test statistic) one must first perform the expected normal scores *transformation* upon the original sample observations, using

Table V of the Appendix, or one of the tables listed in Section 6.1. After making the transformation and calculating the value of the test statistic, the latter's significance can then be tested using one of the tables listed in the present section.

Table VI of the Appendix gives, for $N = n + m = 6\,(1)\,20$ and $n = 2\,(1)\,N/2$, those values of $S = \sum_{i=1}^{n} E_{NR_i}$ whose upper-tail cumulative probabilities are closest to the nominal α's of .001, .005, .01, .025, .05, .075, and .10, and, beneath each such S-value, gives its exact upper-tail cumulative probability. The null distribution of S is symmetric about zero, so the tabled S's are also the absolute values of the (negative) critical or just-beyond-critical values of S corresponding to the *same* numerical α values and exact probability levels for a *lower*-tailed test.

The entire upper half of the distribution of S is given for $N = 2\,(1)\,10$, $n = 2\,(1)\,N/2$ (followed by a short table of critical values and their exact probabilities) by Terry [22] who uses the symbol $c_1(R)$ to denote the test statistic we call S.

The null distribution of S has a mean of zero, a variance of

$$\frac{nm}{N(N-1)} \sum_{R=1}^{N} (E_{NR})^2$$

and approaches a normal distribution as N approaches infinity (provided that n/N does not become virtually zero or 1 in the process) becoming exactly normal when $N = \infty$. So, when N is too large for the exact tables to apply, approximate probability levels can be obtained for S by referring the critical ratio

$$\frac{S}{\sqrt{\dfrac{nm}{N(N-1)} \sum_{R=1}^{N} (E_{NR})^2}}$$

to tables of probabilities for a standard normal deviate with zero mean and unit variance. A better approximation, however, can be obtained by making use of the fact that

$$t_{N-2} \cong \sqrt{\frac{(N-2)S^2}{\dfrac{nm}{N} \sum_{R=1}^{N} (E_{NR})^2 - S^2}}$$

is approximately distributed as Student's t with $N - 2$ degrees of freedom, obtaining probability levels by referring the above statistic to the t tables.

6.2.6 Example

In the example given for the Wilcoxon rank-sum test, three of eight subjects belonging to a certain class were randomly selected and assigned to diet X, experiencing weight increments of 10, 5, and 3, while the five remaining subjects received diet Y and experienced weight increments of 29, 25, 22, 16, and 8. Before obtaining the data the experimenter decided to use a lower-tail rejection region with $\alpha = .05$ to test the null hypothesis that the population distributions of weight increments under the two diets are identical. Thus rejection would call for the conclusion that the populations are nonidentical and would strongly suggest that the median of the X-population lies below that of the Y-population. As it turned out, the Wilcoxon test statistic, calculated from the data, was the critical value for a lower-tail test at the .05 level of significance. Applying the present test to these same data to test the same hypothesis in the same way, we perform the operations indicated in the accompanying table. The X's, their ranks in the pooled sample, and the E_{NR}'s for those ranks are boldface; S is simply the sum of the boldface E_{NR}'s.

X	**3**	**5**		**10**				
Y			8		16	22	25	29
R	**1**	**2**	3	**4**	5	6	7	8
E_{NR}	**−1.4236**	**−.8522**	−.4728	**−.1525**	.1525	.4728	.8522	1.4236

$$S = \sum_{i=1}^{n} E_{NR_i} = -1.4236 + (-.8522) + (-.1525) = -2.4283$$

$$P(S \leq -2.4283) = .03571$$

The E_{NR}'s are obtained from Appendix Table V with $N = 8$, $R = 1, 2, \ldots, 8$. Actually only the E_{NR}'s for the R's associated with X's are needed; the others are listed only to clarify the procedure. The sum of the E_{NR}'s replacing X's in the transformation of observations is S, which has the value -2.4283. Entering Appendix Table VI with $N = 8$ and $n = 3$, the value $S = 2.4283$ is listed under $\alpha = .025$ as the S-value whose upper-tail cumulative probability is closest to .025. Its exact upper-tail cumulative probability is listed below it as .03571, and that is therefore the exact lower-tail cumulative probability of $S = -2.4283$. We therefore reject the H_0 of identical populations at the .05 level of significance and strongly suspect that the median of the X-population lies below that of the Y-population.

6.2.7 Discussion

In testing the hypothesis of identical populations against the alternative of location shift the experimenter may suspect that his populations are essentially normal, but not know it. If he uses the t test, he attains optimum power if he is right but incurs invalidity if he is wrong. By using the Terry-Hoeffding test he attains nearly optimum power (which becomes optimum when sample sizes are infinite and H_0 is only slightly false—the conditions under which the A.R.E. is measured) if he is right but risks no invalidity if he is wrong. True, he incurs the labor of transforming scores, but on the other hand the arithmetical operations in calculating the test statistic are far less involved than is the case for the t test. Indeed, on balance the Terry-Hoeffding test appears easier to perform, at least when N is not large. The only really appreciable advantages of the t test that come to mind (other than optimum power under the unrealistic assumption of exact normality) are that it is more extensively tabled, more information is available about it, such as power functions, etc., and it may by an easier and quicker test when n and m are both very large or when the test is performed by an electronic computer.

The comparison between the Terry-Hoeffding and Wilcoxon tests is less one-sided. Since it requires no transformation for which tables must be consulted, but is otherwise similar, the Wilcoxon test is easier to perform. It is also conceptually simpler and easier to interpret because of the less complicated weights assigned to (i.e., replacing the values of) the observations. And its exact tables are more extensive. It is only in the realm of power that the Wilcoxon test is put on the defensive, and even here the comparison is far from universally unfavorable. When used as tests for unequal locations of otherwise identical populations, the A.R.E. of the Wilcoxon relative to the Terry-Hoeffding test is greater than 1.00 if the populations are logistic, double-exponential, or Cauchy, and is only slightly less than 1.00 if the populations are normal. On the other hand, the A.R.E. can go no higher than 1.910, but can go as low as zero, which it does if the populations are rectangular or exponential. These zero A.R.E.'s count against the Wilcoxon test, but their importance should not be exaggerated. They mean that when very large samples are drawn from (rectangular or exponential) populations differing only very slightly in location, the Wilcoxon test is less powerful than the Terry-Hoeffding test and that very many more observations would have to be included in the Wilcoxon test in order to close the power gap (which may have been very small). To put the matter in perspective, it is well to remember that the A.R.E. of the Wilcoxon test relative to Student's t test when populations are rectangular is 1.00 and that the t test has very good power (although slightly impaired validity at finite sample sizes) when applied to rectangular populations differing only in location. Perhaps even more conducive to a realistic perspective is the fact that if samples are sufficiently small the Wilcoxon and Terry-Hoeffding tests yield identical probability levels when applied to the same data (irrespective of the type of continuous population from which the observations were drawn).

6.3 OTHER EXPECTED NORMAL SCORES TESTS

The concept of an expected normal scores *test* with its own unique test statistic and its own unique tables, based on the Method of Randomization, has been extended to situations other than that dealt with in the preceding section. The basic method can be extended to, and an expected normal scores test can be developed for, all of the testing situations dealt with in the two preceding chapters. Indeed, for every observation-randomization test it is possible to construct a rank-randomization analogue, and for every test of the latter type an "expected-normal-scores-randomization" analogue can be devised. In a variety of situations such analogues *have* been devised. Some of them are described below in order to illustrate the general approach in specific, concrete situations. However, the descriptions will be very brief, since: (a) most such tests are in an immature stage of development, having insufficiently extensive tables (or no tables) and perhaps insufficiently investigated power, etc.; (b) the corresponding rank-randomization test in the preceding chapter will usually be quite adequate.

6.3.1 The Terry-Hoeffding Test for Grouped Data

Terry [21] has published a small table of tail probabilities for the sum of several independent values of S. In so doing, although he seems to have had a somewhat different application in mind, he has extended his own test to the case of data grouped by blocks representing original matching conditions. Such an extension would be completely analogous to Wilcoxon's extension of his rank-sum test to the case of grouped data, both in rationale and procedure (see Section 5.10). The null distribution of $T = \sum_{j=1}^{b} S_j$ has a mean of zero, a variance which is the sum of the variances of the null distributions of the b individual S_j's, and approaches a limiting normal distribution as $\sum_{j=1}^{b} (n_j + m_j)$ approaches infinity [provided that as $n_j + m_j$ approaches infinity $n_j/(n_j + m_j)$ does not become virtually zero or one]. Therefore when large amounts of data are taken, probabilities can be obtained by obtaining the critical ratio of T to its standard deviation and referring it to standard normal tables.

6.3.2 Test for Correlation

An Expected Normal Scores analogue of Pitman's test for correlation (Section 4.2.1) can be obtained as follows. Let X_i be the X-measurement and Y_i be the Y-measurement upon the ith of N units. Let R_{X_i} be the size rank of X_i among the N X's. Similarly, let R_{Y_i} be the size rank of Y_i among the N Y's. Then if we replace each X_i by the expected value of the R_{X_i}th smallest observation in a sample of size N from a standard normal distribution, and perform a similar expected normal scores transformation upon the Y's, the Pitman, observation-randomization, test statistic

$$\sum_{i=1}^{N} (X_i - \bar{X})(Y_i - \bar{Y})$$

becomes

$$\sum_{i=1}^{N} (E_{NR_{X_i}} - 0)(E_{NR_{Y_i}} - 0)$$

Thus we may use

$$B = \sum_{i=1}^{N} (E_{NR_{X_i}})(E_{NR_{Y_i}})$$

as the test statistic for an expected normal scores test of the hypothesis that the X's and Y's are uncorrelated. The null distribution of B can be obtained by the Method of Randomization. (The N constant values of $E_{NR_{X_i}}$ are arranged in ascending order in one row with the corresponding $E_{NR_{Y_i}}$ in a second row beneath them. These $E_{NR_{Y_i}}$ are then permuted in all $N!$ possible ways. B is calculated for each of these ways and the relative frequency distribution of B values is tabulated.) The null frequency distribution of B is symmetric about zero. Its entire upper half has been tabled for the cases $N = 2\,(1)\,6$ in reference [2]. Approximate probabilities for B can be obtained at higher N's by making use of the fact that

$$t_{N-2} \cong B \sqrt{\frac{N-2}{\left[\sum_{R=1}^{N} (E_{NR})^2\right]^2 - B^2}}$$

i.e., that the expression on the right is distributed approximately as Student's t with $N - 2$ degrees of freedom.

6.3.3 Test for Difference-Score Population Symmetry about Zero

Let $D_i = X_i - Y_i$ be the ith difference-score in a sample of N difference-scores; let S_i be its algebraic sign; and let R_i be the size rank of its absolute value, i.e., the size rank of $|D_i|$ among the N different $|D_i|$'s. The Wilcoxon signed-rank test statistic for the null hypothesis that the population of D's is symmetric about zero was

$$W_+ = \sum_{S_i=+} S_i R_i$$

The analogous expected normal scores test statistic is

$$F_+ = \sum_{S_i=+} S_i \mathscr{E}_{NR_i}$$

where $\mathscr{E}_{NR}$ is the expected value of the Rth smallest observation in a random sample of N observations upon a standard *absolute* normal variate, i.e., a variate whose distribution is identical to the upper half of the standard normal distribution except that all ordinates, i.e., probabilities, are doubled. (The square root of chi-square with one degree of freedom is such a variate.) Alternatively $\mathscr{E}_{NR}$ may be defined as the expected *absolute* value of the Rth smallest observation in *absolute* magnitude in a random sample of N observations from a standard normal distribution. Since an expected *absolute* normal score $\mathscr{E}_{NR}$ is not the same as an expected normal score E_{NR}, special tables are needed for the former. Klotz [14] has tabled $\mathscr{E}_{NR}$ for all values of R in the cases $N = 1\,(1)\,10$. From the null, randomization, distribution of F_+, he has also tabled, for $N = 4\,(1)\,10$ and α's of .001, .005, .01, .025, .05, .075, and .10, both F_+ and its exact upper-tail cumulative probability for those F_+ values whose upper-tail probability level lies close to α.

Consider the situation where the D-variate is symmetrically distributed, with variance of 1, about a location parameter μ, and an upper-tail test is to be performed. In that event the null hypothesis of symmetry about zero can be false only if $\mu \neq 0$, tending to be rejected only if $\mu > 0$, and the following statements about the power and efficiency of the expected absolute normal scores test hold true: (a) It is the most powerful rank-type test for the above situation in the case where the D-population is normal and μ is close to its hypothesized value of 0 [5]. (b) If the D population is normal, the test has A.R.E. of 1.00 relative to Student's t test, $\pi/3$, or 1.047, relative to Wilcoxon's signed-rank test (but $3/\pi$, or .955, if the population is logistic), and $\pi/2$, or 1.571, relative to the Sign test (but $2/\pi$, or .637, if the population is double-exponential) [11]. (c) If the population is normal, as $\mu - 0$ increases from 0 to 3 the limiting efficiency of the test relative to Student's t drops from 1.00 to about .60, that relative to the Wilcoxon test drops from 1.047 to about 1.00 (which it reaches at about $\mu = 1$), and that relative to the Sign test drops from 1.571 to about 1.01. If the population is logistic the limiting efficiency relative to the Wilcoxon test improves but stays below 1.00. If the population is double-exponential, the limiting efficiency relative to the Sign test becomes greater than 1.00 at about $\mu = 1.1$ and remains so. When $\mu - 0$ becomes infinite, all of the above A.R.E.'s become 1.00 except that relative to Student's t, under normality, which becomes zero [11]. (d) If the population is normal and N is small the test is less efficient than the t test at all values of μ, the efficiency decreasing as μ increases from 0 to infinity (going, for example when $N = 10$ and $\alpha \cong .05$, from .98 at $\mu = .25$ to .62 at $\mu = \infty$); and although more efficient than the Wilcoxon test when $\mu = 0$, it is less so at $\mu = \infty$ [14]. (e) The test is unbiased [15].

6.4 THE INVERSE NORMAL TRANSFORMATION

Suppose that a random sample of N observations is drawn from a continuous, but otherwise unknown, population of Z's, and let Z_R be the Rth smallest Z-observation in the sample. Then the expected proportion of the Z-population that lies below Z_R is $R/(N + 1)$. But this is also the expected proportion of a standard normal distribution lying below the Rth smallest of N observations randomly drawn from it. The inverse normal transformation is performed by entering a table of lower-tail cumulative probabilities for a standard normal variate, finding the variate-value whose cumulative probability is $R/(N+1)$, and substituting that value for Z_R. Thus one replaces the Rth smallest of N sample observations with the population quantile of which it would be an estimate if the sampled population were standard normal.

More formally, let $p = R/(N+1)$, let η be a standard normal variate, let $\Phi(\eta)$ be its cumulative distribution (i.e., the lower-tail cumulative probability of η in its own distribution), and let η_p be that value of η whose lower-tail cumulative probability is p. Then the inverse normal transformation consists of replacing Z_R by η_p. But since $\Phi(\eta_p) = R/(N+1)$, one can denote η_p as $\eta_p = \Phi^{-1}[R/(N+1)]$, and the expression to the right of the equals sign is often used in place of η_p, since it more clearly implies the operations performed and the inverse character of the transformation.

The inverse normal transformation and expected normal scores transformation are not the same when N is finite. Let η_{NR} be the Rth smallest in a random sample of N observations from a standard normal population. Then E_{NR} is the expected *value* of the *sample* variate η_{NR}, while $R/(N+1)$ is the expected *proportion* of the η *population* lying below η_{NR} and η_p is the population quantile corresponding to this proportion. When N becomes infinite, however, the variance of η_{NR} about E_{NR} becomes zero, so that $\eta_{NR} = E_{NR}$, and the variance of $\Phi(\eta_{NR})$ about $R/(N+1)$ becomes zero, so that $\eta_{NR} = \eta_p$. At infinite N, therefore, $\eta_p = E_{NR}$ and the two transformations are the same.

The inverse normal transformation is a bit more difficult to conceptualize than is the expected normal scores transformation. Perhaps its greatest advantage is that it uses the standard normal tables, which are generally more extensive [17] and easier to locate than are tables of E_{NR}.

6.5 INVERSE NORMAL SCORES TESTS

6.5.1 VAN DER WAERDEN'S TEST

Van der Waerden [23] has proposed a two-sample test for identical populations, sensitive to unequal locations, that is completely analogous to the Terry-Hoeffding test for the same situation, except that the inverse normal scores transformation is used. Van der Waerden's test statistic is

$$V = \sum_{i=1}^{n} \Phi^{-1}\left(\frac{R_i}{N+1}\right)$$

where R_i is the size rank of X_i, the ith of n X-observations, in the combined sample of $N = n + m$ X's and Y's. The null distribution of V has a mean of zero, a variance of

$$\frac{nm}{N(N-1)} \sum_{k=1}^{N} \left[\Phi^{-1}\left(\frac{k}{N+1}\right)\right]^2$$

and a shape that approaches normality as N approaches infinity. Probabilities can therefore be obtained by dividing the obtained value of V by the standard deviation of the null V distribution and referring this critical ratio to standard normal tables. Since the respective transformations used are equivalent at infinite N, this test has the same A.R.E.'s and limiting power properties as the Terry-Hoeffding test.

6.5.2 Klotz's Test

An inverse normal scores analogue of Mood's test for scale (see Section 5.11.3) has been devised by Klotz [12]. The test statistic is

$$K = \sum_{i=1}^{n} \left[\Phi^{-1}\left(\frac{R_i}{N+1}\right)\right]^2$$

where R_i is the size rank of X_i, the ith among n X's, in the combined sample of $N = n + m$ X's and Y's. And the author provides a table of tail probabilities for K in the cases $N = 8$ (1) 20, $n = 1$ (1) $N/2$. The test tests whether or not two populations known to be identical in every respect except the scale aspect of variance are also identical in that regard. Under these conditions the test's A.R.E. relative to either the classical F ratio or to the Siegel-Tukey test can be anything from zero to infinity, depending upon the type of populations; if the population is normal, the A.R.E. relative to the F ratio test is 1.00. The test's A.R.E. relative to Mood's test and also relative to the Siegel-Tukey test is greater than 1.00 if both populations are exponential, rectangular, normal, logistic, or double-exponential (and is infinite in the first two cases), but is less than 1.00 if the populations are Cauchy.

6.6 THE RANDOM NORMAL SCORES TRANSFORMATION

Suppose that instead of replacing the N observations in the pooled sample by the expected (or "inverse normal") values of the observations in a random sample of size N from a standard normal population, we replace them by the random normal observations themselves. This can be accomplished by entering a table [such as 19] of random values for a standard normal variate, randomly selecting N of them, and then, for every value of R, replacing the Rth smallest observation in the original pooled sample by the Rth smallest of the N random values of the standard normal variate. This transformation completely meets the normality assumption of the t and F tests for equal means provided that their null hypothesis is true and the sampled populations have identical shapes. In such cases, the classical, parametric test could, if we wished, be used in the usual way, probability levels being obtained from the classical tables. Furthermore, by changing the null hypothesis of the t or F test from "equal means" to "identical populations," we eliminate all population assumptions except continuity.

Still further advantages can be gleaned. One can, of course, perform an analysis of variance F test in the usual way and obtain probabilities from the F tables. However, a more powerful test can be had. The purpose of the denominator of F is to estimate population variance. However we do not need to estimate it since we know our standard normal population to have a variance of 1. If we substitute 1 for the denominator, the resulting fraction is no longer distributed as F but rather as the numerator of F which, under H_0, is a χ^2/d.f. (i.e., a chi-square-over-degrees-of-freedom) variate. Therefore, we may test the same hypothesis more effectively (and more easily) by calculating the numerator of F and obtaining probabilities for it from tables of χ^2/d.f. Many other parametric test statistics contain the common population variance, or an estimate of it, in their formulas, and they can be similarly simplified under the random normal scores transformation. When the common variance term is replaced by a 1, the resulting, simplified test statistic is generally found to have a familiar, well-tabled classical distribution, often the normal or chi-square. For example, if we replace by a 1 the pooled variance estimate in the two-unmatched-sample t statistic, the expression under the radical becomes simply the sum of the reciprocals of the sample sizes. The statistic therefore reduces to the corresponding Z statistic for the case where both populations have unit variance, and that statistic has a standard normal distribution. Bell and Doksum [1] have devised a whole series of such tests, which they present as analogues of expected normal scores tests and which they show generally to have the same A.R.E.'s as their expected normal scores counterparts, at least for a broad specified class of alternatives.

Tests based on the random normal scores transformation have the unnerving characteristic that subsequent to the collection of data an additional element of pure chance, completely unrelated to the physical experiment, is permitted to influence the test outcome and therefore the decision about the physical experiment. Other flaws, which they share with expected and inverse normal scores tests, are (a) that the "normality" they induce may be restricted to the case where the null hypothesis is true (and perhaps certain other conditions are met, e.g., that populations have identical shapes); and (b) the homogenizing effect of the transformation may considerably reduce the power of the test. Consider, for example, an X-population normally distributed with unit variance about a mean of zero and a Y-population normally distributed with unit variance about a mean of 100. Suppose that we draw a representative sample of n observations from each population and, knowing the population shapes to be the same, wish to test the equality of their means. If we apply Student's t test to the original observations, the numerator of t will be about -100, the variance estimate in the denominator will be about 1, and t will be approximately $-100\sqrt{n/2}$, which would fall extremely far out on the tail of the null distribution of t and would do so at any n value, i.e., at any degrees of freedom. If we perform a random normal scores transformation upon the pooled sample, the gap of probably at least 92 units between the highest X and lowest Y is so reduced that we can only say that the latter lies above the former, although perhaps negligibly so. Furthermore, we are transforming the X's and Y's from completely normal observations to "lower-normal" and "upper-normal" observations respectively, somewhat as though they had been drawn from truncated normal populations. Now if we apply Student's t test to the transformed scores the variance estimate in the denominator becomes about $\frac{1}{4}$, the denominator becomes about $\sqrt{1/2n}$, and the numerator, $\bar{X} - \bar{Y}$, has become about -1.6. The resulting value of t will be approximately $-1.6\sqrt{2n}$ which will be far less significant than the previous t at any common n value. Furthermore, if we increase the distance between population means from 100 to 1,000, say, the (left-tailed) probability level for the classical t test will greatly diminish, as it should; but since there is virtually no overlap between the two populations in either case, the probability level of the random normal scores t test will remain essentially the same.

Had we used Fisher's observation-randomization analogue of the t test we would have found ourselves in the same predicament but for a different reason: the probability level of the test becomes $1/\binom{n+m}{n}$ when all X-observations lie below all Y-observations and cannot sink below that lower bound no matter what the size of the gap between the two groups of observations. Had we used the Terry-Hoeffding expected normal scores test or van der Waerden's inverse normal scores test, both the homogenizing effect of the transformation and the lower bound for the probability level would have insured that the predicament would be the same. If this seems difficult to reconcile with the fact that these tests have A.R.E. $\geqq 1$ relative to Student's t test as tests for unequal locations of otherwise identical populations, it is only necessary to recall that the A.R.E. applies to the case where the true condition is almost identical to the condition described by the null hypothesis.

REFERENCES

1. Bell, C. B., and K. A. Doksum, "Some New Distribution-Free Statistics," *Annals of Mathematical Statistics*, **36** (1965), 203–214.
2. Bhuchongkul, S., "A Class of Nonparametric Tests for Independence in Bivariate Populations," *Annals of Mathematical Statistics*, **35** (1964), 138–149.
3. Chernoff, H., and I. R. Savage, "Asymptotic Normality and Efficiency of

Certain Nonparametric Test Statistics," *Annals of Mathematical Statistics*, **29** (1958), 972–994.

4. Fisher, R. A., and F. Yates, *Statistical Tables for Biological, Agricultural and Medical Research*, 3rd ed. New York: Hafner Publishing Company, 1949, pp. 21–22 and 66–67.
5. Fraser, D. A. S., "Most Powerful Rank-Type Tests," *Annals of Mathematical Statistics*, **28** (1957), 1040–1043.
6. Gibbons, Jean D., "On the Power of Two-Sample Rank Tests on the Equality of Two Distribution Functions," *Journal of the Royal Statistical Society, Series B*, **26** (1964), 293–304.
7. Harter, H. L., "Expected Values of Normal Order Statistics," *Biometrika*, **48** (1961), 151–165.
8. Harter, H. L., *Expected Values of Normal Order Statistics*, Aeronautical Research Laboratories Technical Report 60-292, Wright-Patterson Air Force Base, Ohio, July 1960.
9. Hodges, J. L. and E. L. Lehmann, "Comparison of the Normal Scores and Wilcoxon Tests," in *Proceedings of the Fourth Berkeley Symposium on Mathematical Statistics and Probability*, ed. Jerzy Neyman. Berkeley and Los Angeles: University of California Press, 1961, Vol. 1, pp. 307–317.
10. Hoeffding, W., "'Optimum,' Nonparametric Tests," in *Proceedings of the Second Berkeley Symposium on Mathematical Statistics and Probability*, ed. Jerzy Neyman. Berkeley and Los Angeles: University of California Press, 1951, pp. 83–92.
11. Klotz, J., "Alternative Efficiencies for Signed Rank Tests," *Annals of Mathematical Statistics*, **36** (1965), 1759–1766.
12. Klotz, J., "Nonparametric Tests for Scale," *Annals of Mathematical Statistics*, **33** (1962), 498–512. [See also *Annals of Mathematical Statistics*, **36** (1965), 1306–1307.]
13. Klotz, J., "On the Normal Scores Two-Sample Rank Test," *Journal of the American Statistical Association*, **59** (1964), 652–664.
14. Klotz, J., "Small Sample Power and Efficiency for the One Sample Wilcoxon and Normal Scores Tests," *Annals of Mathematical Statistics*, **34** (1963), 624–632.
15. Lehmann, E. L., *Testing Statistical Hypotheses*. New York: John Wiley & Sons, Inc., 1959, pp. 236–243.
16. Mikulski, P. W., "On the Efficiency of Optimal Nonparametric Procedures in the Two Sample Case," *Annals of Mathematical Statistics*, **34** (1963), 22–32.
17. National Bureau of Standards, *Tables of Normal Probability Functions*. Washington D. C.: U. S. Government Printing Office, 1953.
18. Pratt, J. W., "Robustness of Some Procedures for the Two-Sample Location Problem," *Journal of the American Statistical Association*, **59** (1964), 665–680.

19. Rand Corporation, *A Million Random Digits with 100,000 Normal Deviates.* New York: The Free Press of Glencoe, Inc., 1955.

20. Teichroew, D., "Tables of Expected Values of Order Statistics and Products of Order Statistics for Samples of Size Twenty and Less from the Normal Distribution," *Annals of Mathematical Statistics*, **27** (1956) 410–426.

21. Terry, M. E., "An Optimum Replicated Two-Sample Test Using Ranks," in *Contributions to Probability and Statistics, Essays in Honor of Harold Hotelling*, ed. I. Olkin *et al.* Stanford, Calif.: Stanford University Press, 1960, pp. 444–447.

22. Terry, M. E., "Some Rank Order Tests which are Most Powerful Against Specific Parametric Alternatives," *Annals of Mathematical Statistics*, **23** (1952), 346–366.

23. van der Waerden, B. L., "Order Tests for the Two-Sample Problem and Their Power," *Proceedings Koninklijke Nederlandse Akademie van Wetenschappen, Series A*, **55** (1952), 453–458 and **56** (1953), 80, 303–316.

7

TESTS BASED ON THE BINOMIAL DISTRIBUTION

A number of distribution-free test statistics are binomially distributed. They are among the simplest, safest, most nearly exact, and most extensively tabled nonparametric tests. Although their statistical efficiency is not the highest, it is generally quite reasonable in an absolute sense and is astonishingly good considering the rudimentary nature of the information used by the test. The sample information used by most such tests is simply the direction of the difference between two scores, i.e., the algebraic sign of the difference. Binomial tests are extremely versatile, finding application in testing for location, trend (in either location or dispersion), concentration of population area, differences between correlated proportions, differences among "paired comparisons," and in the setting of confidence limits for quantiles or proportions.

7.1 THE BINOMIAL DISTRIBUTION

Suppose that all of the possible outcomes of an event may be dichotomized into two mutually exclusive categories, arbitrarily labeled "success" and "failure," these two outcomes having probabilities p and $q = 1 - p$ respectively. Then if the event is permitted to occur n times, the probability that r of the n outcomes will be successes is $P(r) = \binom{n}{r} p^r q^{n-r}$ which is the general expression for a term in the expansion of the binomial $(p + q)^n$.

Proof: The probability that r successes and $n - r$ failures will occur in a specified order is $p^r q^{n-r}$. For example, letting subscripts indicate order of appearance, the probability for the order in which all successes occur first, followed by all failures, is the product

$$(p_1)(p_2) \ldots (p_r)(q_{r+1})(q_{r+2}) \ldots (q_n) = p^r q^{n-r}$$

However, since we seek only the probability of a given *frequency* of successes, the probability $p^r q^{n-r}$ of a given frequency of successes occurring in a specified order within the n trials must be multiplied by the number of orders in which r successes and $n - r$ failures can occur. If the n units (p's and q's) were all distinguishable, there would be n distinguishable ways of filling the first position in order (i.e., any one of the n units could be placed at the beginning), $n - 1$ distinguishable ways of placing one of the remaining $n - 1$ units in second place, $n - 2$ distinguishable ways of filling the third position in order, etc., so that *the number of distinguishable patterns of order, i.e., permutations, of n distinguishable things* (see Section 3.6.1) *is* $n(n - 1)(n - 2) \ldots (3)(2)(1) = n!$, and this would be the number of distinguishable patterns of order had there been n possible outcomes rather than just two. Now still supposing that all n outcomes are distinguishable (as, for example, by means of differing subscripts), for any given one of the $n!$ patterns of outcome the r outcomes which are "successes," i.e., the p's, can be permuted in $r!$ ways and the $n - r$ outcomes which are failures, i.e., the q's, can be permuted in $(n - r)!$ ways without changing the pattern of successes and failures, as such. Therefore the number of patterns of outcome in which successes are distinguishable only from failures and vice versa (so that successes are indistinguishable from one another and failures likewise all appear the same) is

$$\frac{n!}{r!\,(n - r)!}$$

which is commonly represented by the notation $\binom{n}{r}$ (see Sections 3.6.2 and 3.6.3). The probability of exactly r successes in n binomial trials is therefore $\binom{n}{r} p^r q^{n-r}$, the product of the number of distinguishable orders of r successes and $n - r$ failures and the probability for a single distinguishable order. The cumulative probability, i.e., the probability of r or fewer successes in n trials is

$$\sum_{i=0}^{r} \binom{n}{i} p^i q^{n-i}$$

The binomial term $\binom{n}{r} p^r q^{n-r}$ expresses the probability for r successes out of n trials only if the following conditions, implicit in its derivation, are met:

(a) Outcomes must be capable of being *dichotomized* (since only two outcome probabilities, p and q, are used in the derivation).
(b) The two outcome categories must be *mutually exclusive* (since $q = 1 - p$).
(c) The outcome of the n events must be completely *independent*. (Since the same value, p, is used to express the probability of success on each of the n trials, the probability of success on a single trial must not change from one trial to another and, therefore, must not be influenced by the outcome of any other trial. The stipulation that the probability of success must not change from trial to trial also implies that either the population yielding the sample "events" consists of an infinite number of successes and failures or that sampling is with replacement. An infinite population may be encountered in practice when the population is conceived of as a population of potential, rather than actual, events. For example, the population of events labeled "heads" and "tails" for a certain coin is not depleted by actually tossing the coin.)
(d) "Events" must be *randomly selected*. (The formula $\binom{n}{r} p^r q^{n-r}$ gives the probability that by *chance* r successes will occur in n trials if the *chance* probability of success in a single trial is p. If events are not randomly selected, then outcomes are susceptible to nonchance influences.) There must therefore be no bias or system in the selection of which n trials, out of an infinite population of potential trials, to test. Specifically, among other things this means that none of the valid data may be systematically excluded from the test.

The above qualifications will appear in modified form as *assumptions* for all tests whose test statistic is binomially distributed. Such tests are outstanding among distribution-free tests for two reasons: First they are extremely simple, both in derivation and in application. Second, exact probabilities for both the point and cumulative binomial have been extensively tabled. Thus, while for most distribution-free tests large n's require probabilities to be calculated approximately from asymptotic formulas, in the case of binomial tests exact probabilities are readily attainable for many large samples.

The distribution of the number r of successes in n binomial trials has mean np, variance npq, and approaches a normal distribution as n increases, the closeness of the normal approximation at any *fixed* value of n worsening as p departs unidirectionally from .5. Thus $(r - np)/\sqrt{npq}$ is approximately a normal deviate with zero mean and unit variance, the degree of approximation depending upon n and p. Probabilities can therefore be obtained from standard normal tables; the approximation is usually improved by making a correction for continuity which consists of reducing the absolute value of $r - np$ by $\frac{1}{2}$. One is seldom forced to use the normal approximation, however, since the binomial distribution is quite adequately tabled.

A number of comprehensive tables of probabilities have been published. In describing the values of a parameter for which probabilities are tabled, the notational format $K = x(y)z$ signifies that the values of K go from x in successive steps of y units each to z. The cumulative probabilities of r or more successes in n binomial trials have been published for $p = .01\ (.01)\ .50$ in (a) the Harvard tables [23] to five decimal places for $n = 1\ (1)\ 50\ (2)\ 100\ (10)\ 200\ (20)\ 500\ (50)\ 1000$ (these tables also apply to the fractional values $p = \frac{1}{16}\ (\frac{1}{16})\ \frac{7}{16}$ and $p = \frac{1}{12}\ (\frac{1}{12})\ \frac{5}{12}$; (b) the Ordnance Corps tables [18] to seven decimal places for $n = 1\ (1)\ 150$; (c) the National Bureau of Standards tables [16] to seven decimal places for $n = 2\ (1)\ 49$ (these tables give point probabilities as well as cumulative probabilities). A much more fine-grained treatment of the tail values of p is given by the following tables: Weintraub's tables [28] give the probability of r or more successes, to ten decimal places, for $p = .00001$ and $.0001\ (.0001)\ .001\ (.001)\ .10$ and $n = 1\ (1)\ 100$. Finally, Robertson's Sandia tables [20] give the probability of r *or fewer* successes, to five decimal places, for $p = .001\ (.001)\ .05$ and n values going from 2 to 1000 in steps of 1 to 50 and thereafter to 1000 in a series of step sizes which vary with p.

7.2 THE SIGN TEST FOR THE MEDIAN

7.2.1 Rationale

Suppose that one hypothesizes, H_0, that the median M of a certain population, known to be continuously distributed, has the value M_0. The outcome of randomly drawing a single observation X from the population can be dichotomized into "X exceeds M_0" and "X is less than M_0," or, equivalently, into "$X - M_0$ is positive" and "$X - M_0$ is negative." If H_0 is true, half the population lies above M_0, half below it, and each member of the dichotomy has probability $\frac{1}{2}$. So the drawing of an observation is a binomial event, and if n observations are drawn, the number r of X's which fall into a given category, e.g., the number of X's $> M_0$, is a binomially distributed variate with point probability

$$P(r) = \binom{n}{r}\ .5^n$$

and cumulative probability

$$\text{Cum } P(r) = \sum_{i=0}^{r} \binom{n}{i}\ .5^n$$

Thus H_0 can be tested by referring the number of $X - M_0$ differences having an algebraic sign of one type to tables of the cumulative binomial distribution.

7.2.2 Null Hypothesis

$P(X > M_0) = P(X < M_0) = .5$, or equivalently, $P(X - M_0 > 0) = P(X - M_0 < 0) = .5$, or still more briefly, for the algebraic sign of the difference $X - M_0$, $P(+) = P(-) = .5$. This will be the case if the median of the X-population is M_0 and all assumptions are true.

7.2.3 Assumptions

(a) $P(X_i = M_0) = 0$; roughly speaking, none of the observations will "ever" fall on the hypothesized median. This will be the case if measurement is precise and the sampled population is continuously distributed in the region of M_0; it will also be the case if measurement is not very imprecise and the sampled population is discretely distributed but contains no variate value equal to M_0. (For proper treatment of obtained difference-scores of zero, see Methods A to E of Sections 3.3 and 3.4, especially Methods A and B.)
(b) The X_i's have been *randomly drawn.*
(c) The X_i's are *independent*, i.e., whether a given observation falls above or below M_0 (or its probability of doing so) is unaffected by the outcome for any other observation. This implies among other things that either the population is an infinite one, which will be the case if it is continuously distributed, or sampling is with replacement.

7.2.4 Efficiency

This test appears to be the most powerful (rank-type) distribution-free test for an hypothesized population median when all that is known about the sampled population is that it is continuous [11, 12].

For a normal distribution, median M and mean μ are equal, so this test and Student's t test are comparable in the sense that they test the same hypothesis $H_0: \mu = \mu_0$. The A.R.E. of the sign test for the median relative to Student's t test under conditions meeting all assumptions of the latter is $2/\pi$, which equals .637. As μ departs increasingly from μ_0, while the variance of the normal population remains constant, the limiting relative efficiency (i.e., at infinite sample sizes), for one-tailed tests at any α, decreases from .637 toward .500, which it reaches when $|\mu - \mu_0| = \infty$ [14]. When the restriction of a normal population is removed, the situation changes. The Sign test is unaffected by the variance of the population, whereas if population variance is infinite the power of the t (or Z) test does not increase with sample size. Therefore, there is no upper bound to the A.R.E. of the Sign test relative to the t (or Z) test when the type of population is unspecified. If the population is rectangular, the A.R.E. relative to t is $\frac{1}{3}$, which is a lower bound for a certain class of unimodal populations to which the rectangular belongs [14].

The finite-sample relative efficiency of the Sign test to the t test increases with decreasing values of n, α, and $|\mu - \mu_0|$ when the population is normal and alternatives involve changes of location only. Values as high as .96 have been found.

The above statements hold under the conventional definition of relative efficiency as the ratio of sample sizes required to attain equal power, but not necessarily under other perfectly reasonable definitions [2, 8]. For example, if relative efficiency is defined in terms of the ratio of powers attained at the same sample size, under the alternative which minimizes the ratio, i.e., under the "worst" alternative, then the relative efficiency of the Sign test (with respect to the t test against normal translation alternatives): (a) tends to remain nearly *constant* for $6 \leqq n \leqq \infty$, at a value of about .74 when $\alpha = .05$ (one-tailed test); and (b) *decreases* with decreasing values of α when n is infinite, going from about .74 to .62 to .49 to 0 as α goes from .05 to .01 to .001 to 0 [2]. For a fixed alternative, of course, the power of the Sign test at a given α increases with increasing n. Thus it is well to bear in mind that efficiency depends strongly upon the precise definition chosen and that it should never be confused with absolute power.

Generally speaking, the Sign test has greatest *relative* power, i.e., efficiency, in those situations (e.g., small n, "close" alternatives) where power tends to be smallest and therefore is badly needed, and has least relative power when absolute power is close to 1.00 anyway, making the matter academic. Power (or efficiency) functions can be found in references [1, 2, 8, 9, 13, 14, 27]. Tables giving the sample size at which the hypothesis $p = .5$ will have a given probability of rejection, i.e., the test will have a given power, for various "true" values of p are given in [10, 24]. The Sign test is consistent provided only that $p \neq q$, i.e., in the present case provided only that H_0 is false and the assumptions are true.

7.2.5 Tables

Table VII of the Appendix gives critical values of r for a two-tailed test, where r is the frequency of the less frequently occurring algebraic sign. The table is based upon the formula

$$P(i \leqq r) = 2 \sum_{i=0}^{r} \binom{n}{i} .5^n$$

and the critical value tabled is the largest value of r for which $P(i \leqq r) \leqq \alpha$. Other tables which contain the necessary information, possibly in different format, are given in references [10, 16, 18, 23].

7.2.6 Example

It is hypothesized that the median time patients spend in a certain doctor's office is exactly 10 minutes. It is decided to test this hypothesis using a two-tailed test with an α of .05. A random sample of 30 observations is taken, of which 8 time periods are less than 10 minutes and 22 are greater. Referring the smaller number, 8, to Table VII, for $n = 30$, it is found to be the critical value corresponding to a two-tailed significance level of .02 so the hypothesis can be rejected at the predetermined .05 level of significance.

7.2.7 Discussion

By adding (or subtracting) a constant C to every X before applying the test, the hypothesis can be tested that the population median has "slipped" a distance C below (or above) a value it is known to have had at some earlier period (and which is substituted for M_0 in the formulas already presented). It is of course assumed that population shape remains constant, i.e., that every possible variate value slips by C units if the median does. If a population is known on a priori grounds to be symmetrical but is not known to be normal, the Sign test is an appropriate test for the population mean, since under symmetry mean and median are the same.

Actually the derivation of the test would permit each observation to be drawn from a different population so that the hypothesis tested could be that all sampled populations have the same median M_0. However, this type of application is more likely in the special case of this test which is taken up next. Its discussion is therefore deferred to that case where it is more appropriate.

The Sign test for the median is, itself, a special case of a more general test which might be termed the Sign test for an hypothesized population quantile. Let Q_p be the value below which exactly a proportion p of a continuous population lies, so that Q_p is the 100pth percentile of the X-population, and let Q_{p0} be the hypothesized value of Q_p. Then if the hypothesis is true, the number r of negative values of $X - Q_{p0}$ in a sample of size n will be binomially distributed with parameter p, and the hypothesis can be tested by referring r to the appropriate binomial distribution. Mathematically, the hypothesis tested is that $P(X < Q_{p0}) = p$ and $P(X > Q_{p0}) = 1 - p$. It is assumed that $P(X = Q_{p0}) = 0$, and this assumption will be met if the X-population is continuously distributed and measurement is precise. This test is the optimal rank-type test for a population quantile when all that is known about the population is that it is continuous [12].

7.3 THE SIGN TEST FOR THE MEDIAN DIFFERENCE

7.3.1 Rationale

A special case of the foregoing test is that in which the sampled population is a population of difference-scores and the tested hypothesis is that its median is zero. Suppose one hypothesizes that a continuously distributed variate X is as likely to exceed as to be exceeded by another continuously distributed variate Y. This is equivalent to hypothesizing that for a randomly drawn pair of observations on the two variates the difference-score $X - Y$ is as likely to exceed zero as to be exceeded by it. And this is equivalent to hypothesizing that the median of a continuously distributed population of difference-scores is zero. Therefore, the previous test applies. If n pairs of observations have been drawn, one member of each pair from the X-population, one from the Y-population, then the number r of $(X_i - Y_i) - 0$ or, more briefly, $X_i - Y_i$ differences having a given algebraic sign is a binomially distributed variate and the parameter p of its distribution has the value .5 only if H_0 is true.

7.3.2 Null Hypothesis

$P(X > Y) = P(X < Y) = .5$, or equivalently, $P(X - Y > 0) = P(X - Y < 0) = .5$. This will be the case if the population of $X - Y$ difference-scores has a median of zero (as it will if X is as likely to exceed Y as not) and all assumptions are true.

7.3.3 Assumptions

(a) $P(X_i = Y_i) = 0$, i.e., roughly speaking, the members of a pair of observations will "never" be exactly equal. Provided that measurement is precise, this will be the case if the difference-score population is *continuously* distributed in the region of zero, or if there is no variate value of zero in that population. It will also be the case (still assuming measurement to be precise) if either the X- or the Y-population is continuously distributed throughout, or if the two populations do not contain a common discrete value. (For proper treatment of obtained difference-scores of zero, see Methods A to E of Sections 3.3 and 3.4, especially Methods A and B.)

(b) The $X - Y$ differences have been *randomly* drawn from the difference-score population.

(c) The $X - Y$ difference-scores are *independent*, thereby implying either that the difference-score population is an infinite one (as it will be if continuous) or that sampling is with replacement.

7.3.4 Efficiency

The matched-pair t test is a special application of the one-sample t test, just as the present test is a special case of the Sign test for the median. Therefore the statements in the "Efficiency" section for the latter apply in the present case so long as the present test is regarded as a test for the median of a population of difference-scores. Specifically, as a test for the median = mean of a normally distributed population of difference-scores against translation alternatives (i.e., changes of location without change of variance or shape), the present test has A.R.E. of $2/\pi = .637$ relative to the matched-pair t test. The test is consistent against all alternatives in which $p \neq q$, i.e., provided only that $H_0 : P(X - Y > 0) = P(X - Y < 0) = .5$ is false and that the assumptions are true.

7.3.5 Tables

The same tables are appropriate as for the previous test.

7.3.6 Example

In every year from 1629 to 1710, more males than females were born in the city of London. What is the a priori probability of this result if in fact the two sexes were equally likely? (Annual birth frequency of male infants and annual birth frequency of female infants may be treated as continuously distributed variates because of the large numbers involved; there were, in fact, no ties.) If males and females are equally likely, P (male frequency > female frequency) = .5 for each year, and the probability that the same (specified) result would occur in 82 or more of the years considered is

$$\sum_{i=82}^{82} \binom{82}{i} .5^{82} = \frac{82!}{82!\,0!} .5^{82} = .5^{82} = 2.068 \times 10^{-25}$$

which is the probability for as extreme or more extreme a result. (Note the astronomical power of this allegedly inefficient test in the present practical application to real data!) John Arbuthnott obtained this probability for these data in 1710. He was not content, however, simply to reject the null hypothesis, but went on to infer that the untested *extent* of the difference in birth frequencies was such as to compensate exactly for a greater mortality rate among males "who must seek their Food with danger" so that at the age of mating "every Male may have its Female."

7.3.7 Discussion

This is a test for the median difference, i.e., the median of a difference-score population, *not* for the difference between medians of the X- and Y-populations. Equality of medians for the X- and Y-populations does not necessarily imply that the median of the $X - Y$ population will be zero. For example, suppose that the X- and Y-populations both have a median of 4 but that almost all of the X-population lies in the intervals between 0 and 2 and between 4 and 6 while nearly all of the Y-population lies in the intervals between 2 and 4 and between 6 and 8. For those pairs of scores whose members come from opposite sides of the common median, $P(X > Y) = P(X < Y)$; but for those coming from the same side, the overwhelming probability is that $X < Y$, so for all possible pairs $P(X > Y) \cong .25$ while $P(X < Y) \cong .75$.

The inference is sometimes made that if the median difference is zero, then the X- and Y-populations are equally "good" in a quantitative sense. However it does not follow that because $P(X > Y) = P(X < Y)$ then $|X - Y|$ has the same average or median value for $X > Y$ as for $X < Y$. For example, if half the X-population lies between 0 and 1 and the other half between 8 and 9, and if all of the Y-population lies between 1 and 3, then $P(X > Y) = P(X < Y)$, but $X - Y$ will always have greater absolute magnitude when $X > Y$ than when $X < Y$. And the X-population will be "better" in a quantitative arithmetic sense than the Y-population. However, if the X- and Y-populations are identical or coaxially symmetric, the $X - Y$ difference-score population will be symmetric about a median of zero, which because of the symmetry will also be its mean. But a *mean* difference of zero implies that the difference between means of the X- and Y-populations is zero. And if the X- and Y-populations are symmetric or have identical shapes (including the variance or "scale" aspect) then a difference of zero between the two means implies a difference of zero between the two medians. Therefore, if the X- and Y-populations are known (i.e., can be "assumed") to have shapes which are either identical or symmetric, the test for a median difference of zero is equally a test for a difference of zero between two population medians. Furthermore, if H_0 is true the *mean* difference will be zero which, together with the symmetry of the difference-score population, implies that $|X - Y|$ has the same average value and the same median value when $X > Y$ as when $X < Y$. And if H_0 is false but the new assumptions are true, then one population will be "better" than the other in a quantitative arithmetic sense.

By adding C to each Y-observation before subtraction from its paired X-observation, one can test the null hypothesis that the median difference is C. If the assumption of identical or symmetric shapes can justifiably be made about the X- and Y-populations, one can test the hypothesis that the mean difference is C, or, equivalently, that the X-population is on the average C units "better" than the Y-population. By multiplying each Y-observation by $(100 + K)/100$ before subtraction, one can, under the same assumptions, test the hypothesis that the X-population is on the average K per cent "better" than the Y-population.

The preceding discussion has assumed that every X-observation has the same parent population and likewise for every Y-observation. Actually, the formula holds good even if every X- or Y-observation comes from a different population so long as each population corresponding to a given $X - Y$ difference has zero median. The null hypothesis tested is that all of the difference-score populations from which the $X - Y$ differences were "drawn" have zero median. Mathematically, the tested hypothesis is H_0: for every i,

$$P(X_i - Y_i > 0) = P(X_i - Y_i < 0) = .5$$

This type of application should be approached with caution, however. Suppose, for example, that half of the pairs represent populations in which X's are truly superior to Y's while the reverse is true for the other half. Although the null hypothesis is entirely false, the probability of its rejection is no greater than if it were true. Again, suppose that for a tenth of the pairs X's are truly superior to Y's, whereas for the remainder there is no real difference. The power of the test would be much greater if that tenth of the data were tested separately. Applications of the type described, therefore, may greatly reduce the power of the test, and even when the null hypothesis is rejected, it is not at all clear what alternative hypothesis is indicated.

It has been stated that the Sign test is particularly appropriate when the members of each pair were obtained under similar conditions, but when conditions differed from one pair to another. This, of course, represents a special case of the application discussed above. Here it is implied that a number of variables may have a real effect upon the absolute values of the X's, the Y's, or even the $X - Y$ differences, but that only one variable, the one in which the experimenter is interested, can have much real effect upon the *direction* of the $X - Y$ differences, i.e., the signs of the differences. This is not necessarily an unrealistic assumption. For example, the X's and Y's might be positions of seismograph needles during, and an hour previous to, an hypothesized tremor. The seismographs being located in widely different parts of the world, the $X - Y$ differences would be expected to vary in size with distance from the source of tremor. Furthermore, the numerical size of the difference might be reported in metric units by some and in British units of measurement by others. These considerations would preclude the direct use of a t test, but not the Sign test since the variable mentioned would affect the size but not the direction of the differences.

It is extremely important, however, that no variable causing differences between pairs shall interact with the variable in which the experimenter is interested, i.e., shall differentially affect the sign of the difference between members of a pair. Suppose, for example, that X and Y are two strains of wheat and that some of the XY pairs were grown in a northeastern county, the rest in a southwestern county. If the former location has a moist climate, the latter a dry one, it may well be that X is superior to Y in one location and inferior in the other. Subjecting pairs to different treatments, therefore, may introduce subtle and spurious interactions between "tested" and "nontested" effects with the result that the power of the test is reduced and the true alternative hypothesis may differ greatly from the alleged one.

7.4 COX AND STUART'S SIGN TESTS FOR TREND IN LOCATION

The following test is a legitimate special application of the preceding one, in which "Y_i" is a subsequent observation upon the "process" which produced

X_i, and the hypothesis tested is that the process has not changed in "location."

7.4.1 Rationale

Suppose one hypothesizes that the distribution of a continuously distributed variate X does not change with time and wishes to test that hypothesis with a test sensitive to monotonic changes in the location of the distribution. Let a sequence of cn observations, $X_1, X_2, \ldots X_i, \ldots, X_n, X_{n+1}, \ldots, X_{(c-1)n+1}, X_{(c-1)n+2}, \ldots, X_j, \ldots, X_{cn}$, be drawn, the subscripts indicating position in the temporal sequence, i.e., the order in which the observations were drawn (so that X_j is subsequent in appearance to X_i). Now for every $i \leqq n$, form the difference-score $X_i - X_{(c-1)n+i}$, and let S_c be the number of those difference-scores that are positive. Then if H_0 is true, each such difference-score will be as likely to be positive as to be negative, and S_c will have the same distribution as the number of successes r in n binomial trials with constant probability $p = .5$ of success on each trial. But if there is a monotonic trend in the location of the X-distribution (but no change in any other aspect), difference-scores of one algebraic sign will be more probable, tending to cause the test statistic to fall in a properly chosen rejection region.

Two statistics are of interest, S_2 and S_3. If we let the symbol T stand for the total number of observations cn in the sequence, then S_2 is based on n difference-scores whose formula is $X_i - X_{(T/2)+i}$, where $i = 1, 2, \ldots, n = T/2$, and S_3 is based on n difference-scores of the type $X_i - X_{(2/3)T+i}$, where $i = 1, 2, \ldots n = T/3$. Thus S_2 breaks the sequence in half and uses both halves, matching corresponding observations in the two halves. But S_3 breaks the sequence into thirds, discards the middle third, and matches corresponding observations in the first and last third. Thus as applied to the observations remaining after discarding the middle third, S_3 is entirely analogous to S_2.

7.4.2 Null Hypothesis

For every $i \leqq n$,

$$P(X_i > X_{(c-1)n+i}) = P(X_i < X_{(c-1)n+i}) = .5$$

Sufficient conditions for the validity of H_0 are that the distributions of the X_i's are continuous and identical, being unaffected by the passage of time, so that the X_i observations are randomly related to sequence.

7.4.3 Assumptions

(a) $P(X_i = X_{(c-1)n+i}) = 0$ for every $i \leqq n$. (Difference-scores of zero should be dealt with by Method B of Section 3.3.2 unless there is a sequentially changing bias in measurement. As explained in the first two paragraphs of Section 3.3.2, tests of randomness, such as this one, virtually guarantee the validity under H_0 of the assumptions, other than unbiased measurement, upon which Method B is based.)

(b) Whether a given X_i falls above or below $X_{(c-1)n+i}$ is independent of the outcome for any other pair.

(c) The sample X's are randomly selected.

7.4.4 Efficiency

Applied to populations known to be normally distributed, the S_2 and S_3 tests for trend in location have A.R.E.'s of .78 and .83, respectively, with respect to the best parametric test, based on the regression coefficient. Under the same conditions, they have respective A.R.E.'s of .79 and .84 compared to the best distribution-free test under these circumstances, i.e., compared to Spearman's or Kendall's rank correlation tests used as one-sample tests of randomness. Power functions of S_3 are given in [5].

Note that the S_3 test which "throws away" the "information" contained in the middle third of the sequence is more efficient than the S_2 test which retains it.

7.4.5 Tables

See Section 7.2.5.

7.4.6 Example

The closing prices of American Telephone and Telegraph Co. (AT & T) on the New York Stock Exchange for the first 36 trading days in 1965 were, in sequence (after subtracting 60 from each price): 9.5, 9.875, 9.25, 9.5, 9.375, 9, 8.75, 8.625, 8, 8.25, 8.25, 8.375, 8.125, 7.875, 7.5, 7.875, 7.875, 7.75, 7.75, 7.75, 8, 7.5, 7.5, 7.125, 7.25, 7.25, 7.125, 6.75, 6.5, 7, 7, 6.75, 6.625, 6.625, 7.125, 7.75. Do these data support the hypothesis that the price sequence is random against the alternative of an essentially downward trend? First, applying the S_2 test, $c = 2$, and since $cn = 36$, $n = 18$ and $(c - 1)\,n + i = (2 - 1)\,18 + i = 18 + i$. So subtracting the nineteenth price from the first, the twentieth from the second, etc., we obtain the following 18 differences: 1.75, 2.125, 1.25, 2, 1.875, 1.875, 1.5, 1.375, .875, 1.5, 1.75, 1.375, 1.125, 1.125, .875, 1.25, .75, 0. Dropping the zero difference, we have 17 difference-scores, all of which have the same, positive, algebraic sign. (The subtractions actually were unnecessary, since only the directions of the differences need be counted.) Therefore $S_2 = 17$, and if H_0 is true the probability of this is the probability of 17 successes out of 17 trials for a binomial variate with $p = .5$. The greater frequency of positive differences, or, equivalently, the smaller frequency of negative ones, tends to support the alternative hypothesis of a downward trend. Entering Table VII with $n = 17$, we find the critical frequency of the less frequent algebraic sign is 1 at $\alpha = .001$ for a two-tailed test and, therefore, at $\alpha = .0005$ for a left-tailed test on the frequency of a specified sign. Therefore, the a priori probability of as few or fewer negative differences as obtained is less than .0005 and the data support the alternative hypothesis of a downward trend rather than an hypothesis of randomness.

Applying the S_3 test, $c = 3$, and since $cn = 36$, $n = 12$ and $(c - 1)n + i = (3 - 1)12 + i = 24 + i$. So subtracting the twenty-fifth price from the first, the twenty-sixth from the second, etc., we obtain the following twelve differences: 2.25, 2.625, 2.125, 2.75, 2.875, 2, 1.75, 1.875, 1.375, 1.625, 1.125, .625. All 12 differences are positive, so $S_3 = 12$. The probability of 12 or more plusses is the probability of $n - 12 = 12 - 12 = 0$ or fewer minuses. Entering Table VII with $n = 12$, we find that zero is the critical value corresponding to $\alpha = .001$ for a two-tailed test, so again the significance level for our one-tailed test is .0005, and consequently the a priori probability of as few or fewer negative differences is less than or equal to .0005 if the sequence is truly random.

7.4.7 Discussion

Since n must be an integer the following procedures should be followed. In applying the S_2 test, if the sequence contains an odd number of observations, discard the middle observation and apply the test to the remaining sequence of observations whose length is now defined to be $2n$. In applying the S_3 test, if the number of observations in the sequence is not divisible by 3, "add" one or two "dummy" observations to the middle of the sequence (where only the presence, not the value, of the observations matter) to make it so; the length of the resulting sequence of observations and dummy observations is now defined to be $3n$.

The S_3 test uses only two-thirds of the raw data employed by the S_2 test; however, the members of each pair of measurements whose difference is taken are one-third farther apart. The net result is an increase in efficiency. If a real monotonic trend exists, then the farther removed two measurements are in sequence, the greater the expected difference in magnitudes and the more likely that the sign of the difference will betray the direction of the trend. The S_2 test, however, has one advantage. Since it uses all of the data, statistical inference can be extended to the entire parent population. Strictly speaking, inferences based on the S_3 test cannot legitimately be extended to the middle third of the sampled sequence, since a temporary trend occupying only this portion could not be detected.

The "distance" between paired observations used by the S_2 test is half the number of observations, in the sequence, and this is the "optimal" separation (in the sense that if the X's are normally distributed, the A.R.E. is maximized) if all observations in the sequence are used in the test and each difference-score is independent and is given equal weight. The "distance" used by the S_3 test, which does not use all observations to obtain difference-scores, is two-thirds the total number of observations, and this is the optimal separation if each difference-score is independent and is given equal weight. And if the equal-weight restriction be removed, still another Sign test [5] is optimal, but its test statistic is not binomially distributed. Let $2n$ be the number of observations in the sequence (where n is an integer) and let $h_{i,\,2n-i+1}$ be 1 if the difference-score $X_i - X_{2n-i+1}$ is positive, and zero if the difference-score is negative. Then

$$\sum_{i=1}^{n} (2n - 2i + 1)h_{i,\,2n-i+1}$$

the test statistic, has under H_0 a distribution with mean $n^2/2$, variance $n(4n^2 - 1)/12$ and shape which approaches normality as n approaches infinity. The test weights each $h_{i,\,2n-i+1}$ by $2n - 2i + 1$, the "distance" between the pair of observations to which the h corresponds. Under the assumption of normally distributed X's, the test has A.R.E. of .86 relative to the best parametric test based on the regression coefficient. A more efficient test than any of the above compares every observation with every other observation and the test statistic is the number of times a later observation exceeds an earlier observation. The pairs, therefore, are not independent and the test is not a binomial one. It is discussed in a later chapter.

7.5 COX AND STUART'S SIGN TESTS FOR TREND IN DISPERSION

The previous test can be modified to test the hypothesis that the dispersion of the sampled population does not vary with time, the modified test being especially sensitive to monotonic changes in dispersion. An additional assumption is required: that the location of the population either does not change or changes linearly (and in the latter case the observations must be equally spaced). Suppose that a sequence of cnk observations has been made on the variate in question. Let w_1 be the range of the first k observations, w_2 be the range of the second k observations, etc., so that w_i is the range of the ith block of k consecutive observations. (In place of the range w, any other dispersion statistic can be used, e.g., the variance.) Then we have a sequence of cn ranges, w's, to which the previous test can be applied, substituting the w_i's for the X_i's in the formulas for the previous test. Thus a test sensitive to monotonic change in the size of observations is converted to a test sensitive to monotonic change in the size of successive ranges of observations. The test's authors [5] recommend that k should not exceed 5 and should be so chosen that the number of ranges in the sequence is at least 16. Under normality, the A.R.E. of the S_3 test for dispersion is .71 compared with the maximum-likelihood test. The corresponding figure for the A.R.E. of the differentially weighted Sign test for dispersion relative to the maximum likelihood test is .74.

The S_3 test for dispersion with $k = 2$ would be applied as follows to the data in the example for the previous test (see Section 7.4.6). The ranges for successive blocks of two observations are, in sequence: .375, .25, .375, .125, .25, .125, .25, .375, .125, 0, .5, .375, 0, .375, .5, .25, 0, .625. This series of values is now dealt with just as in the preceding test. Subtracting values in the final third of the series from corresponding values in the first third and recording only the algebraic signs of the differences, we have, in sequence $+ - - - + -$. So $S_3 = 2$ and Table VII shows that a value of 2 for the less frequently occurring sign when $n = 6$ and $p = .5$ is not significant at any reasonable significance level for a two-tailed test. Clearly the number of observations was insufficient for good power, and the example is illustrative rather than practical.

7.6 NOETHER'S TEST FOR CYCLICAL TREND

The preceding tests of randomness were designed to be sensitive to monotonic trend but not to a trend which periodically reverses direction, especially if the reversals are separated by relatively short periods. The following test is sensitive to the latter type of trend as well as to the existence, but not the direction, of the former.

7.6.1 Rationale

Suppose that $3n$ measurements have been recorded or are available in order of sequence and it is desired to know whether the sequence may contain a fluctuating or cyclical trend. If the measurements are continuously distributed and there is no trend of any kind, no two measurements will be equal, and the measurements will be randomly related to sequence. Any three consecutive measurements will be equally likely to have any of the six sequences represented by the six possible permutations of three things. However, of these six sequences only two are monotonic, i.e., ascend or descend without reversals, while the remaining four change direction in the middle. For example, if the three measurements are ranked, the ranks will be found to have one of the six sequences: **1 2 3**, **3 2 1**, 1 3 2, 2 3 1 , 2 1 3, 3 1 2, the boldface sequences being monotonic. The probability of monotonicity for such a set of three measurements is therefore $\frac{1}{3}$ if the sequence is random and the measurements are continuously distributed. And if the $3n$ measurements are divided into n successive and nonoverlapping sets of three consecutive measurements each, the number of monotonic sets will be binomially distributed with $p = \frac{1}{3}$. On the other hand, if a cyclical or fluctuating trend of any but the shortest possible "wavelength" exists, one would expect more than one-third of the sets to be monotonic.

7.6.2 Null Hypothesis

For every $i \leqq n$,

$$P(X_{3i} > X_{3i-1} > X_{3i-2}) + P(X_{3i} < X_{3i-1} < X_{3i-2}) = \tfrac{1}{3}$$

Sufficient conditions for the validity of H_0 are that the X's are continuously distributed and the size of the X's is unrelated to their position in sequence.

7.6.3 Assumptions

(a) For every $i \leqq n$,

$$P(X_{3i} = X_{3i-1}) = 0 \quad \text{and} \quad P(X_{3i-1} = X_{3i-2}) = 0,$$

i.e., adjacent observations in no set are tied.

(b) Whether or not any given set is monotonic is independent of the monotonicity or nonmonotonicity of any other set, i.e., "outcomes" of the binomial "event" are independent. Among other things, this means that no X is used in more than one set.

(c) The act of sampling is without bias in the sense that if the process generating the X's is random the sample sequence of X's will be random also.

Ties are a practical problem only when adjacent observations in the same set are tied. Interset ties are obviously irrelevant. And if the first and third observations within a given set are tied and the second is not, the set is clearly nonmonotonic and there is no ambiguity in categorizing it.

Adjacent intraset ties may be dealt with by examining the sets in sequence. If the middle observation in a set is tied with an adjacent observation, it is dropped and the series is "closed up" by moving each subsequent observation (in the entire series) backward by one position. This process is continued, moving from earlier toward later sets, until no set remains in which the middle observation is tied with an end observation in the same set. This procedure rules out subjective bias in determining which observations to drop. Its effect should generally be roughly that of taking a smaller sample in which no critical ties occur.

Alternatively, Methods A, C, or D (the midrank method, F, is useless) of Sections 3.3 and 3.4 may be used.

7.6.4 Efficiency

The test's author [17] expresses doubt that it is highly efficient. However, if either (a) data are plentiful, or (b) the trend is pronounced and scedasticity is relatively small, it should prove a good test.

7.6.5 Tables

Since under H_0 the binomial parameter is $p = \frac{1}{3}$, it will be necessary to consult one of the more extensive volumes of tables [16, 18, 23], of which only the Harvard tables [23] apply for the fraction $\frac{1}{3}$ as opposed to the decimal number .33.

7.6.6 Example

If we drop the middle observation in a set of three whenever it is tied with an end observation in the same set, the original 36 AT&T prices of Section 7.4.6. reduce to the following sequence of 30 prices, alternate sets of three being enclosed within parentheses and monotonic sets being boldface: 9.5, 9.875, 9.25, **(9.5, 9.375, 9)**, **8.75, 8.625, 8,** (8.25, 8.375, 8.125), 7.875, 7.5, 7.875, (7.875, 7.75, 8), 7.5, 7.125, 7.25, **(7.25, 7.125, 6.75)**, 6.5, 7, 6.75, **(6.625, 7.125, 7.75)**. Four of the ten sets show monotonic price change. If the H_0 of randomness is true, we have obtained in ten trials four outcomes of a type whose probability of occurrence on a single trial is $\frac{1}{3}$. Entering the Harvard tables [23] with $p = \frac{1}{3}$, $n = 10$, and $r = 4$, we find the cumulative probability for $r \geqq 4$ to be .44, so we cannot reject the hypothesis of randomness using a test sensitive to cyclical trend (although we rejected the same hypothesis on the basis of Cox and Stuart's tests, which are sensitive to monotonic trend).

7.6.7 Discussion

This test should practically always be used as a one-tailed test with rejection occurring for "too many" monotonic sets. In order for cyclic trend to produce "too few" monotonic sets the direction of trend would have to change almost with every observation. Because of the short length of the sets, it is difficult to imagine many types of trend in location to which the test would not be sensitive, especially if the observations used in the test represent all, rather than a sampling of, the variate values generated during a given time period. The test therefore should be valuable as a test for nonrandomness of unspecified type as well as for cyclical trend.

This test (and other relevant ones) is also presented by its author [17] as a sequential probability ratio test.

7.7 OTHER BINOMIAL TESTS

7.7.1 Test for an Hypothesized Proportion

Suppose that all of the units comprising an infinite population can be categorized as definitely possessing or definitely not possessing some characteristic K, and that the proportion of units in the population having the characteristic K is p. Then the number r of units which are "K's" in a randomly and independently drawn sample of size n is a binomially distributed variate with parameters n and p. So the hypothesis $H_0: p = p_0$ can be tested by drawing such a sample and determining the cumulative probability of r in a binomial distribution with parameters n and p_0.

7.7.2 Test for Proportion of Population Above or Below a Specified Value

If the population units have values, X's, on some directional scale (i.e., ordinal, interval, or ratio scale) and if one takes the characteristic K to be that $X < X_K$, the preceding test becomes a test of the hypothesis that a proportion p_0 of the population lies below the specified value X_K. By performing a two-tailed test one tests the hypothesis that the proportion is exactly p_0, whereas a one-tailed test tests whether the proportion is $\geqq p_0$ or $\leqq p_0$, depending upon the tail used for rejection. By letting the characteristic K be that $X > X_K$, one tests that p_0 of the population lies above X_K. The characteristic K can also be that $X \leqq X_K$ or that $X \geqq X_K$, the tested hypothesis being modified accordingly.

7.7.3 Test for an Hypothesized Population Quantile

If the X's are continuously distributed and K is that $X < X_K$, the above test is equivalent to a test of the hypothesis that X_K is the $100p_0$th percentile of the X-population. See Section 7.2.7.

7.7.4 Test for Concentration

If the population units have values, X's, on a directional scale, and if we take K to be the characteristic that $X_{K'} < X < X_{K''}$, the test (in Section 7.7.1) becomes a test of the hypothesis that a proportion p_0 of the population lies between the two specified values $X_{K'}$ and $X_{K''}$. This test is useful when it is of no concern how much of the population lies at or below $X_{K'}$ and how much lies at or above $X_{K''}$, but only how much lies in between. A toy manufacturer, for example, might want to test the hypothesis that 80% of the viewers of a certain television program have ages between 5 years 11 months and 29 days and 8 years 0 months and 0 days because this is the age range to which his toys appeal, the distribution according to age of the remaining 20% being a matter of indifference to him.

7.7.5 McNemar's Test for Equality of Correlated Proportions

Suppose that each unit in the infinite population is categorizable as definitely belonging to one or the other of two mutually exclusive categories K and L and is also categorizable according to another mutually exclusive dichotomy I and II, the second dichotomy being independent of the first. Suppose further that a sample of m units has been drawn from the population and the experimenter wishes to test the hypothesis that in the parent population the proportion of "K's" is the same as the proportion of "I's." The following tables give both the sample and population composition. The lower-case letters stand for the *numbers* of *sample* units in each of the four possible combinations of category membership. Upper-case letters represent the corresponding *proportions* in the *population.*

Sample Frequency Table

	K	L	*Either*
I	a	b	$a + b$
II	c	d	$c + d$
Either	$a + c$	$b + d$	m

Population Proportion Table

	K	L	*Either*
I	A	B	$A + B$
II	C	D	$C + D$
Either	$A + C$	$B + D$	1.00

The proportion of "K's" in the population is $A + C$ and the population proportion of "I's" is $A + B$. If these two proportions are equal, as hypothesized, then $A + C = A + B$, which reduces to $C = B$. So if H_0 is true, then within the *sub*population consisting only of units which are either KII's or LI's (and therefore consisting of a *single* dichotomy) the proportions of these two kinds of units are equal and complementary, i.e., $C = B = .5$. Furthermore, the sample units represented by the northeast-southwest diagonal of the Sample Frequency Table are a sample of size $b + c$ from this subpopulation, of which c are KII's. Thus if H_0 is true, there have been $r = c$ "successes," i.e., KII's, in $n = c + b$ binomial trials when the probability of "success," i.e., drawing a KII unit, on a single trial was .5. Therefore H_0 can be tested by entering binomial tables with $p = .5$ and $n = c + b$, and finding the properly cumulated probability for the test statistic $r = c$. The usual binomial assumptions apply. However it should be emphasized that the K–L and I–II dichotomies apply to the *same* data, i.e., *each* unit is either a K or an L and is *also* either a I or a II, and that these two original dichotomies must have nonoverlapping definitions so that the KII and LI categories will themselves be mutually exclusive. As pointed out by its author [15], this test would be appropriate for testing whether two questions on a test are of equal difficulty: K and L would be, respectively, passing and failing the first question, and I and II would be, respectively, passing and failing the second, the same set of students having answered the two questions. Or K and L could be categories prior to "treatment," and I and II could be the same categories subsequent to treatment, in which case the test would test whether or not the treatment changed the proportionate membership in the first category, i.e., whether or not the proportions of K's and I's are the same.

7.7.6 Tests Appropriate to the Technique of Paired Comparisons

It was pointed out in Section 7.3.7 that the Sign test for the median difference does not require that the various $X - Y$ differences used in the test come from the same population and that each difference may in fact be drawn from a different population, the tested hypothesis being simply that each of the contributing difference-score populations has a median of zero. An experimental technique known as the Method of Paired Comparisons often produces precisely these conditions, so that this special case of the Sign test is appropriate for statistical analysis of results. Sometimes it is difficult to arrange several objects in order of preference, even if they do not differ in a complex way, simply because they must be perceived one at a time and because an intervening perception tends to cloud a judgment. This is the case, for example, in taste-testing of hams from swine raised on different diets, e.g., peanuts versus corn versus swill. In other cases, the various objects may differ along several different dimensions, in which case it does not necessarily follow that because A is preferred over B and B over C that A must be preferred over C. In such cases one may have to be content with a direct comparison between A and C, abandoning any attempt to obtain a meaningful rank ordering of more than two objects. Suppose that there are K objects, $O_1, O_2, \ldots, O_i, \ldots, O_K$ and J judges to compare them, each judge making all of the required comparisons exactly once. Now let d_{hij} be a variable which equals one if O_i is preferred over O_j by the hth judge and equals zero in the opposite case, i.e.,

$$\left.\begin{array}{lll} d_{hij} = 1 & \text{if} & O_i > O_j \\ d_{hij} = 0 & \text{if} & O_i < O_j \end{array}\right\} \quad \text{according to } h\text{th judge}$$

If we wish to test the general null hypothesis that the difference-score populations corresponding to every possible comparison between different objects have medians of zero, then in mathematical terms we have H_0:

$$P(O_i - O_j > 0) = P(O_i - O_j < 0) = .5$$

for every *ij* pair in which $i \neq j$ (where O_i now represents the "value" of object O_i on some strictly theoretical, perhaps crude, unspecified, directional scale). The number of possible paired comparisons between different objects in a set of K objects is simply the number of combinations of K things taken two at a time, $\binom{K}{2} = K!/[2!\,(K-2)!] = K(K-1)/2$. And if all of these possible different comparisons are made by each of the J judges, there will be $JK(K-1)/2$ comparisons in all (actually, J is simply the number of times the set of all different comparisons is performed, so J could be the number of replications by a single judge or the total number of replications by various judges, provided, of course, that replication does not violate the assumption of independence). Therefore, if H_0 is true,

$$r = \sum_{h=1}^{J} \sum_{i \neq j} d_{hij}$$

is a binomially distributed variate with parameters $n = JK(K-1)/2$ and $p = .5$, and H_0 can be tested by referring r to cumulative binomial tables and performing a two-tailed test. If interest is confined to the relationship between a *single* object, say O_K, and all others, we may wish to test $H_0: P(O_i - O_K > 0) = P(O_i - O_K < 0) = .5$ for every $i < K$. In that case there are $K - 1$ possible different comparisons between O_K and a different object and $J(K - 1)$ comparisons in all. So if H_0 is true,

$$r = \sum_{h=1}^{J} \sum_{i=1}^{K-1} d_{hiK}$$

is a binomially distributed variate with parameters $n = J(K-1)$ and $p = .5$ and cumulative binomial tables can be used to make either a one- or two-tailed test of H_0. If instead of p being .5 for every comparison it has an average value of .5 over all comparisons, it has been shown [7] that the test is a conservative one in the sense that r will be less likely to fall in any given conventional rejection region. Therefore, the above test is an approximate and conservative test of the modified H_0 that the average probability that O_i exceeds O_K is .5, i.e., that

$$\sum_{i=1}^{K-1} \frac{P(O_i - O_K > 0)}{K-1} = .5$$

The usual binomial assumptions apply. The assumptions that outcome categories be mutually exclusive members of a dichotomy can be satisfied by requiring judges to guess when they are uncertain, thereby preventing ties and forcing judgments to be of only two types, e.g., "better" and "worse." The assumption of independent event outcomes requires that no judgment be influenced by any other judgment. This means, among other things, that there are no carryover effects, e.g., that a taste judgment is uninfluenced by the lingering taste from the previous judgment or by the memory of it, and that judges do not compare notes during the experiment. (Carryover effects are almost inevitable, but their influence can be mitigated by proper counter-balancing and other design features.) The assumption of random sampling means that whatever the true probabilities of the judgments "better" and "worse" from a particular judge making a particular comparison, the *order* in which judgments of better or worse are "generated" is a random one, i.e., *potential* judgments of the two types fluctuate randomly. Practically speaking, this means simply an absence of bias in the selection of judgments to be tested. The assumption of randomness may also apply to the selection of judges or to the selection of objects, depending upon the populations to which the experimenter wishes to extend statistical inference.

7.8 CONFIDENCE LIMITS FOR QUANTILES

Suppose that a random sample of n independent observations is to be drawn from an infinite but otherwise unknown population and that it is desired to establish a confidence interval for Q_p, the smallest value of the variate X below which at least a specified proportion p of the X-distribution lies. Thus Q_p is the unknown population quantile corresponding to the specified proportion p; it is the $100p$th percentile, and if $p = .5$ it is the median.

Let X_i designate the observation in the ith position in a sample of size n, not yet drawn, when the observations are arranged in a line in order of increasing magnitude. Thus i is an integer which assumes all values from 1 to n and $X_i \geqq X_{i-1}$. Then the following statement gives an inequality for the a priori probability that X_r and X_s will form an open or closed interval containing Q_p:

$$P(X_r < Q_p < X_s) \leqq \sum_{i=r}^{s-1} \binom{n}{i} p^i(1-p)^{n-i} \leqq P(X_r \leqq Q_p \leqq X_s)$$

Thus, after the sample is drawn, the a posteriori statement that $X_r < Q_p < X_s$ can be made at a confidence level no greater than the middle term above, i.e., the summation, and the statement that $X_r \leqq Q_p \leqq X_s$ can be made at a level of confidence at least as great as the middle term. (As usual, in order to avoid bias, r and s should be chosen prior to sampling.)

If the sampled population is continuous, there will be zero probability for X_r or X_s equaling any other X_i or equaling Q_p, so the first and third terms will be equal and the $\leqq$ signs on either side of the middle term, i.e., the confidence level, can be replaced by $=$ signs. If the population is discrete, the first and last probabilities are unequal, and the inequality involving the middle and last terms should be used.

Proof: Consider first the case in which X is continuously distributed so that Q_p is the value *below* which *exactly* a proportion p of the population lies. When the sample is finally drawn, the n observations, X's, can be arranged from smallest to largest, as follows, with subscripts indicating rank position in order of increasing magnitude:

$$X_1, X_2, \ldots, X_r, X_{r+1}, \ldots, X_{s-1}, X_s, \ldots, X_{n-1}, X_n$$

Now if i of these observations come from below Q_p, then it necessarily follows that $X_i < Q_p$, and if *exactly* i of the observations, no more no less, come from below Q_p, then $X_i < Q_p \leqq X_{i+1}$. In fact, *if and only if* exactly i of the n observations are drawn from below Q_p will it be true that $X_i < Q_p \leqq X_{i+1}$, in which case it *must* be true. Because of this one-to-one relationship, i.e., because the one condition necessitates the other, the a priori probability that X_i and X_{i+1} will take values such that Q_p lies in the interval between them is the same as the a priori probability of drawing exactly i observations from below Q_p, and we already know that probability to be $\binom{n}{i}p^i(1-p)^{n-i}$. Bearing in mind that Q_p is a constant and that the probabilities refer to the positions that the ordered X's will assume in a sample not yet drawn, it follows that

$$\begin{aligned} P(X_r < Q_p \leqq X_s) &= P(X_r < Q_p \leqq X_{r+1}) + P(X_{r+1} < Q_p \leqq X_{r+2}) \\ &\quad + \ldots + P(X_{s-1} < Q_p \leqq X_s) \\ &= \binom{n}{r} p^r(1-p)^{n-r} + \binom{n}{r+1} p^{r+1}(1-p)^{n-r-1} \\ &\quad + \ldots + \binom{n}{s-1} p^{s-1}(1-p)^{n-s+1} \\ &= \sum_{i=r}^{s-1} \binom{n}{i} p^i (1-p)^{n-i} \end{aligned}$$

The $\leqq$ sign is used above simply for logical completeness. Actually since the population is continuous $P(X_i = Q_p) = 0$, so that the $<$ sign could have been used throughout. Specifically, $P(Q_p = X_s) = 0$, so

$$P(X_r < Q_p < X_s) = \sum_{i=r}^{s-1} \binom{n}{i} p^i (1-p)^{n-i}$$

The above derivation tells us that if every time we draw a sample of size n from a continuous population we allege that $X_r < Q_p < X_s$, we shall be right a proportion of the time equal to $\sum_{i=r}^{s-1} \binom{n}{i} p^i (1-p)^{n-i}$, which is therefore the confidence level for the allegation. The confidence level can also be written as

$$1 - \sum_{i=0}^{r-1} \binom{n}{i} p^i (1-p)^{n-i} - \sum_{i=s}^{n} \binom{n}{i} p^i (1-p)^{n-i}$$

i.e., as 1 minus the probability of $r - 1$ or fewer observations falling below Q_p minus the probability of $n - s + 1$ or fewer observations falling above Q_p.

Now consider the case where X may be discretely distributed [22]. Every unit comprising the infinite X-population can be associated with, i.e., can be regarded as represented by, a different point on the graph giving the cumulative probability distribution of X. (For discrete values of X, the point will lie somewhere on a vertical line segment above the discrete value of X; it is assumed that there are no horizontal "connecting" lines, i.e., that the graph is just as discontinuous as the actual X's in the population.) Every point on the graph is represented by two values, an abscissa, X, which it may share with other points and an ordinate which is unique. Since the population is assumed infinite, ordinates are continuously distributed. Thus we may concentrate upon ordinates rather than abscissas. By definition, Q_p is the abscissa corresponding to an ordinate of p, and p is the proportion of population units represented by ordinates less than p. Let p_i be the ith ordinate in order of increasing magnitude in a sample of n points drawn from the graph and let X_i be the abscissa of the point whose ordinate is p_i (so that the size of the ordinate determines the subscript for tied X's). Then the probability that exactly i of the n sample points will have ordinates less than p is $\binom{n}{i} p^i (1 - p)^{n-i}$, which must also be the probability that $p_i < p < p_{i+1}$. Thus, in complete analogy with the derivation for the continuous case, we can show that

$$P(p_r < p < p_s) = \sum_{i=r}^{s-1} \binom{n}{i} p^i (1 - p)^{n-i}$$

Now let r' and r'' be the smallest and largest subscripts, respectively, for which it is true that $X_{r'} = X_r = X_{r''}$, i.e., the smallest and largest subscripts for sample X's tied with X_r; and let s' and s'' be the smallest and largest subscripts, respectively, for which it is true that $X_{s'} = X_s = X_{s''}$. Because of its use of the $<$ sign, rather than $\leqq$, the statement that $X_r < Q_p < X_s$ implies that Q_p is not equal to X_r or X_s and is therefore not equal to any X_i's tied with them, i.e., since X_r and X_s are excluded from the interval, so also are any X_i's equal to them. Therefore

$$P(X_r < Q_p < X_s) = P(X_{r''} < Q_p < X_{s'}) = P(p_{r''} < p < p_{s'})$$

$$= \sum_{i=r''}^{s'-1} \binom{n}{i} p^i (1 - p)^{n-i} \leqq \sum_{i=r}^{s-1} \binom{n}{i} p^i (1 - p)^{n-i}$$

which proves the first two-thirds of the original inequality.

Now let p' and p'' be the smallest and largest population ordinates, respectively, corresponding to the abscissa Q_p. The statement that $X_r \leqq Q_p \leqq X_s$ implies an interval which includes X_r and X_s and which therefore includes all X_i tied with them. It therefore implies that $X_{r'} \leqq Q_p \leqq X_{s''}$. Furthermore, its probability must take account of occasions in which Q_p equals one of the interval endpoints, since they are part of the interval. Now the probability that $p_{r'} < p < p_{s''}$ is

$$\sum_{i=r'}^{s''-1} \binom{n}{i} p^i (1 - p)^{n-i}$$

and whenever it is true that $p_{r'} < p < p_{s''}$, it must also be true that $X_{r'} \leqq Q_p \leqq X_{s''}$. But $X_{r'}$ and $X_{s''}$ can be equal to Q_p not only when $p_{r'} < p$ or $p_{s''} > p$ as taken account of in the above summation, but also when $p < p_{r'} < p''$ or $p' < p_{s''} < p$ as not taken into account. (And similar remarks hold for X_i's tied with $X_{r'}$ or $X_{s''}$.) Therefore

$$P(X_r \leqq Q_p \leqq X_s) = P(X_{r'} \leqq Q_p \leqq X_{s''}) > P(p_{r'} < p < p_{s''})$$
$$= \sum_{i=r'}^{s''-1} \binom{n}{i} p^i (1-p)^{n-i}$$

So

$$P(X_r \leqq Q_p \leqq X_s) > \sum_{i=r'}^{s''-1} \binom{n}{i} p^i (1-p)^{n-i} \geqq \sum_{i=r}^{s-1} \binom{n}{i} p^i (1-p)^{n-i}$$

and this proves the last two-thirds of the original inequality. (The $>$ becomes a $\geqq$ in order to include the continuous case.)

7.8.1 Example

To illustrate the method, suppose that we wish confidence limits for the first quartile, $Q_{.25}$, corresponding to a confidence level of $1 - \alpha \geqq .95$, based on a sample of $n = 20$ observations. Entering binomial tables with $p = .25$ and $n = 20$, we find a probability of .0139 for ten or more successes and a probability of .9757 for two or more successes. Therefore, the probability of from two to nine successes is $.9757 - .0139 = .9618$. So we can state that

$$P(X_{r=2} < Q_{p=.25} < X_{s=10}) \leqq .9618 = \sum_{i=r=2}^{s-1=9} \binom{20}{i} .25^i (.75)^{20-i}$$
$$\leqq P(X_{r=2} \leqq Q_{.25} \leqq X_{s=10})$$

with respect to the a priori probability for the X-values in the sample not yet drawn. We now draw a sample of 20 observations and arrange them, as follows, in order of increasing magnitude: 31, 47, 55, 81, 90, 90, 96, 104, 111, 125, 142, 153, 159, 165, 180, 204, 227, 239, 254, 278. Since $r = 2$, $X_r = X_2 = 47$ and for $s = 10$, $X_s = X_{10} = 125$, we can state at a confidence level of at least .9618 that $47 \leqq Q_{.25} \leqq 125$ and at most .9618 that $47 < Q_{.25} < 125$. Because of ties (the two 90's) in the obtained data we conclude that we are dealing with the "discrete" case so that the latter statement is fairly useless. We therefore state at the .9618 level of confidence that $47 \leqq Q_{.25} \leqq 125$.

7.9 CONFIDENCE LIMITS FOR PROPORTIONS

The previous section concerned confidence limits for an unknown population *value* corresponding to a known and therefore "fixed" proportion. The present section concerns the much different problem of establishing confidence limits for a proportion itself, the proportion being unknown.

For fixed values of n and p, we can find, using binomial formulas already presented, two values of r, r' and r'', where r' is the largest value of r having a lower-tail cumulative probability $\leqq \alpha/2$ and r'' is the smallest value of r having an upper-tail cumulative probability $\leqq \alpha/2$. Thus

$$P(r \leqq r') \leqq \alpha/2 \qquad \text{and} \qquad P(r \geqq r'') \leqq \alpha/2$$

from which it follows that $P(r' < r < r'') \geqq 1 - \alpha$.

Now suppose that for a fixed value of n we construct a graph with p plotted on the horizontal axis, r on the vertical axis. For every value of p (practically speaking we could use the p's corresponding to exact hundredths), we could calculate r' and r'' and plot them as ordinates for the abscissa corresponding to p. We could then draw lines connecting all the adjacent r''s and all the adjacent r'''s. The area between the r' and r'' lines would be shapled like a football whose long axis extends from southwest to northeast.

Since the interval between the *points* r' and r'' encloses at least $1 - \alpha$ of the univariate distribution of r for *any given* p, the *lines* r' and r'' must enclose at least $1 - \alpha$ of the bivariate distribution of r and p. It follows therefore that we shall be right at least $1 - \alpha$ of the time if every time we draw a sample of the fixed size n we allege that the point whose ordinate is the observed number of successes r, and whose abscissa is the true probability of success on a single trial p, lies in the football-shaped area enclosed between the r' and r'' lines. Let a horizontal line be drawn through the observed number of successes, r, and let p_1 and p_2 be the abscissas of the intersections of the horizontal line with the line connecting r'''s and the line connecting r''s, respectively. Then, since r is known but p is not, the allegation amounts to the assertion that $p_1 < p < p_2$. Whatever the proportion of correct assertions when r is fixed, the proportion is at least $1 - \alpha$ over all values of r, i.e., in the general case where the allegation is to be made irrespective of the actual value observed for r.

The above approach is that used by Clopper and Pearson [4] who actually performed the calculations and plotted the lines yielding the confidence contours at various values of n. One can read the critical values of p_1 and p_2 off of their published curves. However, one can also obtain them by letting $\alpha_1 = \alpha_2 = \alpha/2$ and solving the following equations for p_1 and p_2:

$$\alpha_1 = \sum_{i=r}^{n} \binom{n}{i} p_1^i (1 - p_1)^{n-i}$$

$$\alpha_2 = \sum_{i=0}^{r} \binom{n}{i} p_2^i (1 - p_2)^{n-i}$$

In practice this amounts to entering the binomial tables with n and r and finding the largest p satisfying the first equation (i.e., yielding an $\alpha_1 \leqq \alpha/2$) and the smallest p satisfying the second. One then states at confidence level $1 - \alpha_1 - \alpha_2$ that $p_1 < p < p_2$.

Although Clopper and Pearson took $\alpha_1 \leqq \alpha/2$ and $\alpha_2 \leqq \alpha/2$, the method requires only that $\alpha_1 + \alpha_2 \leqq \alpha$. There are various techniques for choosing α_1 and α_2 [3, 4, 6, 19] and each has its advantages and drawbacks. One can choose them so as to minimize the interval, so as to minimize $\alpha - \alpha_1 - \alpha_2$, so as to make α_1 most nearly equal to α_2, etc. The important thing, however, is to decide upon the method prior to sampling.

To illustrate the method, suppose we decide that α_1 and α_2 are each to be the largest possible values $\leqq .025$, so that $\alpha \leqq .05$ and $1 - \alpha \geqq .95$. We make $n = 40$ observations of which $r = 7$ turn out to be "successes," the rest "failures." What is the confidence interval for p at confidence level $\geqq .95$? Entering tables of the cumulative binomial probability distribution [23] with $n = 40$, $r = 7$, we find the probability of 7 or more successes to be .0199 for $p = .07$ and .0376 for $p = .08$, so $\alpha_1 = .0199$ and $p_1 = .07$. The probability of 7 or fewer successes will be 1 minus the probability of 8 or more. For $n = 40$, the probability for 8 or more successes is .9685 for $p = .32$, and .9766 for $p = .33$. So $\alpha_2 = .0234$ and $p_2 = .33$. Therefore, we can state that $.07 < p < .33$ at a level of confidence $\geqq .95$ and in fact equal to $1 - .0199 - .0234$. Note that the obtained proportion of successes was $7/40 = .175$.

REFERENCES

1. Bahadur, R. R., "Simultaneous Comparison of the Optimum and Sign Tests of a Normal Mean," in *Contributions to Probability and Statistics, Essays in Honor of Harold Hotelling*, ed. I. Olkin *et al.* Stanford, Calif.: Stanford University Press, 1960, pp. 79–88.
2. Blyth, C. R., "Note on Relative Efficiency of Tests," *Annals of Mathematical Statistics*, **29** (1958), 898–903.
3. Blyth, C. R., and D. W. Hutchinson, "Table of Neyman-Shortest Unbiased Confidence Intervals for the Binomial Parameter," *Biometrika*, **47** (1960), 381–391.
4. Clopper, C. J., and E. S. Pearson, "The Use of Confidence or Fiducial Limits Illustrated in the Case of the Binomial," *Biometrika*, **26** (1934), 404–413.
5. Cox, D. R., and A. Stuart, "Some Quick Sign Tests for Trend in Location and Dispersion," *Biometrika*, **42** (1955), 80–95.
6. Crow, E. L., "Confidence Intervals for a Proportion," *Biometrika*, **43** (1956), 423–435.
7. David, H. A., "A Conservative Property of Binomial Tests," *Annals of Mathematical Statistics*, **31** (1960), 1205–1207.
8. David, H. A., and C. A. Perez, "On Comparing Different Tests of the Same Hypothesis," *Biometrika*, **47** (1960), 297–306.
9. Dixon, W. J., "Power Functions of the Sign Test and Power Efficiency for Normal Alternatives," *Annals of Mathematical Statistics*, **24** (1953), 467–473.
10. Dixon, W. J., and A. M. Mood, "The Statistical Sign Test," *Journal of the American Statistical Association*, **41** (1946), 557–566.
11. Fraser, D. A. S., "Most Powerful Rank-Type Tests," *Annals of Mathematical Statistics*, **28** (1957), 1040–1043.

12. Fraser, D. A. S., "Non-Parametric Theory: Scale and Location Parameters," *Canadian Journal of Mathematics*, **6** (1954), 46–68.

13. Gibbons, Jean D., "Effect of Non-Normality on the Power Function of the Sign Test," *Journal of the American Statistical Association*, **59** (1964), 142–148.

14. Hodges, J. L., and E. L. Lehmann, "The Efficiency of Some Nonparametric Competitors of the *t*-Test," *Annals of Mathematical Statistics*, **27** (1956), 324–335.

15. McNemar, Q., "Note on the Sampling Error of the Difference between Correlated Proportions or Percentages," *Psychometrika*, **12** (1947), 153–157.

16. National Bureau of Standards, *Tables of the Binomial Probability Distribution.* Washington, D. C.: U. S. Government Printing Office, 1949.

17. Noether, G. E., "Two Sequential Tests against Trend," *Journal of the American Statistical Association*, **51** (1956), 440–450.

18. Ordnance Corps, *Tables of the Cumulative Binomial Probabilities*, Ordnance Corps Pamphlet ORDP 20-1, Washington, D. C., September 1952. (Distributed by Office of Technical Services, U. S. Dept. of Commerce, Washington, D. C.)

19. Pachares, J., "Tables of Confidence Limits for the Binomial Distribution," *Journal of the American Statistical Association*, **55** (1960), 521–533.

20. Robertson, W. H., *Tables of the Binomial Distribution Function for Small Values of P*, Sandia Corporation Monograph SCR-143, January 1960. (Distributed by Office of Technical Services, U. S. Dept. of Commerce, Washington, D. C.)

21. Savur, S. R., "The Use of the Median in Tests of Significance," *Proceedings of the Indian Academy of Science*, **5** (1937), 564–576.

22. Scheffe, H., and J. W. Tukey, "Non-Parametric Estimation: I. Validation of Order Statistics," *Annals of Mathematical Statistics*, **16** (1945), 187–192.

23. Staff of the Computation Laboratory, *Tables of the Cumulative Binomial Probability Distribution.* Cambridge, Mass.: Harvard University Press, 1955.

24. Stewart, W. M., "A Note on the Power of the Sign Test," *Annals of Mathematical Statistics*, **12** (1941), 236–239.

25. Thompson, W. R., "On Confidence Ranges for the Median and Other Expectation Distributions for Populations of Unknown Distribution Form," *Annals of Mathematical Statistics*, **7** (1936), 122–128.

26. Walsh, J. E., "Some Bounded Significance Level Properties of the Equal-Tail Sign Test," *Annals of Mathematical Statistics*, **22** (1951), 408–417.

27. Walsh, J. E., "On the Power Function of the Sign Test for Slippage of Means," *Annals of Mathematical Statistics*, **17** (1946), 358–362.

28. Weintraub, S., *Tables of the Cumulative Binomial Probability Distribution for Small Values of p.* New York: The Free Press of Glencoe, Inc., 1963.

8

TESTS BASED UPON THE HYPERGEOMETRIC DISTRIBUTION

In both the preceding and the present chapter the test statistic is the number r of successes in a sample of size n drawn from a dichotomous population consisting only of mutually exclusive "successes" and "failures." If, prior to sampling, the proportion of successes in the population is p, then the probability that the first-drawn sample observation will be a success is also p. If the population is an infinite one, or if sampling is with replacement of each observation before the next observation is drawn, then the probability that an observation will be a success will be constant from one drawing to the next and the distribution of r will be the binomial. However, if the population is finite and sampling is without individual replacements, the act of drawing the observations will deplete the population, and the ratio of number-of-remaining-successes to number-of-remaining-failures in the population will vary from the drawing of one sample observation to that of the next. In this case, where the binomial assumption of a constant p is not met, the distribution of r is not binomial—rather it is hypergeometric.

8.1 THE HYPERGEOMETRIC DISTRIBUTION

Suppose that a random sample of size n is drawn without individual replacements from a finite, dichotomous population consisting of N units belonging to one of two mutually exclusive categories arbitrarily labeled "success" and "failure." Suppose further that of the N units comprising the population, R are "successes" and $N - R$ are "failures." Then the probability that exactly r of the n sample units will be successes is

$$P(r) = \frac{\binom{R}{r}\binom{N-R}{n-r}}{\binom{N}{n}} = \frac{R!\,(N-R)!\,n!\,(N-n)!}{r!\,(R-r)!\,(n-r)!\,(N-R-n+r)!\,N!}$$

which is the formula for a hypergeometrically distributed variate.

Proof: The number of different ways of drawing (without replacement) the r sample successes from the R population successes is simply $\binom{R}{r}$, the number of combinations of R things taken r at a time. For each such "way," there are $\binom{N-R}{n-r}$ different ways in which the $n - r$ sample failures can be drawn from the $N - R$ population failures. So the number of different ways of drawing a sample of size n consisting of r successes and $n - r$ failures is $\binom{R}{r}\binom{N-R}{n-r}$. But the number of different ways of drawing a sample of size n, without restriction as to the number of successes or failures, from a population of size N is $\binom{N}{n}$. Thus, of the $\binom{N}{n}$ possible different samples of size n, $\binom{R}{r}\binom{N-R}{n-r}$ of them contain exactly r successes and $n - r$ failures. Consequently

$$\frac{\binom{R}{r}\binom{N-R}{n-r}}{\binom{N}{n}}$$

is the proportion of samples of size n having the specified makeup and is therefore the a priori probability that a single such sample will have it.

If the sample size n is greater than the number of failures $N - R$ in the population, there *must* be at least $n - (N - R)$ successes in the sample. So the cumulative probability of r or fewer successes in the sample must be

$$\text{Cum } P(r) = \sum_{i=\max(0,\, n-N+R)}^{r} \frac{\binom{R}{i}\binom{N-R}{n-i}}{\binom{N}{n}}$$

i.e., the cumulation starts with zero unless $n - N + R$ is positive, in which case it starts with the latter.

The following assumptions are implicit in the derivation of the hypergeometric distribution:

(a) All population units belong to a *dichotomy* and the two categories comprising the dichotomy are *mutually exclusive*. (This is implied by the fact that the number of successes R plus the number of failures $N - R$ exactly equals the size N of the population. Thus no unit belongs to both categories.)

(b) In order to use the "chance" combinatorial formulas used in the derivation, it was necessary that each drawing of an observation be a chance event whose outcome depended only upon the relative frequencies of population successes and failures at that moment. Therefore sampling must be *random*, and the *act* (as contrasted with the outcome) of drawing any observation must be *independent* of the act of drawing any other observation.

The hypergeometric distribution has been extensively tabled, to six decimal places, by Lieberman and Owen [12] who use the symbol x for our r and k for our R. They give both the point probability for exactly x successes,

$$\frac{\binom{k}{x}\binom{N-k}{n-x}}{\binom{N}{n}}$$

and the cumulative probability of x or fewer successes. Their tables yield (either directly or by the use of equivalence relationships) any probability desired (i.e., corresponding to any possible combination of n, k, and x) for $N = 2\,(1)\,50\,(10)\,100$. Two additional tables are for $N = 1000$, $n = 500$ (all cases of k and x), and for $N = 100\,(100)\,2000$, $n = N/2$, $k = n - 1$ or n, and all values of x. The mean and variance of the hypergeometric distribution are $(nR)/N$ and $[nR(N - R)(N - n)]/[N^2(N - 1)]$, respectively. The distribution is approximated by a number of other distributions, including the normal [see 12].

The hypergeometric distribution is one of the most important distributions in the theory of probability and one of the most useful in applications. This is particularly true in distribution-free statistics. The tests presented in this chapter all have a hypergeometric derivation, i.e., the point probabilities of the test statistic are point hypergeometric probabilities. For certain other tests, discussed in the next chapter, the test statistic does not have exactly univariate hypergeometric point probabilities but, due to certain equivalences, does have cumulative probabilities which are exactly the cumulative probabilities for some univariate hypergeometric distribution and can therefore be obtained from cumulative hypergeometric tables such as [12].

8.2 FISHER'S EXACT METHOD FOR ANALYZING FOURFOLD (2 × 2) CONTINGENCY TABLES

One of the fundamental concepts in distribution-free statistics was developed by R. A. Fisher. It is that if one sample is drawn from each of two identical populations, the overall, combined sample will be homogeneous, so that the two samples may be regarded as having been "drawn," randomly and without replacement, from a finite, miniature population consisting entirely of the combined sample. By testing the hypothesis that the "first" sample is a *random* sample from the miniature population one tests, in effect, the hypothesis that (aside from sample size) it is chance alone that distinguishes the two samples. This is equivalent to testing the hypothesis that the two samples are "homogeneous," and that is equivalent to testing the hypothesis that the two original populations are identical. Data analyses based upon this Fisherian concept are known as the Method of Randomization when observation magnitudes are involved and as Fisher's exact method when category frequencies in a fourfold contingency table are involved.

8.2.1 Rationale

Suppose that two infinite populations each consist entirely of units which belong to one or the other of two mutually exclusive categories, S and F (standing for "success" and "failure"). Suppose further that a sample has been drawn from each population and the experimenter wishes to test the hypothesis that the proportion of S's in Population I is the same as that in Population II (i.e., that, with respect to the dichotomization of their units, the two populations are identical). Let the frequency data be represented by the accompanying table. If the two populations are identical and infinite,

	Category		
	Success	*Failure*	*Total*
Sample I	r	$n - r$	n
Sample II	$R - r$	$(N - R) - (n - r)$	$N - n$
Total: Combined Sample	R	$N - R$	N

they may be regarded as a single, common, infinite population. Sample I and Sample II may be pooled and the combined sample may be regarded as a single, homogeneous sample drawn from the common population. The two original samples may therefore be regarded as having been obtained by randomly assigning the label "Sample I" to n of the N units in the combined sample, the $N - n$ remaining units being labeled "Sample II." The degree to which Samples I and II differ in any respect other than size is therefore a matter of chance. On the other hand, if the two populations are not identical with regard to the proportions of S's and F's, the combined sample will not be homogeneous and Sample I will be a biased, rather than a random, selection of units from it. Thus a test of the hypothesis that Sample I is a random sample drawn, without replacements, from the combined sample represented by the row total is equivalent to a test of the hypothesis that Populations I and II are identical with respect to the S–F dichotomy. The point probability of randomly drawing the r successes and $n - r$ failures of Sample I, without replacements, from the R successes and $N - R$ failures of the combined sample is the hypergeometric probability

$$\frac{\binom{R}{r}\binom{N-R}{n-r}}{\binom{N}{n}}$$

which equals

$$\frac{R!\,(N-R)!\,n!\,(N-n)!}{r!\,(R-r)!\,(n-r)!\,(N-R-n+r)!\,N!}$$

Note that since Sample II is simply what is left after Sample I has been drawn, this is also the probability of "drawing" both samples (without replacements), i.e., it is the probability for the entire 2×2 table, given fixed marginal frequencies. Note further that if instead of considering the hypergeometric probability of "drawing" the first row, i.e., Sample I, from the row total, i.e., the combined sample, we had considered the probability of "drawing" the first column from the column total, we would have obtained

$$\frac{\binom{n}{r}\binom{N-n}{R-r}}{\binom{N}{R}}$$

which equals

$$\frac{n!\,(N-n)!\,R!\,(N-R)!}{r!\,(n-r)!\,(R-r)!\,(N-n-R+r)!\,N!}$$

the same probability obtained earlier. Clearly we are dealing with the probability for an entire 2×2 table (with fixed marginal frequencies) based upon membership in two dichotomies and, mathematically speaking, it makes no difference which dichotomy is called S–F and which is designated I–II (i.e., "Sample I" may simply be all sampled units belonging to a given member of a dichotomy *other than* success–failure, Sample II consisting of units belonging to the opposite member). Thus we may conceive of a single infinite population in which every unit is either an S or an F and is also either a I or a II, i.e., "Populations I and II" are simply the portions of a single population belonging to the respective members of the I–II dichotomy. Finally, our original hypothesis that the proportion of successes in Population I is the same as in Population II is now seen to amount to the hypothesis that the two dichotomies are independent, i.e., that there is no association between membership in one dichotomy and membership in the other (e.g., knowing that a unit is a II yields no additional information on whether it is an S or an F).

The derivation obtained only a point probability. If the null hypothesis of equal population proportions of successes is true, one would expect the proportion r/n of successes in Sample I to tend to equal the proportion $(R - r)/(N - n)$ of successes in Sample II. Therefore to test the hypothesis one should take as rejection region those values of r that cause r/n to differ from $(R - r)/(N - n)$ by the greatest amount (in a single, predesignated direction for a one-tailed test or in either direction for a two-tailed test). The probability level for the test is therefore (see above table) the cumulation of

$$\frac{\binom{R}{i}\binom{N-R}{n-i}}{\binom{N}{n}}$$

over all values of i that cause

$$\frac{i}{n} - \frac{R-i}{N-n} \quad \left(\text{or } \left|\frac{i}{n} - \frac{R-i}{N-n}\right| \text{ in the case of a two-tailed test}\right)$$

to assume values more extreme (in the predicted direction) than

$$\frac{r}{n} - \frac{R-r}{N-n} \quad \left(\text{or } \left|\frac{r}{n} - \frac{R-r}{N-n}\right| \text{ in the case of a two-tailed test}\right)$$

i.e., more extreme than is the case in the table actually obtained.

In order to perform a two-tailed test it is necessary to know the value of r for as extreme a difference in proportions in the opposite direction. Call this value x. Then if for the obtained value of r the difference in proportions is

$$\frac{r}{n} - \frac{R-r}{N-n} = d$$

it follows that

$$\frac{x}{n} - \frac{R-x}{N-n} = -d$$

so that

$$\frac{x}{n} - \frac{R-x}{N-n} = -\frac{r}{n} + \frac{R-r}{N-n}$$

Solving the latter equation for x we obtain $x = (2nR - Nr)/N$. If $2n = N$ (in which case each row marginal frequency equals $N/2$), this reduces to $x = R - r$ and the 2×2 table for x is the same as that for the obtained value of r except that the two rows are interchanged, so the probability of the two tables is the same. Likewise, if $2R = N$ (in which case each column marginal frequency equals $N/2$), we get $x = n - r$, so that the 2×2 table for x is the same as that for r except that the two columns are interchanged, and, again, the probabilities for the two tables are equal. Therefore, if either the two row marginals, n and $N - n$, or the two column marginals, R and $N - R$, are equal, the probability level for a two-tailed test is simply double that for a one-tailed test. Otherwise, the probability for a two-tailed test is the cumulation of

$$\frac{\binom{R}{i}\binom{N-R}{n-i}}{\binom{N}{n}}$$

over all values of i except those lying between x and r, e.g., if $x < r$, it is the cumulation over all values of $i \leqq x$ plus the cumulation over all values of $i \geqq r$. Neither this nor any other method of performing a two-tailed test is entirely satisfactory, since the asymmetry of the probability distribution of i makes it difficult to choose an a priori rejection region that is both simple and sensible (see [1]).

8.2.2 Null Hypothesis

Sample I is, in effect, a random sample drawn, without replacements, from the combined sample, so that

$$P(r) = \frac{\binom{R}{r}\binom{N-R}{n-r}}{\binom{N}{n}}$$

This will be the case if the proportion of successes in Population I is the same as in Population II and all assumptions are met.

8.2.3 Assumptions

(a) Each of the two sampled populations is *infinite* in size, or else sampling is with individual replacements. (The reason for this assumption can be most easily seen in the extreme case where Sample I is equal in size to Population I and is drawn without replacements from it. In that case r would be a constant rather than a variable and Sample I could not be regarded as a *random* sample drawn from the combined sample, as required by the hypergeometric derivation. The reason for this assumption, however, is much more apparent in the "binomial" derivation presented in the *Discussion* section.)
(b) The members of each dichotomy are *mutually exclusive* and *exhaustive*. That is, every unit in each population is either a success or a failure but not both, and every unit in the table belongs either to Sample I or to Sample II but not to both.
(c) The outcome of drawing one observation is *independent* of the outcome of drawing any other observation, i.e., to what categories a unit will belong is uninfluenced by the categories to which any other unit belongs. (This assumption applies to the generation of the table and its marginal frequencies, and therefore is not in conflict with the fact that the table is completely specified by its marginal frequencies and a single cell frequency.)
(d) Sampling is *random*.

8.2.4 Efficiency

In a sense the test is perfectly "efficient," since it is an exact method which uses all of the "information" in the sample; parametric tests for the same problem merely substitute the normal approximation for the true binomial distribution of frequencies within a cell and therefore use the same "information" but use it somewhat inaccurately. In the practical, computational sense, the test becomes slightly inefficient for moderate and large samples, when these same approximations are resorted to because exact tables do not extend far enough. Power functions are given in [3, 10, 15].

8.2.5 TABLES

Table VIII of the Appendix was expressly designed for this test. Let A be the larger of the two sample sizes, and let B be the smaller. Let a and b be two cell frequencies chosen so as to fulfill the following requirements: a is in the same row as A; b is in the same row as B; a and b are in the same column; and that column is so chosen that $a/A \geqq b/B$ (or $aB \geqq Ab$). One then has the accompanying table. One enters Table VIII with A, B, a, and a

Table

a	$A - a$	A
b	$B - b$	B
$a + b$	$(A + B) - (a + b)$	$A + B$

Qualifying Conditions

$A \geqq B$

$\frac{a}{A} \geqq \frac{b}{B}$

one-tailed significance level α. These locate a cell which gives the (largest) value of b such that a/A is just significantly greater than b/B for a one-tailed test at the α level of significance, i.e., the cell gives the critical value of b. Alongside the critical value is its exact one-tailed cumulative probability (which is often considerably smaller than α). This probability is the probability of a 2×2 table whose left column marginal frequency is "a + critical value" and it will be the probability of the actually obtained table only if the critical value equals the obtained value of b.

Despite the fact that, depending on the particular 2×2 table, a may turn out to be any one of the four cell frequencies, Table VIII can be used for a one-tailed test involving any pair of cell frequencies in the same column. Since marginal frequencies are fixed, if a/A is significantly greater than b/B, then $(A - a)/A$ is significantly smaller than $(B - b)/B$. And since a was chosen to make a/A greater than b/B, one knows without looking it up that a/A cannot be significantly smaller than b/B, since it is not smaller to any degree; and likewise $(A - a)/A$ cannot be significantly larger than $(B - b)/B$ since it is not larger at all.

For a two-tailed test, if $A = B$, or if the two column marginal frequencies are equal, one need only perform a one-tailed test and double the obtained probability level. Otherwise one must perform two one-tailed tests and add the resulting probability levels [see last paragraph of Section 8.2.1 (*Rationale*)].

For cases not covered by Table VIII, the reader may consult the tables of Finney, Latscha, Bennett, and Hsu [8] from which Table VIII was adapted and which give critical values of b for all cases up to $A = 40$ (and exact probabilities for all cases up to $A = 30$). For cases not covered by the latter consult Lieberman and Owen's extensive hypergeometric tables [12]. For cases not covered by these tables, all "expected" cell frequencies will be large enough to permit use of the classical chi-square test for contingency tables. The result will be an approximation to the true, exact probability. Quite a large number of tables have been published expressly for Fisher's exact test [such as 1, 8, 14]. However, they tend to be highly idiosyncratic as to format, method of entry, symbolic notation, etc., so that considerable introductory material must be read before one is ready to use the table.

8.2.6 Example

Suppose that the accompanying table gives all the known data on the fatality of a rare disease, and it is desired to test at the .01 level whether the

	Survived	*Died*	
Men	9	1	10
Women	4	10	14
	13	11	24

chance of survival is the same for men as for women. We regard *Men* as Sample I, *Women* as Sample II. Since $14 > 10$, we set $A = 14$ and $B = 10$; and since $10/14 > 1/10$ while $4/14 < 9/10$, we set $a = 10$, which makes $b = 1$. Entering Table VIII with these values of A, B, and a, we find that $b = 1$ is the critical value for one-tailed tests at $\alpha = .01$ or $\alpha = .005$ and that the exact one-tailed probability is .004. Had our hypothesis been that the death rate is greater for women than for men (or that the survival rate is greater for men) we could reject at any $\alpha \geqq .005$. For a two-tailed test, however, we must consider extreme outcomes in the opposite direction. Solving the equation

$$\frac{x}{14} - \frac{11 - x}{10} = -d = -\left(\frac{10}{14} - \frac{1}{10}\right)$$

for x, we obtain $x = 2.83$ indicating that for as large a difference in the opposite direction in the proportion dying, x must be equal to or less than 2. Substituting 2 for 10 as the number of women who died, and adjusting the other three cell frequencies accordingly, we get the accompanying table. Note

	Survived	*Died*	
Men	$1 = b$	9	$10 = B$
Women	$12 = a$	2	$14 = A$
	13	11	24

that the "a" cell is now in a different column. Entering Table VIII with $A = 14$, $B = 10$, and $a = 12$, we find that a "b" of 2 would have an exact cumulative probability of .002; so since $b = 1$ must have a smaller cumulative probability, the two-tailed probability for the test must be less than $.004 + .002 = .006$. We are therefore able to reject at the two-tailed .01 level. (Of course in practice our conclusions would be tempered by the ages, physical conditions, etc., of the men and women involved and rejection would be in order only if such factors did not favor the survival of one sex.)

8.2.7 Discussion

The derivation given in Section 8.2.1 makes it easy to see why the obtained probability is a hypergeometric one and was given for that reason. The following classical derivation however, makes it easier to see why the original sampled populations are assumed to be infinite and emphasizes the fixity of all marginal frequencies. The null hypothesis states that the proportion of successes in Population I is the same as in Population II, and if the hypothesis is true, this is also the proportion of successes in both populations combined. Let p be this common, but unknown proportion of successes, $1 - p$ therefore being the common proportion of failures. Finally let the populations be infinite (or sampling be with individual replacements) so that p remains constant from one drawing to the next and let the other binomial assumptions of random sampling and independent observations be met. Then, given that there are n observations in Sample I, the probability that r of them will be successes is the binomial probability

$$\binom{n}{r}p^r(1 - p)^{n-r}$$

Given that there are $N - n$ observations in Sample II, the probability that $R - r$ of them will be successes is

$$\binom{N-n}{R-r}p^{R-r}(1 - p)^{N-n-R+r}$$

and the probability of both events, if Samples I and II are independent, is the product of the two probabilities, which reduces to

$$\binom{n}{r}\binom{N-n}{R-r}p^R(1 - p)^{N-R}$$

The probability that of the N units in Samples I and II combined R will be successes is

$$\binom{N}{R}p^R(1-p)^{N-R}$$

Therefore the conditional probability of the obtained sample frequencies of success and failure, given that there are a total of R successes and $N - R$ failures in the combined sample and that there are n and $N - n$ observations in Samples I and II, respectively (i.e., given the four marginal frequencies), is

$$\frac{\binom{n}{r}\binom{N-n}{R-r}p^R(1-p)^{N-R}}{\binom{N}{R}p^R(1-p)^{N-R}} = \frac{\binom{n}{r}\binom{N-n}{R-r}}{\binom{N}{R}}$$

which reduces to the same probability obtained earlier.

8.3 THE WESTENBERG-MOOD MEDIAN TEST

Suppose that, in the previous test, Populations I and II are infinite populations of variate-values, X's, and the two column categories "success" and "failure" are replaced by "$X < Q$" and "$X \geqq Q$." Then if the value Q is specified prior to sampling, Fisher's exact method becomes an exact and uncomplicated test of the hypothesis that equal proportions of the two populations lie below Q. It does not tell us what that common proportion is, but we could estimate it from the combined sample, the estimate being R/N. Then if H_0 is true and the populations are continuous, Q is the same quantile in the two populations; it is the pth quantile Q_p, where the estimated value of p is R/N. An experimenter, however, may be more interested in controlling (i.e., specifying a priori) the proportion p than the value of Q. It is tempting, therefore, to decide upon a value of p, estimate the corresponding value of Q_p from the combined sample, and then use this value as the basis for the column categories, despite the fact that a posteriori dichotomization of an infinite population on the basis of obtained sample values may introduce both statistical and logical complications. The following test is based upon this approach.

8.3.1 RATIONALE

Suppose that Populations I and II are both continuously distributed populations of variate-values, that the median variate-value of the combined sample has the value $\hat{M}$, and that the two column categories are $<\hat{M}$ and $>\hat{M}$ respectively. Then (if when the number of observations in the combined sample is odd, the single observation equal to $\hat{M}$ is excluded from all frequencies in the 2×2 table so that N is an even number), Fisher's exact method becomes a test of the hypothesis that equal but inexactly known proportions of Populations I and II lie below $\hat{M}$ (or, if the original combined sample contained an even number of observations, below some population value lying between the $[N/2]$th and $[(N/2) + 1]$th observations in order of increasing size in the combined sample). Thus it tests the hypothesis that $\hat{M}$ (or some value close to it) is the same, but unknown, quantile in the two populations; and if H_0 is true and N is large, the common proportion of the two populations lying below $\hat{M}$ will tend to be close to .5, so $\hat{M}$ will tend to be a population quantile which is nearly a median. For this reason the test is sometimes stated to be a test for identical population medians; however, it is quite possible for two populations to have unequal medians and equal "$\hat{M}$-quantiles" or vice versa, especially when both samples are small. Instead of a median test, therefore, it might be better described as a quasi-median test or as a test for a common, probably more or less centrally located, quantile.

8.3.2 Null Hypothesis

For the basic H_0 see Section 8.2.2. However, we may substitute a more precisely meaningful H_0. Let X and Y represent the respective variates of Populations I and II. Then if the original combined sample contained an odd number of observations, so that $\hat{M}$ corresponds to the value of one of them, we may consider H_0 to be that $P(X < \hat{M}) = P(Y < \hat{M})$ and $P(X > \hat{M}) = P(Y > \hat{M})$, i.e., that $\hat{M}$ is the same quantile in the two populations. If the original combined sample contained an even number of observations and $\hat{M}$ falls between two adjacent combined-sample values, Z' and Z'', so that $\hat{M} = (Z' + Z'')/2$, H_0 is the same as above, except that $\hat{M}$ must be replaced by D, where all we know about the value of D is that $Z' < D < Z''$.

Since the populations are assumed to be infinite and since the value of $\hat{M}$ is determined by chance, it is *possible* for $\hat{M}$ to be *any* value in either population (although, if the populations are identical, as N increases the variance of $\hat{M}$ about the population median M diminishes toward zero, so that the probability that $\hat{M}$ will stray far from the vicinity of the latter eventually becomes negligible). Thus, prior to sampling, the test is, in a sense, a test for identical populations. Once samples are drawn and the value of $\hat{M}$ is identified, however, it becomes a test of whether or not equal proportions of the two populations lie below $\hat{M}$.

8.3.3 Assumptions

The test makes the same assumptions as Fisher's exact method. As described under *Rationale* (Section 8.3.1), it also assumes that both populations are *continuously distributed*, and under this assumption all observations equal to $\hat{M}$ (other than the single observation which must be so when the original combined sample is odd) should be included in the table frequencies and dealt with according to Method A of Chapter 3. However, certain modifications of the test do not require a continuity assumption (see Section 8.3.7). Finally, if when H_0 is true $\hat{M}$ is to be regarded as a very centrally located population quantile (i.e., as a quasi-median), *N must be large.*

8.3.4 Efficiency

The A.R.E. of the median test for location relative to Student's t test, when both tests are applied to normal populations with equal variances, is $2/\pi$, or .637. Relative to the likelihood ratio test, when both tests are used as tests for a shift in location of an exponential population, the median test has an A.R.E. of zero, as does a test based on dichotomization by the first quartile, rather than the median, of the combined sample. The "first quartile" test has, relative to the median test as a test for shift in location, an A.R.E. of .846 when both populations are normal with equal variances and an A.R.E. of 3 when both populations are exponential [5].

8.3.5 Tables

See Section 8.2.5.

8.3.6 Example

An experimenter, content to let the exact nature of "location" be defined a posteriori, wishes a rough test (at the .05 level) of whether or not two continuous populations have a common location. From Population I he draws a sample consisting of the values 2, 3, 5, 6, 8, 9, 13, 15, 16, 18, 31, 54, 89, and from Population II he obtains the sample observations 10, 11, 17, 19, 20, 22, 23, 27, 30, 33, 34, 35, 39, 40, 42. Of the $N = 28$ observations in both samples combined, the fourteenth and fifteenth in increasing order of size are 19 and 20, so $\hat{M} = (19 + 20)/2 = 19.5$. We therefore have the accompanying 2×2 table. Entering Table VIII of the Appendix with

	$<\hat{M}$	$>\hat{M}$	*Total*
Sample I	10	3	13
Sample II	4	11	15
Total: Combined Sample	14	14	28

$A = 15$, $B = 13$, and $a = 11$, we find the one-tailed probability for $b = 3$ to be .011. Since the two column marginal frequencies are equal, the two-tailed probability is simply twice the one-tailed probability or .022. We may reject, therefore, at the predesignated .05 level the hypothesis that equal proportions of the two populations lie below $\hat{M} = 19.5$ (or at least below some point between 19 and 20—see Section 8.3.7, following).

8.3.7 Discussion

Suppose that, due either to intrinsic discreteness or to imprecision of measurement, more than one sample observation equals $\hat{M}$. Then we have a trichotomy, $<\hat{M}$, $=\hat{M}$, and $>\hat{M}$, of outcomes. If, after determining $\hat{M}$, we discard all observations equal to $\hat{M}$ and base the table only on the dichotomy $<\hat{M}$ and $>\hat{M}$ (so that N is the total number of observations $\neq\hat{M}$), we are testing whether or not equal proportions lie below $\hat{M}$ in what is left of Populations I and II after removing from them all units measurable as $\hat{M}$, i.e., in the subpopulations of "not-$\hat{M}$'s." This is equivalent to testing whether or not the ratio of units below $\hat{M}$ to units above $\hat{M}$ in the original Population I (which includes units equal to $\hat{M}$) is the same as the corresponding ratio in the original Population II. But unless the proportion of units equal to $\hat{M}$ in the two populations is the same, it is *not* equivalent to testing whether or not the proportion of units below $\hat{M}$ is the same in the two original populations. Note that the test outlined under *Rationale* is the limiting case of the above test as the proportion of units measurable as $\hat{M}$'s, in both populations, diminishes to zero, i.e., practically speaking, as discrete populations become sufficiently "fine-grained" and as measurement becomes sufficiently precise that no sample observations equal $\hat{M}$ when the size of the combined sample is an even number and only one observation equals $\hat{M}$ when combined sample size is odd.

Another approach to the problem of multiple sample observations equal to $\hat{M}$ would be to dichotomize observations as $<\hat{M}$ or $\geq\hat{M}$. If this were valid, then surely in the case of continuous populations and odd N, where only a single observation equals $\hat{M}$, it should make no difference whether we use the dichotomy $<\hat{M}$ and $\geq\hat{M}$ or the dichotomy $\leq\hat{M}$ and $>\hat{M}$. It does though, and we get two different tables having two different probabilities under the two dichotomizations. The flaw is in claiming $\hat{M}$ to be the point of dichotomization when, for example, for the dichotomy $<\hat{M}$ versus $\geq\hat{M}$ the same table would have resulted for a dichotomization at any point between, but not including, $\hat{M}$ and the observation just smaller than $\hat{M}$. The point halfway between these two values would be a reasonable estimate of the point of dichotomization, but the point infinitesimally different from one of these two values, $\hat{M}$, is not. The use of a biased estimate thus affects the probability of the test.

We could, of course, rectify this particular difficulty by letting H be the point halfway between $\hat{M}$ and the next smaller observation and using $<H$ versus $\geq H$ as the dichotomy. However, H would still be only an estimate of the "true" point of dichotomization (a shortcoming it shares with $\hat{M}$, in the test outlined under *Rationale*, when the original size of the combined sample is an even number; this is one of the complications, mentioned in Section 8.3, due to a posteriori dichotomization). Furthermore, if the two popu-

lations had a common median, H would not be an unbiased estimate of it, although $\hat{M}$ would be.

This test is a special case of a more general test which, in analogy, might be called the Quantile test. Let $\hat{Q}_p$ be any (unique, i.e., untied) value, based on the combined sample, which is an unbiased estimate of the pth quantile Q_p in the population from which the combined sample was drawn. Then, by using $<\hat{Q}_p$ and $>\hat{Q}_p$ as the column categories and applying Fisher's exact method, under the continuity assumption, we are testing whether or not equal proportions of the two populations lie below $\hat{Q}_p$. To the extent that $\hat{Q}_p$ is a precise estimate of Q_p, this is roughly equivalent to testing whether or not the two populations have the same pth quantile Q_p. (The kth observation in order of increasing size in a sample of size N from a continuously distributed population is an estimate of the population value below which a proportion $k/(N + 1)$ of the population lies.)

8.4 THE MEDIAN TEST FOR TREND

If Sample I is taken to be the first half, and Sample II the second half, of a series of observations taken sequentially, the median test can be used to test for trend. (Populations I and II are therefore the sampled population as it existed in the interval from the beginning of sampling until half the observations had been drawn and in the interval from that point until the end of sampling, respectively.) The only types of trend to which the test can be sensitive, of course, are those which affect the proportion of the population lying below $\hat{M}$. One such type of trend is simple slippage of location. If, as time passes, the population distribution, without changing in shape, simply "slides" upward or slides downward, unidirectionally on the "X-axis," then the proportion of values below $\hat{M}$ (and, indeed, below any constant which is an abscissa of the population) will change from Population I to Population II. As just implied, the statement will be equally valid whatever population quantile $\hat{M}$ represents when the null hypothesis is true. Therefore, if it can be legitimately assumed that the sampled population may change in location but not in the shape of its distribution, the test will be sensitive to "slippage" of any location parameter, and the question of how closely $\hat{M}$ represents the common population median (under H_0), or of what the actually dichotomizing value is, will not be a problem. So, in order to maximize power, it is probably best to use all observations by dichotomizing units as $<\hat{M}$ versus $\geqq \hat{M}$. The test has A.R.E. of .78 relative to the best parametric test against normal regression, based on the regression coefficient b, under conditions satisfying all assumptions of the latter. The test was suggested by Cox and Stuart (referenced in the previous chapter).

8.5 WESTENBERG'S INTERQUARTILE RANGE TEST

This test is an application of Fisher's exact method to test for dispersion in the same way that the median test applies it to test for location. Let Z stand for a unit in either population and let q_1 and q_3 be the first and third quartiles, respectively, of the combined sample. Then the two column categories are "$q_1 < Z < q_3$" and "$Z \leqq q_1$ or $Z \geqq q_3$." In analogy with the median test, the null hypothesis tested is that identical proportions of Populations I and II lie within the interquartile range of the combined sample, not that the two populations have identical interquartile ranges. The test suffers from the same type of imprecision discussed in connection with the median test.

8.6 BLOMQVIST'S DOUBLE MEDIAN TEST FOR ASSOCIATION

8.6.1 RATIONALE

Consider a sample of units or individuals upon each of which an X-measurement and a Y-measurement have been made, and let $\hat{M}_X$ and $\hat{M}_Y$ be the respective medians of the sample X-values and Y-values. Now suppose it is known that the sample was drawn from an infinite (and, for the present, continuously distributed) bivariate population, and that it is desired to test whether or not in that population the location of an X as above or below $\hat{M}_X$ is independent of whether its paired Y-value is located above or below $\hat{M}_Y$. The alternative hypothesis, of course, is that the locations of the paired variate-values (relative to $\hat{M}_X$ and $\hat{M}_Y$) tend to be associated, i.e., correlated. We may imagine the infinite bivariate population divided into two infinite populations, Population I consisting of units whose Y-value is $<\hat{M}_Y$ and Population II containing all units whose Y-value is $>\hat{M}_Y$. Then, if the hypothesis of independence is true, the proportion of units whose X-value is $<\hat{M}_X$ is the same in "Population I" as in "Population II." Therefore, the hypothesis can be tested by applying Fisher's exact method to the accompanying 2×2 table, where it is understood that a cell frequency refers to the number of sample units whose X-value falls into the cell's column category and whose paired Y-value falls into the cell's row category. Notice that this

	$X < \hat{M}_X$	$X > \hat{M}_X$	*Total*
$Y < \hat{M}_Y$	r	$n - r$	n
$Y > \hat{M}_Y$	$R - r$	$(N - R) - (n - r)$	$N - n$
Total	R	$N - R$	N

test can be regarded as a special case of the median test in which Populations I and II are as defined above. It therefore suffers from the same technical shortcomings due to a posteriori dichotomizations as pointed out and discussed in connection with the median test (and these shortcomings apply now to both row and column dichotomizations).

8.6.2 Null Hypothesis

For the basic H_0 see Section 8.2.2. However, if the original sample contained an odd number of observations, so that $\hat{M}_X$ and $\hat{M}_Y$ correspond to the values of (discarded) sample observations, we may consider H_0 to be that

$$P(X < \hat{M}_X \mid Y < \hat{M}_Y) = P(X < \hat{M}_X \mid Y > \hat{M}_Y)$$

as well as the same equation with $X < \hat{M}_X$ replaced by $X > \hat{M}_X$. (The vertical line inside the parentheses stands for "given that" the condition expressed to the right of it is true.) In words, this H_0 means that, in the population, whether an X will be above or below the constant $\hat{M}_X$ is *independent* of whether its paired Y-value is above or below the constant $\hat{M}_Y$. If the original sample size was an even number, so that $X' < \hat{M}_X < X''$ and $Y' < \hat{M}_Y < Y''$, where X' and X'' are sample X-values that are adjacent in size and Y' and Y'' are analogously defined, then H_0 is the same as above, except that $\hat{M}_X$ and $\hat{M}_Y$ must be replaced (in the above equations and statement of H_0) by D_X and D_Y, where all we know about the latter is that $X' < D_X < X''$ and $Y' < D_Y < Y''$.

8.6.3 Assumptions

The test assumes that observations have been drawn randomly and independently from an infinite bivariate population. Since none of the four categories of association in the fourfold table includes equality with a sample median, the infinite bivariate population being "tested" excludes all units whose X equals $\hat{M}_X$ or whose Y equals $\hat{M}_Y$. Therefore this test does *not* assume continuously distributed variates, despite the fact that continuity was introduced in *Rationale* as a descriptive convenience.

8.6.4 Efficiency

When the sampled population is bivariate normal, this test has A.R.E. of $(2/\pi)^2 = .405$ relative to the parametric test based on the product moment correlation coefficient. A.R.E.'s under other circumstances are given in [11]. Power functions can be found in [7].

8.6.5 Tables

See Section 8.2.5.

8.6.6 Example

It is hypothesized that in a certain bivariate population categorization of a unit's X-value as above or below $\hat{M}_X$ is uncorrelated with categorization of its Y-value as above or below $\hat{M}_Y$, where $\hat{M}_X$ and $\hat{M}_Y$ are the X- and Y-medians in a sample not yet drawn. The following sample is then drawn, each sample unit being represented by the format (X, Y), and the sample units being arranged, for convenience, in order of increasing X-values: (5, 62), (7, 65), **(7, 15)**, (7, 13), (11, 17), **(13, 15)**, (14, 35), (14, 21), **(17, 44)**, (23, 39), (30, 21), (41, 11), (44, 10), (57, 11), (63, 6), (81, 11), (95, 8). The median X-observation is $\hat{M}_X = 17$ and the Y-median is $\hat{M}_Y = 15$. Discarding the three boldface observations whose X falls on $\hat{M}_X$ or whose Y corresponds to $\hat{M}_Y$, the original sample of 17 units is reduced to 14 whose values are categorized in the accompanying table. Entering Table VIII of the Appen-

	Number of units whose $X < \hat{M}_X$	$X > \hat{M}_X$	*Sum*
Number of units whose $Y < \hat{M}_Y$	1	6	7
Number of units whose $Y > \hat{M}_Y$	5	2	7
Sum	6	8	14

dix with $A = 7$, $B = 7$, and $a = 6$, we find that $b = 1$ is the critical value for a one-tailed $\alpha = .025$, and therefore for a two-tailed α of .05 when a pair of marginal frequencies are equal, as is the case here. Since our obtained value of b is 2, we are unable to reject the hypothesis on the basis of the obtained data.

8.6.7 Discussion

If we are willing to accept an inexactly defined point of population dichotomization (as may be necessary anyway if there are an even number of observations—see Section 8.6.2), we may include the medians in the test, i.e., we may use $<\hat{M}_X$ and $\geq\hat{M}_X$ as the column categories and $<\hat{M}_Y$ and $\geq\hat{M}_Y$ as the row categories. In that case, of course, we are testing a correspondingly altered hypothesis of independence between row and column category memberships where the population categories do not correspond exactly with the sample categories (see Section 8.3.7). The test would be an exact test of an inexactly defined association.

There is of course no compulsion to dichotomize at the medians, nor to dichotomize the X's in the same way as the Y's. By dichotomizing at the tenth X, X_{10}, and seventy-fifth Y, Y_{75}, in respective order of increasing size, in a sample of 100 units, we could test whether, in the population, the proportion of units whose $X < X_{10}$ is the same when $Y < Y_{75}$ as when $Y > Y_{75}$ (and likewise for units whose $X > X_{10}$).

8.7 WILKS' EMPTY CELL TEST FOR IDENTICAL POPULATIONS

8.7.1 Rationale

Suppose that a sample of n observations, X's, has been drawn from one continuously distributed population, that a sample of m observations, Y's, has been drawn from another, and that it is desired to test whether or not the two populations are identical. If they are, then X's and Y's should be well intermingled when the $n + m$ observations in the combined sample are arranged in order of increasing size; but if they are not, one would expect observations from the same original sample, e.g., X's, to tend to cluster together to a degree unlikely by chance alone. The n different X-values, arranged in order of increasing value, may be considered to divide the continuum from minus to plus infinity into $n + 1$ intervals or "cells." Let e be the number of these cells that contain no Y-observations, i.e., that are "empty." If the populations differ, this fact should be reflected in a tendency for like-sample observations to cluster which, if it occurs, will cause e to assume large values. On the other hand, if the populations are identical, X's and Y's are merely arbitrary labels for observations considered as drawn from a common continuous population or drawn, without replacements, from a homogeneous combined sample, in which case the distribution of e can be derived as follows.

The number of ways one can select from among the $n + 1$ cells created by the X's the e cells to be empty is $\binom{n+1}{e}$. If e cells are empty, then $n + 1 - e$ cells contain groups of one or more Y's. Imagine the m different Y's arranged in order of increasing value and imagine a vertical line or partition between each two adjacent Y's. The number of ways of separating the ordered series of m Y's into $n + 1 - e$ groups is the number of ways of choosing $n - e$, i.e., the number of groups minus 1, of the $m - 1$ partitioning lines between adjacent Y's and removing those not chosen, which is $\binom{m-1}{n-e}$. Thus the number of ways of choosing the e X-cells to be empty and the $n + 1 - e$ groups of Y's to occupy the nonempty cells is $\binom{n+1}{e}\binom{m-1}{n-e}$. This in fact is the number of ways of arranging n X's and m Y's in such a way as to obtain e empty X-cells. The number of ways of dividing the combined sample of $n + m$ ordered observations into n X's and m Y's without restriction as to arrangement is $\binom{n+m}{n}$. Therefore the proportion of all possible arrangements which result in e empty X-cells, and therefore the point probability of e, is

$$\frac{\binom{n+1}{e}\binom{m-1}{n-e}}{\binom{n+m}{n}}$$

And the cumulation of this probability over all possible values of e equal to or greater than the obtained value of e is the probability level of the test.

The derived point probability is hypergeometric. Furthermore, the data can be cast into a 2×2 table in which e is a cell entry and all marginal frequencies are fixed, so cumulative probabilities can be obtained in the same way as with Fisher's exact method. This is shown in the accompanying table. The "hypergeometric" relationship between the table frequencies is unequivocal. It is less immediately apparent that the two column categories form a mutually exclusive dichotomy, and still less obvious that each unit in the first row can be identified with an X-observation and that each unit in the second row can be related to an observation from the Y-sample. Nevertheless reflection will reveal that this is indeed the case. It is clear from the table that the

	Cells formed by X's	*Partitions between Y's*	*Sample size*
X-sample	e (empty)	$n - e$ ("used," i.e., replaced by groups of X's)	n
Y-sample	$n + 1 - e$ (occupied by groups of Y's)	$m - 1 - n + e$ ("removed," thus forming the Y-groups mentioned at left)	m
Total	$n + 1$	$m - 1$	$n + m$

fact that there must be at least one nonempty cell does not restrict the cumulation of e. The number of cells, $n + 1$, minus 1 leaves the row marginal frequency n, above which e could not be cumulated anyway. The test, of course, is one-tailed, and, for reasons given in the *Discussion* (Section 8.7.7), it must be decided prior to sampling which sample is to be the X-sample, i.e., the sample whose number of empty cells is to be the test statistic.

8.7.2 Null Hypothesis

The null hypothesis is, in effect, that the distribution of e is the same as if the label "X" had been randomly assigned to n of the $n + m$ observations in the combined sample, the remaining m observations receiving the label "Y." This will be the case if the X- and Y-populations are identical and all assumptions are met.

8.7.3 Assumptions

It is assumed that both sampled populations are *continuously distributed*, that sampling is *random* and that observations are *independent*. Tied observations should be dealt with according to Method A of Chapter 3.

8.7.4 Efficiency

This test differs only slightly from the Wald-Wolfowitz total-number-of-runs test, which is one of the least efficient distribution-free tests in comparison either with parametric tests or with other nonparametric tests. Presumably, then, its A.R.E. and relative power are generally very low. Its absolute power may be large, however, when both sample sizes are large, since it has been shown to be consistent under conditions given in [18]. Considerations by Wilks in [18] suggest that power is optimal when m/n equals 2. However it is undesirable for the m/n ratio to be extremely large, especially when n is not large, since in that case if the two populations cover the same range, *all* X-cells will tend to be occupied irrespective of whether or not the two populations are identical.

8.7.5 Tables

Table VIII in the Appendix, tables for Fisher's exact method, and hypergeometric tables are all appropriate. Tables specifically prepared for this test can be found in [6].

8.7.6 Example

It is desired to test at the .05 level the hypothesis that two continuously distributed populations are identical against the general alternative hypothesis that they are not. Prior to sampling one population is labeled "X," the other "Y." Then 10 observations are drawn from the X-population and 15 observations from the Y-population. These 25 observations are then arranged in order of increasing value irrespective of which sample they originally came from, but the observations from the X-sample are each enclosed within parentheses, yielding the following series. (5) 7 9 11 14 19 21 27 (28) (30) (36) 41 55 (67) (68) 75 (79) (81) (83) (85) 88 89 91 94 95. There are seven empty X-cells: the one to the left of the first X and the six between adjacent X's. We can cast these data into the form of a fourfold table, as shown here. We want the probability of seven or more empty X-cells, given the marginal

Spaces between, above, or below X's	*Spaces between Y's*	*Sample size*
Empty $= 7$	Occupied $= 3$ (By groups of X's)	X-Sample $= 10$
Occupied $= 4$ (By groups of Y's)	Empty $= 11$	Y-Sample $= 15$
Total $= 11$	Total $= 14$	Combined sample $= 25$

frequencies, which, with fixed marginals, is the same as the probability that the lower-right-cell frequency will be 11 or more. Entering Table VIII with $A = 15$, $B = 10$, and $a = 11$, we find this cumulative probability to be .042, since this is the cumulative probability listed for b (our upper-right-cell frequency, in this case) equal to 3. Since the test is intrinsically one-tailed, this probability is all we need, and we reject the hypothesis of identical populations at our predesignated .05 level of significance.

8.7.7 Discussion

An unbroken series of like objects is called a run. Clearly each of the $n + 1 - e$ groups of Y's which occupy a nonempty X-cell represents a run of Y's. So we know that there are exactly $n + 1 - e$ runs of Y's. Each of these groups of Y's is separated from its neighboring Y-groups by one or more X's, so there must be at least as many runs of X's as there are spaces between Y-groups, i.e., at least $n - e$. And there may or may not be an X-run below the lowest Y-group or above the highest Y-group, so the number of X-runs may be $n - e$, $n - e + 1$, or $n - e + 2$, depending upon whether both, one, or neither of the end X-cells are occupied by Y's. Thus, knowing the number of empty X-cells, we know that the total number of runs of X's and Y's is within ± 1 of $2(n + 1 - e)$. The empty cell test therefore uses very nearly the same information as does a test whose test statistic is the total number of runs, the Wald-Wolfowitz runs test, and the two tests are quite similar. The primary advantage of the empty cell test is the availability of fairly exhaustive tables, since its test statistic has a hypergeometric distribution, whereas that of the Wald-Wolfowitz test does not.

REFERENCES

1. Armsen, P., "Tables for Significance Tests of 2 × 2 Contingency Tables," *Biometrika*, **42** (1955), 494–511.

2. Barnard, G. A., "Significance Tests for 2 × 2 Tables," *Biometrika*, **34** (1947), 123–138.

3. Bennett, B. M., and P. Hsu, "On the Power Function of the Exact Test for the 2 × 2 Contingency Table," *Biometrika*, **47** (1960), 393–397.

4. Blomqvist, N., "On a Measure of Dependence between Two Random Variables," *Annals of Mathematical Statistics*, **21** (1950), 593–600.

5. Chakravarti, I. M., F. C. Leone, and J. D. Alanen, "Asymptotic Relative Efficiency of Mood's and Massey's Two Sample Tests against Some Parametric Alternatives," *Annals of Mathematical Statistics*, **33** (1962), 1375–1383.

6. Csorgo, M., and I. Guttman, "On the Empty Cell Test," *Technometrics*, **4** (1962), 235–247.

7. Elandt, Regina C., "Exact and Approximate Power Function of the Non-Parametric Test of Tendency," *Annals of Mathematical Statistics*, **33** (1962), 471–481.

8. Finney, D. J., R. Latscha, B. M. Bennett, and P. Hsu, *Tables for Testing Significance in a 2 × 2 Contingency Table*. London and New York: Cambridge University Press, 1963. (Extension of tables in *Biometrika*, **35** (1948), 145–156 and **40** (1953), 74–86.)

9. Fisher, R. A., *Statistical Methods for Research Workers*, 12th ed. New York: Hafner Publishing Co., 1954, pp. 96–97.

10. Harkness, W. L., and L. Katz, "Comparison of the Power Functions for the Test of Independence in 2 × 2 Contingency Tables," *Annals of Mathematical Statistics*, **35** (1964), 1115–1127.

11. Konijn, H. S., "On the Power of Certain Tests for Independence in Bivariate Populations," *Annals of Mathematical Statistics*, **27** (1956), 300–323.

12. Lieberman, G. J., and D. B. Owen, *Tables of the Hypergeometric Probability Distribution*. Stanford, Calif.: Stanford University Press, 1961.

13. Mood, A. M., *Introduction to the Theory of Statistics*. New York: McGraw-Hill Book Co., 1950, pp. 394–395.

14. Owen, D. B., *Handbook of Statistical Tables*. Reading, Mass.: Addison-Wesley Publishing Co., Inc., 1962, pp. 458–508.

15. Pearson, E. S., and Maxine Merrington, "2 × 2 Tables; The Power Function of the Test on a Randomized Experiment," *Biometrika*, **35** (1948), 331–345.

16. Pratt, J. W., "Robustness of Some Procedures for the Two-Sample Location Problem," *Journal of the American Statistical Association*, **59** (1964), 665–680.

17. Westenberg, J., "Significance Test for Median and Interquartile Range in Samples from Continuous Populations of Any Form," *Proceedings Koninklijke Nederlandse Akademie van Wetenschappen*, **51** (1948), 252–261.

18. Wilks, S. S., "A Combinatorial Test for the Problem of Two Samples from Continuous Distributions," in *Proceedings of the Fourth Berkeley Symposium on Mathematical Statistics and Probability*, ed. Jerzy Neyman. Berkeley and Los Angeles: University of California Press, 1961, Vol. I, pp. 707–717.

19. Yates, F., "Contingency Tables Involving Small Numbers and the χ^2 Test," *Journal of the Royal Statistical Society* (*Series B*), **1** (1934), 217–235.

9

TESTS BASED UPON THE MULTIVARIATE HYPERGEOMETRIC DISTRIBUTION

The previous chapter concerned tests whose point and cumulative probabilities were those of the univariate hypergeometric distribution. The point probabilities of the tests discussed in the present chapter are those of a multivariate hypergeometric distribution. But in every case the cumulative probabilities can be expressed as cumulative probabilities of some univariate hypergeometric distribution, so tables for the latter can be used to determine probability levels for tests in the present chapter.

9.1 THE k-VARIATE HYPERGEOMETRIC DISTRIBUTION

Suppose that a population consisting of N units can be divided into $k + 1$ mutually exclusive categories, so that each unit belongs to exactly one category, and let R_i be the number of population units belonging to the ith category, so that $\sum_{i=1}^{k+1} R_i = N$. Now suppose that a random sample of n units is drawn, without individual replacements, from the population, and that the number of sample units falling in the ith category is r_i, as depicted in the accompanying table.

	Category							
	1	2		i		k	$k+1$	*All*
Sample withdrawn	r_1	r_2	...	r_i	...	r_k	$r_{k+1} = n - \sum_{i=1}^{k} r_i$	n
Units remaining	$R_1 - r_1$	$R_2 - r_2$	...	$R_i - r_i$	...	$R_k - r_k$	$R_{k+1} - r_{k+1}$	$N - n$
Population	R_1	R_2	...	R_i	...	R_k	$R_{k+1} = N - \sum_{i=1}^{k} R_i$	N

There are $\binom{R_1}{r_1}$ ways of drawing the r_1 sample units from the R_1 population units in Category 1; for each such way, there are $\binom{R_2}{r_2}$ ways of drawing the sample units in Category 2; and for each of those ways there are $\binom{R_i}{r_i}$ ways of obtaining the sample units in the ith category, etc., etc.; whereas there are $\binom{N}{n}$ ways of obtaining the sample without the above restrictions as to category membership. Thus the proportion of unrestricted ways which, however, happen to meet the restricting conditions is

$$\frac{\binom{R_1}{r_1}\binom{R_2}{r_2}\cdots\binom{R_i}{r_i}\cdots\binom{R_k}{r_k}\binom{R_{k+1}}{r_{k+1}}}{\binom{N}{n}} = \frac{\prod_{i=1}^{k+1}\binom{R_i}{r_i}}{\binom{N}{n}}$$

and this is therefore the point probability of the obtained sample, i.e., of the obtained set of category "frequencies." Since, if k of the $k + 1$ cell frequencies are designated, the remaining frequency can be obtained by subtraction of their sum from n, one may regard only k of the r_i as variables. Clearly, therefore, the above probability is a generalization to the k-variate case of the univariate hypergeometric probability derived at the beginning of the previous chapter.

9.2 EXCEEDANCES: BASIC FORMULAS

The basic exceedance problem is as follows. Two random and independent samples are to be drawn from the same continuously distributed population. The first sample is to consist of n observations, designated X's, and the second sample is to contain m observations, designated Y's. What is the probability that e or more of the Y's will exceed (or, equivalently, that $b = m - e$ or fewer of the Y's will be less than) X_r, the rth observation in order of increasing value in the first sample?

Since the two samples are defined to be from the same population, the first sample can be regarded as a random sample of n observations drawn, without individual replacements, from the $n + m$ observations comprising the two samples, the second sample being the m observations remaining. Consider both samples to have been drawn and the $n + m$ observations in the two samples to have been arranged in order of increasing value, irrespective of sample, and labeled Z's, with subscripts indicating rank:

$$Z_1, Z_2, \ldots, Z_{r-1+b}, Z_{r+b}, \ldots, Z_{n+m-1}, Z_{n+m}$$

Consider now the probability of drawing an "X-sample" of n observations from these Z's so as to leave a remaining "Y-sample" of m observations, exactly b of which are less than X_r and $e = m - b$ of which exceed it. If exactly b of the Y's are less than X_r, then the $r + b$th observation in the combined sample must be X_r, i.e., Z_{r+b} must be X_r. Therefore, in order to obtain the required "X-sample": we must draw $r - 1$ of the $r - 1 + b$ Z's below Z_{r+b}, which become the $r - 1$ X's less than X_r; we must draw Z_{r+b}, which becomes X_r; and we must draw $n - r$ of the $(n + m) - (r + b)$ Z's above Z_{r+b}, which become the $n - r$ X's above X_r. There are $\binom{r-1+b}{r-1}$ ways of meeting the first condition, $\binom{1}{1}$ ways of meeting the second, $\binom{n+m-r-b}{n-r}$ ways of meeting the third, and $\binom{n+m}{n}$ ways of drawing the X-sample without these restrictions. So

$$P(\text{exactly } b \text{ of the } Y\text{'s will be} < X_r) = \frac{\binom{r-1+b}{r-1}\binom{1}{1}\binom{n+m-r-b}{n-r}}{\binom{n+m}{n}}$$

which is a point probability of a bivariate hypergeometric distribution. The sampling situation is depicted in the accompanying table. The point prob-

	$< Z_{r+b}$ $(< X_r)$	$= Z_{r+b}$ $(= X_r)$	$> Z_{r+b}$ $(> X_r)$	
X-sample	$r - 1$	1	$n - r$	n
Y-sample	b		$m - b$	m
Z-sample	$r - 1 + b$	1	$n + m - r - b$	$n + m$

ability obtained above can be expressed as a constant times a univariate hypergeometric probability, the latter appearing within square brackets:

$$P\,(\text{exactly } b \text{ of the } Y\text{'s will be} < X_r) = \frac{n}{n+m}\left[\frac{\binom{r-1+b}{r-1}\binom{n+m-r-b}{n-r}}{\binom{n+m-1}{n-1}}\right]$$

So although point probabilities are *not* univariate hypergeometric point probabilities, they can be obtained from tables of the latter by a simple multiplication.

The probability we seek, however, is the probability of b or fewer Y's less than X_r, and this involves summing the point probabilities of tables with different column marginal frequencies but the same first-row cell frequencies. In other words, we must cumulate point probabilities of *different* bivariate hypergeometric distributions, since a single bivariate hypergeometric distribution corresponds to a table with fixed marginal frequencies. Fortunately, however, we do not have to take this approach, as the whole problem can be reduced to that of obtaining the cumulative probability for a fourfold table with fixed marginal frequencies: Let i be the exact number of Y's with values below X_r. Then Z_{r+i} is X_r, and exactly $n - r$ of the X's lie above it. For any value of i less than or equal to b, the space above Z_{r+b} is a subset of the space above Z_{r+i}, so no more than $n - r$ of the X's can lie above Z_{r+b}. For any value of i greater than b, Z_{r+i}, which is X_r, must lie above Z_{r+b}, so at least $n - r + 1$ of the X's (the $n - r$ X's above X_r plus X_r itself) must lie above Z_{r+b}. Thus, letting f stand for the number of X's that lie above Z_{r+b}, we have shown that

$$f \leqq n - r \text{ whenever } i \leqq b$$
$$f > n - r \text{ whenever } i > b$$

i.e., we have established a one-to-one relationship between the (respectively, mutually exclusive, and exhaustive) categories of two dichotomies encompassing the same total set of events. It follows therefore that

$$P(i \leqq b) = P(f \leqq n - r)$$

But the latter is simply the probability of the accompanying fourfold table cumulated over all positive integral values of Δ from zero up to the value that first causes a cell frequency to become zero. Thus we have shown that the

	$\leqq Z_{r+b}$	$> Z_{r+b}$	
X-sample	$r + \Delta$	$n - r - \Delta$	n
Y-sample	$b - \Delta$	$m - b + \Delta$	m
Z-sample	$r + b$	$n + m - r - b$	$n + m$

problem of finding the cumulative probability, over all values of $i \leqq b$, for the first table below reduces to the same problem (over realistic values of $i \leqq b$) for the second table below.

	$< X_r$	$= X_r$	$> X_r$	
X-Sample	$r-1$	1	$n-r$	n
Y-Sample	i	0	$m-i$	m
Z-Sample	$r-1+i$	1	$n+m-r-i$	$n+m$

	$\leqq Z_{r+b}$	$> Z_{r+b}$	
X-Sample	$r+b-i$	$n-r-b+i$	n
Y-Sample	i	$m-i$	m
Z-Sample	$r+b$	$n+m-r-b$	$n+m$

That is, we have shown that

$$P\,(b \text{ or fewer } Y\text{'s} < X_r) = \sum_{i=0}^{b} \frac{\binom{r-1+i}{i}\binom{1}{0}\binom{n+m-r-i}{m-i}}{\binom{n+m}{m}}$$

$$= \sum_{i \leqq b} \frac{\binom{r+b}{i}\binom{n+m-r-b}{m-i}}{\binom{n+m}{m}}$$

which we shall call E, since it is also $P\,(e \text{ or more } Y\text{'s} > X_r)$. And the latter summation can be obtained directly from tables of cumulative probabilities for the univariate hypergeometric distribution. (Note that when $i = b$, i.e., for the actually obtained set of data, the frequencies in the "derived" fourfold table are simply those of the original table with its first two columns combined into a single column representing frequencies $\leqq X_r$.)

By making use of the following relationships the probability for any one of four possible cases of interest may be obtained:

$$E = P\,(b \text{ or fewer } Y\text{'s} < X_r) = P\,(e \text{ or more } Y\text{'s} > X_r)$$
$$= 1 - P\,(b+1 \text{ or more } Y\text{'s} < X_r) = 1 - P\,(e-1 \text{ or fewer } Y\text{'s} > X_r)$$

Thus $P\,(b+1 \text{ or more } Y\text{'s} < X_r)$ is obtained by taking the complement, $1 - E$, of the last summation given, i.e., by summing over values of $i \geqq b+1$. Or, if, to avoid possible confusion, we let $b' = b+1$ and $e' = e-1$, we have

$$P\,(b' \text{ or more } Y\text{'s} < X_r) = P\,(e' \text{ or fewer } Y\text{'s} > X_r)$$

$$= \sum_{i \geqq b'} \frac{\binom{r+b'-1}{i}\binom{n+m-r-b'+1}{m-i}}{\binom{n+m}{m}}$$

which, of course, is also the probability of a fourfold table with fixed marginals, although the table and the direction of summation differ from the case previously treated.

Two particularly interesting special cases occur when X_r is an extreme value of the X-sample. When $r = 1$, the summation in the last equation above reduces to a single and very simple term:

$$P(b' \text{ or more } Y\text{'s} < X_1) = \sum_{i \geq b'} \frac{\binom{b'}{i}\binom{n+m-b'}{m-i}}{\binom{n+m}{m}} = \frac{\binom{n+m-b'}{n}}{\binom{n+m}{n}}$$

And when $r = n$, an analogous simplification occurs:

$$P(b \text{ or fewer } Y\text{'s} < X_n) = \sum_{i \leq b} \frac{\binom{n+b}{i}\binom{m-b}{m-i}}{\binom{n+m}{m}}$$

$$= \sum_{i \leq b} \frac{\binom{n+b}{n+b-i}\binom{m-b}{i-b}}{\binom{n+m}{n}} = \frac{\binom{n+b}{n}}{\binom{n+m}{n}}$$

9.3 EXCEEDANCES: TESTS OF HYPOTHESES

9.3.1 RATIONALE

Since E is the a priori probability that b or fewer of the Y's will fall below X_r, provided that the two samples are drawn randomly and independently from the same continuously distributed population, it can also serve as the probability level for a test upon actually obtained data which takes one of the qualifying conditions as the null hypothesis, specifically the condition of identical populations. The latter, however, is a sufficient condition which becomes decreasingly critical as n increases. Rather than identical populations, it is better to think of the test as testing the hypothesis that equal proportions of the two populations lie below whatever value becomes X_r and that equal proportions of the two populations are concentrated at the point X_r. That is, letting $F_V(A)$ and $f_V(A)$ stand for ordinates in the cumulative distribution and in the density function, respectively, of the variate V at the abscissa A, we take H_0 to be that $F_X(X_r) = F_Y(X_r) = p$ and that $f_X(X_r) = f_Y(X_r)$. The following alternative derivation shows that this is indeed what the test is actually testing.

If the two populations have identical ordinates (i.e., densities) at the point Z_{r+b} (as, in effect, hypothesized, since Z_{r+b} will become X_r), the probability that the obtained Z_{r+b} will be an X is simply $n/(n + m)$, the proportion of

X's in the combined sample. If equal proportions p of the two populations lie below the obtained value of Z_{r+b}, then (in analogy with the derivation in Section 8.2.7) the conditional probability that $r - 1$ of the $n - 1$ remaining X's and b of the m Y's will fall below Z_{r+b} (thereby causing Z_{r+b} to become X_r) given that $r - 1 + b$ observations from both samples lie below Z_{r+b} (as implied by its subscripts) is

$$\frac{\left[\binom{n-1}{r-1}p^{r-1}(1-p)^{n-r}\right]\left[\binom{m}{b}p^{b}(1-p)^{m-b}\right]}{\binom{n-1+m}{r-1+b}p^{r-1+b}(1-p)^{n+m-r-b}} = \frac{\binom{n-1}{r-1}\binom{m}{b}}{\binom{n-1+m}{r-1+b}}$$

Therefore, the probability that Z_{r+b} will be an X and that $r - 1$ X's and b Y's will fall below it, under the respective hypothesized conditions, is the product of the two probabilities

$$\frac{n}{n+m}\frac{\binom{n-1}{r-1}\binom{m}{b}}{\binom{n-1+m}{r-1+b}} = \frac{\binom{r-1+b}{r-1}\binom{n+m-r-b}{n-r}}{\binom{n+m}{n}}$$

and the latter expression is the point probability, P (exactly b of the Y's will be $< X_r$), derived in Section 9.2.

Since the populations are assumed to be infinite and since the actual value of X_r is determined by chance, it is *possible* for X_r to be any value in the X-population. Thus to the extent that X_r is a variable which is free to assume all X-values, i.e., prior to sampling, a test of whether or not $F_X(X_r) = F_Y(X_r)$ and $f_X(X_r) = f_Y(X_r)$ is a test for identical populations. However, whereas X_r is relatively likely to be any X-value when n is extremely small—for example when n equals 1, 2, or 3—as n increases toward infinity, the variance [see 16] of $F_X(X_r)$ about an expected value of $r/(n + 1)$ diminishes toward zero for either constant r or constant r/n. And this tends to imply that the variance of X_r about its expected value, i.e., about whatever population quantile it estimates at that value of n, diminishes accordingly. Therefore, when n is of moderate size or larger, the test, prior to sampling, will be sensitive primarily to nonidentity of F_X and F_Y (and f_X and f_Y) in the vicinity of $F_X = r/(n + 1)$. After sampling, it is sensitive exclusively to nonidentity of F_X and F_Y (and f_X and f_Y) at the common abscissa value X_r.

Exceedance tests have considerable versatility. By proper selection of r one can tailor the test in such a way that it will probably be sensitive to certain specified portions of the cumulative distributions of X and Y. (The expected proportion of the X-population lying below X_r is $r/(n + 1)$ [see 16]. Thus by letting the X-sample be a control sample with $n = 2r - 1$, so that X_r is the sample median, one has a test of whether or not the same proportions of treatment and control populations lie below (and lie at) the median of the control sample, i.e., below (and at) an estimate of the control population median. By letting $r = n$, one has the classic exceedance test of whether or not the same proportions of the X- and Y- populations lie below (and at) the upper extremity of the X-sample; by letting $r = 1$, one has the analogous test with reference to the lower extremity.

9.3.2 Null Hypothesis

H_0 is that $F_X(X_r) = F_Y(X_r)$ and $f_X(X_r) = f_Y(X_r)$, where prior to sampling $F_X(X_r)$ is a variable whose expected value is $r/(n + 1)$ and whose variance

$$\frac{r(n - r + 1)}{(n + 1)^2(n + 2)}$$

about that expected value diminishes rapidly with increasing n. At reasonably large n's, $F_X(X_r)$ is exceedingly unlikely to depart very much from its expected value, so the test will be almost exclusively sensitive to alternatives of non-identity between F_X and F_Y or between f_X and f_Y in the region where $F_X \cong r/(n + 1)$. After sampling, of course, X_r is a constant and the test will reject (more than α of the time) only if unequal proportions of the two populations lie below, or if unequal proportions lie at, the obtained value of X_r.

9.3.3 Assumptions

(a) In order to avoid bias, the question of which sample is to be the "X-sample" and the values of r, n, m, and α must all be determined *prior* to viewing the data, and preferably in advance of sampling.
(b) Sampling is *random*.
(c) Observations are *independent*.
(d) The sampled populations are *infinite*, or each observation is replaced before the next observation is drawn. Otherwise the proportion of each population lying below X_r (hypothesized to be equal) varies during, and is probably altered by, sampling.
(e) X_r can be tied with no other observation in either sample. This will be the case if both populations are *continuously distributed*.

If other observations, either X's or Y's, are tied with X_r due to imprecision of measurement upon a continuously distributed variate (so that the assumption of an infinite population is met) the following inequality can be used:

$$P\,(b \text{ or fewer } Y\text{'s} < X_r) \leqq E \leqq P\,(b \text{ or fewer } Y\text{'s} \leqq X_r)$$

for reasons analogous to those given in Section 7.8. (Or, see reference [14]; cf. [6].)

9.3.4 Efficiency

If r is the middle of n integers so that X_r is the median of the X-sample, and if the sampled populations have normal distributions and equal variances, the test has A.R.E. of $2/\pi$ or .637 relative to Student's t test as a test of equal population means under the condition of infinite sample sizes [3]. Thus when X_r is the median of the X-sample, the test has the same A.R.E. as the Westenberg-Mood median test which is based upon the median of the two samples pooled together. The above statements apply to the case where all m of the Y-observations are actually drawn. However, if the observations become available in order of increasing value as is often the case in life testing, and if one is testing a one-sided hypothesis which is to be rejected if "too many" Y's fall on a specified side of X_r, it is only necessary to continue the "X-experiment" long enough to identify X_r and the "Y-experiment" long enough to meet the rejection criterion (or until m observations are recorded, whichever happens first). To the extent that observations "initiated" but not "consummated" can be regarded as observations not "drawn," sample size is reduced by the number of observations eliminated, and there is a consequent increase in efficiency. This increase may be spectacular, e.g., from .637 to .966 [see 3]. This and the resulting savings in time and materials constitute a particularly attractive feature of the test. If X_r is the median of the X-sample, efficiency is optimal when $n = m$ [see 3].

If used as a test for identical populations against general alternatives, the test is not consistent. The two populations can assume widely differing forms and, so long as they have equal amounts of area below, and at, X_r the test will not reject more than α of the time.

However, the test is consistent against slippage alternatives, i.e., as a test of whether two populations identical in shape differ only in location (as would be the case if the "second" population were the first population after it had slipped a certain distance along the abscissa axis). And, of course the test is unbiased and consistent as a test of whether or not equal areas of the two populations lie below, and lie at, an abscissa of X_r.

9.3.5 TABLES

Since cumulative probabilities are equivalent to those of a fourfold table with fixed marginal frequencies (see Section 9.2), any tables appropriate to Fisher's exact method can be used. This includes Table VIII and Lieberman and Owen's tables of the univariate hypergeometric cumulative distribution, referenced in the previous chapter. A number of tables have been especially prepared [2, 11, 13], and these include tables for the probability of e or more Y's above X_n [13] and for b' *or more* Y's below X_1 [11], mathematically equivalent special cases of particular interest. The latter case maximizes economy in life testing (when the test begins at the same moment for all units in a given sample), since it requires the minimum number of units to fail in order for the null hypothesis to be rejected.

9.3.6 Example

A manufacturer of television picture tubes is considering the adoption of a new manufacturing process. He is convinced that customers are strongly and unfavorably impressed by tubes that expire too quickly but that they fail to take note of the fact when a tube lasts inordinately long; also, when a tube expires too quickly he must replace it under his guarantee. He decides to take 10 tubes manufactured under the old process and 15 manufactured under the new process, place them under simultaneous life test, and abandon consideration of the new process if, at the .05 level, significantly many new process (Y) tubes expire before the first old process (X) tube expires. Thus $n = 10$, $m = 15$, $r = 1$, and the null hypothesis (that the proportion of tubes that would expire before X_1 is the same in both populations) will be rejected if b' is such that

$$P(b' \text{ or more } Y\text{'s} < X_r) = \sum_{i \geqq b'} \frac{\binom{r+b'-1}{i}\binom{n+m-r-b'+1}{m-i}}{\binom{n+m}{m}}$$

$$= \sum_{i \geqq b'} \frac{\binom{b'}{i}\binom{25-b'}{15-i}}{\binom{25}{15}} \leqq .05$$

The summation is simply the cumulative probability of the accompanying 2×2 table and all tables with the same marginal frequencies but larger frequencies in the lower-left cell (of which there are none). Entering Table

0	10	10
b'	$15 - b'$	15
b'	$25 - b'$	25

VIII in the Appendix with $A = 15$, $B = 10$, it is found that the smallest value b' can have (if the upper-left cell frequency is 0) for $\alpha \leqq .05$ is 6 (at which value $\alpha = .028$). The 10 X's and 15 Y's are now placed simultaneously on life test. The first six tubes to expire are all new process tubes, i.e., Y's, so testing is terminated, H_0 is rejected, and consideration of the new process is abandoned. Notice that it was not even necessary to learn the value of X_1. Nelson [11] has called this highly economical type of test a precedence test.

9.3.7 Discussion

The event, "b or fewer Y's fall below X_r," and the event, "b or fewer Y's fall above X_r," are mutually exclusive so long as $b < m/2$. Therefore, their probabilities can be added to yield the probability for the event, "b or fewer Y's fall on one side of X_r," i.e., to produce the probability level for a two-sided test of the hypothesis that equal proportions of the X- and Y-populations lie below, and lie at, X_r.

If any Y's at all fall below X_1, then no X's have fallen below Y_1, and vice versa. Therefore if c' and d' are both integers greater than zero, the events, "c' or more Y's fall below X_1" and "d' or more X's fall below Y_1" are mutually exclusive and their probabilities can be added to yield the probability level for a two-tailed test which does not require that the "reference" sample be specified in advance. But a test of what? The conditional nature of the probabilities involved makes it difficult to pinpoint the absolutely necessary hypothesis that is actually being tested. What is certain, however, is that the obtained probability levels will be appropriate to the sufficient null hypothesis of identical populations. But the conventional exceedance test is inconsistent as a test for identical populations unless identical shapes can be safely assumed. Therefore if the reference sample cannot be specified in advance, it would seem wise to use the test only as a test for "slippage," i.e., unequal locations for populations of identical shape.

We may generalize the above modification. Let c and d be integers such that $c + 1 \leqq r$ and $d + 1 \leqq r$.

If c or fewer Y's fall below X_r, then $X_r < Y_{c+1} \leqq Y_r$ so that $X_r < Y_r$

If d or fewer X's fall below Y_r, then $Y_r < X_{d+1} \leqq X_r$ so that $Y_r < X_r$

Obviously the two events are mutually exclusive. Let α_c and α_d be their respective probabilities, and let $\alpha = \alpha_c + \alpha_d$. Then a two-tailed test at the α level of significance is performed as follows: if $X_r < Y_r$, reject if the number of Y's $< X_r$ is $\leqq c$; if $Y_r < X_r$, reject if the number of X's $< Y_r$ is $\leqq d$. For a test sensitive to too *many* observations in one sample below the rth in the other, let c' and d' be integers such that $r \leqq c'$ and $r \leqq d'$.

If c' or more Y's fall below X_r, then $Y_r \leqq Y_{c'} < X_r$, so that $Y_r < X_r$

If d' or more X's fall below Y_r, then $X_r \leqq X_{d'} < Y_r$, so that $X_r < Y_r$

Again the two events are mutually exclusive. Let their respective probabilities be $\alpha_{c'}$ and $\alpha_{d'}$ whose sum is α. Then to perform a two-tailed test at the α level: if $Y_r < X_r$, reject if the number of Y's below X_r is $\geqq c'$; or, if $X_r < Y_r$, reject if the number of X's below Y_r is $\geqq d'$.

9.4 EXCEEDANCES: PREDICTION

In Section 9.2 it was shown that if $n + m$ observations are to be drawn at random from the same continuously distributed population, the first n observations in temporal sequence being called X's, the last m, Y's, the a priori probability that b or fewer of the Y's will be smaller than X_r is

$$E = P\,(b \text{ or fewer } Y\text{'s} < X_r) = \sum_{i \leq b} \frac{\binom{r+b}{i}\binom{n+m-r-b}{m-i}}{\binom{n+m}{m}}$$

Thus if an infinite number of pairs of X- and Y-samples, of sizes n and m respectively, were drawn from a common continuously distributed population, the *proportion of pairs* for which it is true that b or fewer Y's are $<X_r$ (where r is a fixed, predesignated, integer) would be E. So if, after drawing each X-sample but before drawing the accompanying Y-sample, we had predicted that b or fewer observations in the latter would be less than the obtained value of X_r in the former, the proportion of predictions that later proved to be correct would be exactly E.

Therefore, in a single, particular case, if we have an already obtained sample of size n (and the integer r was selected prior to viewing the obtained data) we may predict that b or fewer observations in a single future sample of size m will be less than the magnitude of the rth observation in order of increasing size in the already obtained sample, and we may have *confidence* of *level* E in the validity of our prediction.

It is important to realize that while E is the probability of b or fewer Y's $< X_r$ before either sample is drawn (i.e., while X_r is a variable), it is not the *probability*, but only the confidence level for the statement, that b or fewer Y's in a future Y-sample will fall below the constant value of X_r in an already obtained X-sample. Consider again the infinite number of pairs of X- and Y-samples drawn from the common, continuous population, and let v be the value of X_r and u be the number of Y's in the paired Y-sample falling below v. Then u and v are variables having a bivariate distribution, a proportion E of which corresponds to values $u \leqq b$. Only if u and v are independent will $P(u \leqq b)$ always be the same for v fixed as for v variable. But u and v are not independent. Let $F(v)$ be the cumulative distribution of the common population up to the value v, so that $F(v)$ is the proportion of the sampled population lying below v. Then the probability that $u \leqq b$ is simply the cumulative binomial probability

$$\sum_{i=0}^{b} \binom{m}{i}[F(v)]^i[1 - F(v)]^{m-i}$$

which varies with $F(v)$, which in turn varies with v. (The larger the value of v, the greater the proportion of the population lying below it, $F(v)$, and the larger the expected value of the number of randomly drawn Y's which will come from that region.) Since the population was assumed to be continuously distributed, it is possible for X_r to be any value in it and consequently it is possible, after X_r is obtained, for $P(b$ or fewer Y's $< X_r)$ in a still undrawn Y-sample to be anything from zero to 1.00, depending upon the obtained value of X_r. The critical fact which explains everything is that the probability E concerns an event determined by *both* members of a *pair* of samples, and cannot be regarded as applying to just one member either before or after the other member is drawn. By invoking the concept of confidence level we are, in fact, changing our frame of reference from a single obtained X-sample and future Y-sample to an infinite number of "obtained" X-samples and future Y-samples, which, in effect, treats X_r as a variable. In this context E is simply what it has always been, an a priori probability for an event involving two samples and therefore the proportion of times the event occurs in an infinite number of trials.

If we let m become infinite while the ratio b/m remains constant, say equal to p, then the future sample becomes, in effect, the entire population. So the allegation that b or fewer Y's will be $<X_r$ is equivalent to the statement that the proportion of the population below X_r is less than or equal to p, and therefore to the statement that $X_r \leqq Q_p$ (where Q_p is the 100pth population percentile, or pth population quantile). Thus, in the case of an infinite future sample, the problem becomes one of attaching confidence levels to statements about population quantiles, and this problem has already been solved in Section 7.8. Using the formula given in that section, and letting $F(X_r)$, as usual, stand for the proportion of the population $\leqq X_r$, we obtain the relationships shown immediately below. These confidence levels can be obtained

Statement	*Equivalent statement*	*Confidence level is* ***at least***
$F(X_r) \leqq p$	$X_r \leqq Q_p$	$\sum_{i=r}^{n} \binom{n}{i} p^i(1-p)^{n-i}$
$1 - F(X_r) \geqq 1 - p$	$X_r \leqq Q_p$	$\sum_{i=r}^{n} \binom{n}{i} p^i(1-p)^{n-i}$
$F(X_r) \geqq p$	$X_r \geqq Q_p$	$\sum_{i=0}^{r-1} \binom{n}{i} p^i(1-p)^{n-i}$
$1 - F(X_r) \leqq 1 - p$	$X_r \geqq Q_p$	$\sum_{i=0}^{r-1} \binom{n}{i} p^i(1-p)^{n-i}$

from tables of cumulative binomial probabilities or from tables [10, 15] especially prepared for this particular application.

If we view X_r as a boundary in a probability statement about the location of the *parameter p* of a binomial distribution (i.e., about the proportion of the population lying to one side of X_r), then X_r is a *confidence* limit. If we view it as a boundary in a probability statement about the probable location of the *variate X*, then X_r is a *tolerance* limit. The probability statements already discussed in this section carry both implications and one can speak of either confidence limits and confidence levels or of tolerance limits and levels, depending upon whether one's emphasis is upon p or upon X, i.e., upon the population or a randomly selected unit from it. In the context of exceedances, convention favors the terms tolerance limit and tolerance level. However, in order to maintain consistency with other sections of this book, the "confidence" terminology will be used.

Sometimes the experimenter may wish to know the minimum number of future observations required in order that he may have confidence at level C that *at least* b of the Y's will be $<X_r$. The *minimum* sample size for the number of Y's $< X_r$ to become *at least* b is, of course, the minimum sample size for the number to become exactly b. Now if m is the minimum number of Y's that must be drawn to attain the criterion of b Y's $< X_r$, then the criterion must have been attained on the mth draw, i.e., the first $m - 1$ draws of Y's result in exactly $b - 1$ observations $<X_r$ and the mth draw results in one more $Y < X_r$. Thus we may regard C as the probability that if $n + m$ observations, Z's, are drawn in sequence from a common, continuous population: of the first n obesrvations drawn, $r - 1$ will be less than Z_{r+b}, one will be Z_{r+b}, and $n - r$ will be greater than Z_{r+b}; of the next $m - 1$ observations, $b - 1$ will fall below, and $(m - 1) - (b - 1)$ will fall above, Z_{r+b}; and the final observation will fall below Z_{r+b}. The sampling situation is depicted in the accompanying table.

	$<Z_{r+b}$	Z_{r+b}	$>Z_{r+b}$	*Total*
X's	$r - 1$	1	$n - r$	n
First $m - 1$ *Y's*	$b - 1$	0	$m - b$	$m - 1$
Last Y	1	0	0	1
Combined sample	$r + b - 1$	1	$n + m - r - b$	$n + m$

The probability of the table is

$$C = \frac{\dfrac{(r + b - 1)!}{(r - 1)!\,(b - 1)!\,1!}\;\dfrac{1!}{1!\,0!\,0!}\;\dfrac{(n + m - r - b)!}{(n - r)!\,(m - b)!\,0!}}{\dfrac{(n + m)!}{n!\,(m - 1)!\,1!}}$$

where the denominator is the number of ways of obtaining the row margins from the total frequency and the numerator is the number of ways of obtaining the cell frequencies under the restrictions imposed by both row and column margins. (Each of the three terms in the numerator is, for a given column, the number of ways of obtaining the column cell frequencies from the column total.) Thus the minimum sample size at which one may have a confidence level of C that at least b of the Y's will be $<X_r$ is obtained by solving the above equation for m. It will, of course, usually be much more convenient simply to make successive guesses at m, substitute them, and solve for C until an acceptable confidence level is obtained.

9.5 INCLUDANCES: BASIC FORMULAS

Consider again the sampling situation in which $n + m$ observations are to be randomly drawn from the same continuously distributed population, the first n observations in drawing sequence to be called X's (with subscripts indicating rank in order of increasing value), the last m observations to be called Y's. The probability that e or more of the Y's would exceed X_r (or, equivalently, that $b = m - e$ or fewer would fall below it) was derived in the section on Exceedances (see Section 9.2). Instead of the above probability, however, the experimenter may desire the probability that e or more of the Y's will be enclosed, i.e., included, within the interval between X_{r_1} and X_{r_2}, where $r_1 < r_2$ (or, equivalently that $b = m - e$ or fewer of the Y's will fall either below X_{r_1} or beyond X_{r_2}, i.e., will lie outside of the interval). If we let $r = n - (r_2 - 1 - r_1)$, it turns out that the two cumulative probabilities are the same, i.e., the probability that e or more Y's will be included in the interval between X_{r_1} and X_{r_2} is the same as the probability that e or more of the Y's will exceed $X_{n-r_2+1+r_1}$. Thus not only can cumulative probabilities for includances be equated to those for some case involving exceedances, but the includance probability is seen to depend upon the numerical *difference* between r_2 and r_1 rather than upon their individual values.

Consider both samples to be drawn and let the $n + m$ observations in the combined sample be labeled Z's, with subscripts indicating rank in order of increasing value. Then the a priori probability that of the m Y's, exactly j will be less than X_{r_1}, exactly e will lie between X_{r_1} and X_{r_2}, and $m - e - j$ will lie above X_{r_2} is the probability of drawing exactly j of the $r_1 - 1 + j$ Z's below Z_{r_1+j}, not drawing Z_{r_1+j} (which must become X_{r_1}), drawing e of the $r_2 - 1 - r_1 + e$ Z's between Z_{r_1+j} and Z_{r_2+j+e}, not drawing Z_{r_2+j+e} (which must become X_{r_2}), and drawing $m - j - e$ of the $n + m - r_2 - j - e$ Z's above Z_{r_2+j+e}, when a sample of m observations is drawn randomly and without individual replacements from the combined sample. The sampling situation is depicted in the accompanying table. Obviously e Y's can lie be-

	$<X_{r_1}$	$=X_{r_1}$	$X_{r_1} < Z < X_{r_2}$	$=X_{r_2}$	$>X_{r_2}$	
X-sample	$r_1 - 1$	1	$r_2 - 1 - r_1$	1	$n - r_2$	n
Y-sample	j	0	e	0	$m - j - e$	m
Z-sample	$r_1 - 1 + j$	1	$r_2 - 1 - r_1 + e$	1	$n + m - r_2 - j - e$	$n + m$

tween X_{r_1} and X_{r_2} while j assumes any one of the values from zero to $m - e$. Therefore the *point* probability of exactly e Y's between X_{r_1} and X_{r_2} is

$$P(e) = \sum_{j=0}^{m-e} \frac{\binom{r_1 - 1 + j}{j}\binom{1}{0}\binom{r_2 - 1 - r_1 + e}{e}\binom{1}{0}\binom{n + m - r_2 - j - e}{m - j - e}}{\binom{n+m}{m}}$$

The terms containing no j's can be brought outside the summation, and we may multiply them by a constant and divide the terms inside the summation by the same constant.

$$P(e) = \frac{\binom{r_2 - 1 - r_1 + e}{e}\binom{1}{0}\binom{n + m - r_2 - e + r_1}{m - e}}{\binom{n+m}{m}} \cdot \sum_{j=0}^{m-e} \frac{\binom{r_1 - 1 + j}{j}\binom{1}{0}\binom{n + m - r_2 - j - e}{m - j - e}}{\binom{n + m - r_2 - e + r_1}{m - e}}$$

Note that inside the summation the sum of all "upper-level" values in the numerator equals the "upper-level" expression in the denominator, and likewise for the "lower-level" values; so the fraction is a multivariate hypergeometric point probability. In fact, by comparing the summation term with the first equation containing a summation in Section 9.2, it becomes clear that it is equivalent to a summation of point probabilities for a fourfold table with fixed marginal frequencies, the summation being taken over all possible cell frequencies (since j takes all values from 0 to $m - e$, which is a marginal frequency) and therefore equaling 1. Since e and b are complementary their point probabilities are equal, therefore substituting $m - b$ for e and substituting 1 for the summation, we have

$$P(b) = \frac{\binom{r_2 - 1 - r_1 + m - b}{m - b}\binom{1}{0}\binom{n - r_2 + r_1 + b}{b}}{\binom{n+m}{m}}$$

Now letting $r = n - r_2 + 1 + r_1$ and making the appropriate substitutions, then taking the cumulative probability of b or less, we have

$$P(b \text{ or fewer } Y\text{'s outside interval}) = \sum_{i=0}^{b} \frac{\binom{n - r + m - i}{m - i}\binom{1}{0}\binom{r - 1 + i}{i}}{\binom{n+m}{m}}$$

But this summation is identical to one of the summations giving the probability of b or fewer Y's below X_r (see first formula containing a summation in Section 9.2). It is therefore also equal to the other and we have

$$\begin{aligned}
&P\,(b \text{ or fewer } Y\text{'s outside interval from } X_{r_1} \text{ to } X_{r_2}) \\
&\quad = P\,(b \text{ or fewer } Y\text{'s less than } X_{r=n-r_2+1+r_1}) \\
&\quad = \sum_{i \leq b} \frac{\binom{r+b}{i}\binom{n+m-r-b}{m-i}}{\binom{n+m}{m}} \qquad \text{where } r = n - r_2 + 1 + r_1 \\
&\quad = P\,(e = m - b \text{ or more } Y\text{'s included in interval from } X_{r_1} \text{ to } X_{r_2}) \\
&\quad = E
\end{aligned}$$

Likewise, letting $b' = b + 1$, and $e' = m - b' = e - 1$,

P (b' or more Y's outside interval from X_{r_1} to X_{r_2})

$= P$ (e' or fewer Y's included in interval from X_{r_1} to X_{r_2})

$= 1 - E$

$$= \sum_{i \geqq b'} \frac{\binom{r + b' - 1}{i}\binom{n + m - r - b' + 1}{m - i}}{\binom{n + m}{m}} \qquad \text{where } r = n - r_2 + 1 + r_1$$

Notice that $r = n - [(r_2 - 1) - r_1]$ is n minus the number of X's that are less than X_{r_2} but greater than X_{r_1}, so that r is the number of X's that are not actually inside the interval. One or the other of the above cumulative probabilities reduces to a fairly simple combinatorial expression, in two interesting special cases: (a) when X_{r_1} and X_{r_2} are adjacent in size so that $r_2 = r_1 + 1$ and $r = n$; (b) when r_1 and r_2 are the endpoints of the range of the X-sample so that $r_1 = 1$, $r_2 = n$, and $r = 2$.

9.6 INCLUDANCES: TESTS OF HYPOTHESES

9.6.1 RATIONALE

For reasons (and with qualifications) analogous to those stated in Section 9.3.1, the includance probability E may serve as the probability level for a test of whether or not the X- and Y-populations have equal proportions, i.e., areas, between the two points X_{r_1} and X_{r_2} and equal densities at each of the two points. (The point probability for includances, derived in Section 9.5, can also be derived on the basis of the above-stated conditions plus the assumption of continuity. It is the product of the probability $(\frac{n}{n+m})(\frac{n-1}{n+m-1})$ that both Z_{r_1+j} and Z_{r_2+j+e} will be X's times a fraction which is the "multinomial analogue" of the fraction obtained in Section 9.3.1 whose numerator was the product of two binomial probabilities and whose denominator was a third binomial probability.)

The proportions of the two populations lying between, and the densities at, X_{r_1} and X_{r_2} can be equal despite differences between the two populations in location, dispersion, shape, or combinations thereof, so one must be cautious in drawing inferences beyond the simple equality or inequality of proportions. If it can be assumed, i.e., if it is known, that the two populations can only differ as to "scale parameter" (so that a properly chosen transformation of scale only, i.e., leaving location unaffected, would render the two populations identical in shape as well as location) the includances test becomes a test for identical dispersions. However, such assumption-laden, parameter-conscious tests are a throwback to "parametric" modes of thought.

By letting r_1, r_2, and n be integers such that $r_1 = (n + 1)/4$ and $r_2 = 3(n + 1)/4$, one has a test of whether or not equal proportions of the two populations lie between (and equal densities lie at) estimates of the first and third quartiles of the X-population. (If the populations are known to have identical first or third quartiles, and sample sizes are large, the test becomes more or less a test for identical population interquartile ranges.) Or, in general, by selecting the integers r_1, r_2, and n so that $r_1 = p(n + 1)$ and $r_2 = (1 - p)(n + 1)$, one has a test of whether or not equal proportions of the X- and Y-populations lie between (and equal densities lie at) the estimated $100p$th and $100(1 - p)$th percentiles of the X-population. Another approach is typified by letting $r_1 = 1$, $r_2 = n$, and simply stating that the test is of whether or not equal proportions of the X- and Y-populations lie between (and equal densities lie at) the endpoints of the X-sample. As in the case with exceedances, one can tailor the test to his needs. Fundamentally, includances tests are tests of equality of concentration of two populations between and at two limits which, prior to sampling can be identified as having specifiable ranks in the X-sample or as estimating specifiable percentiles in the X-population, and subsequent to sampling have specifiable values.

9.6.2 Null Hypothesis

H_0 is that $F_X(X_{r_2}) - F_X(X_{r_1}) = F_Y(X_{r_2}) - F_Y(X_{r_1})$ and $f_X(X_{r_1}) = f_Y(X_{r_1})$ and $f_X(X_{r_2}) = f_Y(X_{r_2})$, i.e., that equal proportions of the two populations lie between whatever values turn out to be X_{r_1} and X_{r_2} and that the two populations have equal densities at each of the points X_{r_1} and X_{r_2}.

Prior to sampling it is *possible* for X_{r_1} and X_{r_2} to be virtually any values in the infinite X-population, so in a technical sense the test is, before sampling, a test for identical populations. However, it would be dangerous to use it as such, since for $n > 1$ the likelihood that X_{r_1}, say, will assume a certain value is far different from the likelihood that a randomly drawn X will assume that value. The larger the value of n, the narrower becomes the range of likely values for X_{r_1} or X_{r_2}. The populations could differ in many ways while still having equal proportions between, and equal densities at, any X_{r_1} and any X_{r_2} within their ranges of likely values.

9.6.3 Assumptions

If is assumed: that r_1, r_2, n, m, α, and which sample is to be the "X" sample are all determined *prior* to sampling, or at least prior to viewing the data, that sampling is *random*, observations are *independent*, the sampled populations are *infinite*, and that *no observations* are *tied* with either X_{r_1} or X_{r_2}.

If ties with X_{r_1} or X_{r_2} occur, due to imprecision of measurement upon a continuously distributed variate, the following inequality can be used:

$$P[(\text{number of } Y\text{'s for which } X_{r_1} < Y < X_{r_2}) \geq e] \leq E$$
$$\leq P[(\text{number of } Y\text{'s for which } X_{r_1} \leq Y \leq X_{r_2}) \geq e]$$

See Sections 9.3.3 and 7.8.

9.6.4 Efficiency

If observations become available in order of increasing value, it may be unnecessary to obtain all $n + m$ observation values in order to conduct the test (see Section 9.3.4), in which case efficiency benefits. Includances are poorly adapted to testing for identical populations and should never be used for this purpose when n is moderate or large since the test for identical populations is not consistent against general alternatives.

9.6.5 Tables

Cumulative probabilities can be obtained from univariate hypergeometric tables or from any tables appropriate to Fisher's exact method. Tables for the special case where $r_1 = 1$, $r_2 = n$ have been published by Rosenbaum [12], and a variety of special tables exist for tests of dispersion which are modifications of the includances test (see Discussion).

9.6.6 Example

It is a matter of acute concern to infantrymen that the time interval between activation and explosion of their hand grenades should be neither so short that they explode prematurely nor so long that the enemy has time to hurl them back. A front line infantry captain becomes suspicious that the time interval for a newly received shipment of grenades differs dangerously from that to which his soldiers are accustomed. Having no specifications for either "old" or "new" grenades, he decides to test 12 "old" and 15 "new" grenades, using the .05 level of significance, and to reject the shipment if "too many" of the new grenades have time intervals outside the range of the 12 old grenades. Thus X's are observations on "old" grenades, Y's are observations on "new" ones, $n = 12$, $m = 15$, $r_1 = 1$, $r_2 = n = 12$. Now,

$$P(b' \text{ or more } Y\text{'s outside interval from } X_1 \text{ to } X_n)$$
$$= P(b' \text{ or more } Y\text{'s less than } X_{r=n-r_2+1+r_1} = X_2)$$
$$= \sum_{i \geq b'} \frac{\binom{r+b'-1}{i}\binom{n+m-r-b'+1}{m-i}}{\binom{n+m}{m}}$$
$$= \sum_{i \geq b'} \frac{\binom{b'+1}{i}\binom{26-b'}{15-i}}{\binom{27}{15}}$$

which is the probability that the lower left cell frequency in the accompanying fourfold table with fixed marginal frequencies will be b' or larger.

1	11	12
b'	$15 - b'$	15
$b' + 1$	$26 - b'$	27

Entering Table VIII in the Appendix with $A = 15$, $B = 12$, and $\alpha = .05$, it is seen that for a table with a frequency of 1 in the upper-left cell and significant at the .05 level, the smallest value b' can have is 7 (in which case the cumulative probability that $b' \geqq 7$ in a table with a first-column marginal frequency of $7 + 1 = 8$ is .038). All 12 of the "X" grenades are tested and the values of X_1 and X_{12} are determined. Testing of the "Y" grenades is then begun. After only 10 grenades have been tested 4 of the Y's have been $< X_1$ and 3 have been $> X_{12}$. So the rejection criterion of 7 Y's outside the interval from X_1 to X_{12} is met, the experiment is terminated (at a savings in time and materials), and the shipment of new grenades is rejected.

9.6.7 Discussion

Tests by Rosenbaum [12] and Moses [9] are special cases of the test outlined above. Different but grossly related tests can be found in [7, 5, 17].

By combining the categories "below X_{r_1}" and "above X_{r_2}" into the single category "outside the interval from X_{r_1} to X_{r_2}," we were able to equate cumulative probabilities in the multivariate hypergeometric case to those for a univariate hypergeometric case. Although this has the advantage of giving the cumulative probabilities an already well-tabled distribution, it curtails the variety of tests that can be devised. Therefore let us return to the multivariate case. Let b_Y be the number of Y's falling below X_1, a_Y be the number of Y's above X_n, b_X be the number of X's below Y_1, and a_X be the number of X's above Y_m. Then the following four categories of events are mutually exclusive, so the probabilities of events falling into different categories can be added. (Conditions enclosed within parentheses are a necessary consequence of the immediately preceding condition.)

$$\begin{aligned}
&\text{(i)} \quad b_Y = 0, (b_X > 0); \quad a_Y = 0, (a_X > 0) \\
&\text{(ii)} \quad b_Y = 0, (b_X > 0); \quad a_Y > 0, (a_X = 0) \\
&\text{(iii)} \quad b_Y > 0, (b_X = 0); \quad a_Y = 0, (a_X > 0) \\
&\text{(iv)} \quad b_Y > 0, (b_X = 0); \quad a_Y > 0, (a_X = 0)
\end{aligned}$$

Tukey [17] has devised a "pocket test" (by which he means one simple enough to be performed in one's head) which is limited to cases (ii) and (iii) above, i.e., which requires that one sample contain the highest value, the other the lowest. The test statistic is "the number of individuals in the one sample above all individuals in the other" plus "the number of individuals in the other below all those in the one," i.e., the test statistic is $(b_Y + a_X) + (b_X + a_Y)$, where the sum inside one of the parentheses is zero. It can therefore serve as a test for slippage of location under the assumption of identical, continuous populations. Critical values for the test statistic are remarkably insensitive to either absolute or relative sample sizes. If the sample sizes do not differ by more than 10, the critical values corresponding to the two-tailed .05, .01, and .001 levels of significance are *roughly* 7, 10, and 13, respectively. Haga [5] has published a test for slippage which takes $(a_Y - a_X) - (b_Y - b_X)$ as the test statistic, which therefore tends to become large and positive when the Y-population slips to the right of the X-population and tends to become large in absolute value and negative in the opposite event. The test has the advantage of being rather insensitive to differences between the variances of the two populations when they are both normally distributed. Haga provides tables for a second statistic $(a_Y - a_X) + (b_Y - b_X)$, which therefore tends to become large and positive when a larger portion of the Y-population than of the X-population lies outside the extremes of the X-sample and tends to assume extreme negative values when the extremes of the Y-sample are overflowed more by the X- than by the Y-population. This second test is equivalent to one presented by Kamat [7]. Mathisen [8] bases a test statistic upon the number of Y's falling into each of the four intervals whose boundaries are $-\infty$, X_{k+1}, X_{2k+2}, X_{3k+3}, and $+\infty$, where $n = 4k + 3$, so that the three X's are sample estimates of the first quartile, median, and third quartile, respectively, in the X-population. The test therefore involves generalization from the classic "includances" case of three intervals to more than three. A number of such "interval" tests have been developed, but the test statistic is seldom based simply upon the "frequencies" of Y-observations within the various intervals, so the distribution of the test statistic is seldom a multivariate hypergeometric one. Mathisen, for example, subtracts $m/4$ from the number of Y's in each interval, squares each difference, and bases his test statistic upon the sum of the squares. The departure from simple frequency counts renders the extensive hypergeometric tables virtually useless and necessitates specially constructed tables for each such test (at least when sample sizes are small). Tukey's and Haga's tests have a basically multivariate hypergeometric derivation, but because the test statistic is a *composite* of hypergeometrically distributed variates, probability levels cannot be obtained by a single swift reference to general hypergeometric tables. Fortunately both authors have provided tables for their tests. All three tests are sensitive only to certain forms of nonidentity between the two populations and therefore are presumably not consistent as tests for identical populations against general alternatives (a characteristic they share with most tests discussed in the last two chapters).

On the other hand, *exactly* what they actually test is not always easy to identify in a strict mathematical sense, although the general function of each test is fairly clear on an intuitional basis.

9.7 INCLUDANCES: PREDICTION

Let X_{r_1} and X_{r_2} be the r_1th and r_2th observations, in order of increasing value, in an already obtained sample of size n from a continuously distributed population. Also, let the integers r_1 and r_2 be selected prior to viewing the data and let $r = n - r_2 + 1 + r_1$. Then we may predict that b or fewer observations in a single future sample of size m from the same population will lie outside the interval from X_{r_1} to X_{r_2}, and we may have confidence of level

$$E = \sum_{i \leq b} \frac{\binom{r+b}{i}\binom{n+m-r-b}{m-i}}{\binom{n+m}{m}}$$

in the validity of our prediction. This follows from considerations analogous to those examined in Section 9.4 and from formulas derived in Section 9.5. Thus by letting $r = n - r_2 + 1 + r_1$, tables appropriate to the prediction of exceedances [1, 2, 10, 15] become equally appropriate to the prediction of includances, and the formula in Section 9.4 implicitly giving the minimum size of the future sample required for a confidence level of C that at least b of the Y's will be $< X_r$ applies equally if b is the number of Y's outside the interval from X_{r_1} to X_{r_2}.

9.8 THE BROWN-MOOD MULTI-SAMPLE MEDIAN TEST

This test is a generalization to the multi-sample (and multi-population) case of the two-sample Westenberg-Mood median test, which was outlined in the preceding chapter.

9.8.1 RATIONALE

Suppose that observations have been drawn from each of C continuous populations, and let $\hat{M}$ stand for the grand median of all these observations. If the total number of observations is odd so that one observation exactly equals $\hat{M}$, let that observation be discarded. Of the remaining observations, let n_i be the number of observations drawn from the ith population, let $n = \sum_{i=1}^{C} n_i$ be the total number of observations, let b the total number of observations that lie below the grand median $\hat{M}$, and let b_i be the number of observations drawn from the ith population that lie below $\hat{M}$. The frequencies with which observations fall above or below $\hat{M}$ are therefore expressed in the contingency table shown here.

Number of Observations	*Drawn from Population* 1	2	...	i	...	$C-1$	C	*Total*
Below $\hat{M}$	b_1	b_2	...	b_i	...	b_{C-1}	b_C	b
Above $\hat{M}$	$n_1 - b_1$	$n_2 - b_2$	...	$n_i - b_i$	...	$n_{C-1} - b_{C-1}$	$n_C - b_C$	$n - b$
Total	n_1	n_2	...	n_i	...	n_{C-1}	n_C	n

Now suppose that prior to sampling it had been hypothesized that $\hat{M}$ would be the same quantile in every population, i.e., that the same proportion p of the C populations lies below whatever value $\hat{M}$ turns out to be. Since the value of $\hat{M}$ is not known prior to sampling and is determined by chance, the above hypothesis is, strictly speaking, tantamount to the hypothesis that the C populations have identical distributions, for only in that case is it assured that the same proportion of all C populations falls below any value named. However, if the populations have identical quantiles over a broad central set of quantiles, such as the thirtieth to seventieth percentiles, $\hat{M}$ will become increasingly likely to fall within the set, thus confirming the hypothesis, as the n_i increase.

In any case, if the hypothesis is true the a priori point probability that b_i of the n_i observations drawn from the ith population will fall below $\hat{M}$ is

$$\binom{n_i}{b_i} p^{b_i}(1-p)^{n_i-b_i}$$

And the a priori point probability that b_1, b_2, b_3, etc., observations from the first, second, third, etc., populations will fall below $\hat{M}$, *given that* the numbers of observations drawn (i.e., given that the column totals) are n_1, n_2, n_3, etc., respectively, is the product

$$\prod_{i=1}^{C} \binom{n_i}{b_i} p^{b_i}(1-p)^{n_i-b_i}$$

Finally the same a priori point probability given that of the grand total of n observations b fall below and $n - b$ fall above $\hat{M}$ (i.e., given the row totals) is obtained by dividing the previous product by $\binom{n}{b} p^b (1-p)^{n-b}$, the a priori point probability of the restricting condition just stated. In the resulting fraction, the terms containing p cancel out (since $\Pi_{i=1}^{C} p^{b_i} = p^{\Sigma b_i} = p^b$) as do those containing $1 - p$, leaving

$$\frac{\prod_{i=1}^{C} \binom{n_i}{b_i}}{\binom{n}{b}}$$

as the a priori point probability for the obtained table under the stated hypothesis. This is, of course, a multivariate hypergeometric point probability.

The significance level for a given table can be obtained by cumulating the point probabilities for all tables as extreme or more so with the same row and column marginal frequencies. However, with increasing values of n_i or C, calculations are likely to become prohibitively laborious. Fortunately, when $n \geq 20$ and all $n_i \geq 5$ a fairly good approximation is available. If we apply the usual chi-square test for contingency tables to the table given at the end of the first paragraph of this section, we obtain

$$\begin{aligned}\chi^2_{C-1} &= \sum_{i=1}^{C} \frac{(b_i - [bn_i/n])^2}{bn_i/n} + \frac{\{(n_i - b_i) - [(n-b)n_i/n]\}^2}{(n-b)n_i/n} \\ &= n \sum_{i=1}^{C} \frac{(n-b)n_i[b_i - (bn_i/n)]^2 + bn_i[n_i - b_i - n_i + (bn_i/n)]^2}{bn_i(n-b)n_i} \\ &= n \sum_{i=1}^{C} \frac{nn_i[b_i - (bn_i/n)]^2}{bn_i(n-b)n_i} \\ &= \frac{n^2}{b(n-b)} \sum_{i=1}^{C} \frac{[b_i - (bn_i/n)]^2}{n_i}\end{aligned}$$

The chi-square test for contingency tables is, of course, only an approximate test, and according to Mood the approximation can be improved by substituting $n(n-1)$ for the n^2 preceding the summation in the above formula, obtaining

$$\chi^2_{C-1} = \frac{n(n-1)}{b(n-b)} \sum_{i=1}^{C} \frac{[b_i - (bn_i/n)]^2}{n_i}$$

as the appropriate test statistic for a good approximate test. When H_0 is true, χ^2_{C-1} is distributed approximately as chi-square with $C - 1$ degrees of freedom.

9.8.2 Null Hypothesis

For each of the C populations the proportion of the population lying below $\hat{M}$ is the same, i.e., $\hat{M}$ is the same quantile in all C populations. (If the total number of observations originally drawn is even, $\hat{M}$ is not precisely defined and the null hypothesis is simply that the C populations have a common quantile whose value lies somewhere between the $[n/2]$th and $[(n/2)+1]$th smallest observations in the pooled sample.) Technically, this hypothesis is equivalent to the H_0 that the C populations have identical distributions, since $\hat{M}$ (or the observation values that bracket it) is not known until after the C samples are drawn and since it is *possible* for $\hat{M}$ to be *any* quantile of a common population distribution. However, if the C populations have a common median M, then $\hat{M}$ is an estimate of it. And as all C samples increase proportionately in size, $\hat{M}$ will vary less and less about M. So, in the large-sample case, the hypothesis that the C populations have a common quantile whose value is $\hat{M}$ will become very nearly equivalent to the hypothesis that the C populations have a common median.

9.8.3 Assumptions

Sampling is *random*. The sampled populations are *continuous*, at least in the region of $\hat{M}$ (but, for exceptions, see Sections 8.3.3 and 8.3.7). And, if the test is to be regarded as a test of whether or not the C populations have a common, very *centrally* located, quantile, all n_i must be *large*.

9.8.4 Efficiency

When both tests are applied to populations having normal distributions differing only in location (and therefore having equal variances) the median test has an A.R.E. of $2/\pi$ or .637 relative to the analysis-of-variance F test. Under the same circumstances it has an A.R.E. of $\frac{2}{3}$ relative to Kruskal and Wallis' H test. If the populations have uniform distributions differing only in location, the median test has an A.R.E. of $\frac{1}{3}$ relative to the F test and also relative to the H test. For other types of distribution the A.R.E. of the median test relative to either the F or H tests can be less than, equal to, or greater than, 1, depending upon the particular distribution involved.

The median test is consistent against translation alternatives, i.e., against the class of alternatives in which the populations have distributions that are identical except for location.

9.8.5 Tables

There appear to be no exact tables for the median test in the general case. When $n \geq 20$ and all $n_i \geq 5$, approximate probability levels can be obtained by referring

$$\chi^2_{C-1} = \frac{n(n-1)}{b(n-b)} \sum_{i=1}^{C} \frac{[b_i - (n_i b/n)]^2}{n_i}$$

with $C - 1$ degrees of freedom to chi-square tables such as Table XVI of the Appendix.

If the n observations upon which the test is based include no observations equal to the grand median (i.e., if the assumptions are met and the test is performed exactly as described up to now), then $b = n/2$ and the formula for the approximate test statistic reduces to

$$\chi^2_{C-1} = \frac{4(n-1)}{n} \sum_{i=1}^{C} \frac{[b_i - (n_i/2)]^2}{n_i}$$

9.8.6 Example

An experimenter wishes to test, at the .05 level, the hypothesis that the respective cumulative distributions of three continuous populations have identical ordinates at a common, but as yet unspecified, abscissa value (which will be $\hat{M}$, the grand median of the pooled sample, as yet undrawn). Thus, he wishes to test whether or not the as yet unspecified value of $\hat{M}$ is the same quantile in the three populations. He draws random samples of 6, 7, and 8 observations, respectively, from the three populations, obtaining the observation values given in the following table, in which the grand median $\hat{M}$ of all 21 observations is in boldface.

Sample from Population I	215	252	255	271	**283**	302		
Sample from Population II	231	264	267	269	277	340	611	
Sample from Population III	233	285	288	299	315	369	394	457

Discarding the pooled sample median $\hat{M} = 283$ from the sample from Population I, the frequencies of observations above and below $\hat{M}$ in the three samples are given in the accompanying contingency table. From the values

Number of Observations	*In Sample from* Population I	Population II	Population III	*Total*
Below $\hat{M}$	4	5	1	10
Above $\hat{M}$	1	2	7	10
Total	5	7	8	20

given in the contingency table, the experimenter computes

$$\frac{[b_1 - (n_1/2)]^2}{n_1} = \frac{(4 - 2.5)^2}{5} = \frac{2.25}{5} = .450$$

$$\frac{[b_2 - (n_2/2)]^2}{n_2} = \frac{(5 - 3.5)^2}{7} = \frac{2.25}{7} = .321$$

$$\frac{[b_3 - (n_3/2)]^2}{n_3} = \frac{(1 - 4)^2}{8} = \frac{9}{8} = 1.125$$

which he then uses to obtain

$$\begin{aligned}\chi^2_{C-1} &= \frac{4(n-1)}{n} \sum_{i=1}^{C} \frac{[b_i - (n_i/2)]^2}{n_i} \\ &= \frac{4(19)}{20}(.450 + .321 + 1.125) \\ &= 3.8(1.896) = 7.2048\end{aligned}$$

The obtained value, $\chi^2_{C-1} = 7.205$, exceeds the critical value, 5.991, of chi-square with $C - 1 = 2$ degrees of freedom for $\alpha = .05$, obtained from Table XVI of the Appendix. Therefore, the experimenter rejects the H_0 that the three populations have a common quantile whose value is $\hat{M} = 283$, i.e., that the proportions of the three populations lying below 283 are equal. The three populations cannot have different proportions lying below a specified value and still be identical. However, they could have equal proportions lying below a specified value and still be nonidentical. Therefore rejection is equivalent to a verdict of nonidentical populations, but acceptance of H_0 is *not* equivalent to the verdict that the populations are identical.

9.8.7 Discussion

The disadvantages of including $\hat{M}$ in one of the dichotomies, i.e., of using dichotomies such as $\leq \hat{M}$ and $> \hat{M}$, or $< \hat{M}$ and $\geq \hat{M}$, instead of $< \hat{M}$ and $> \hat{M}$, have been taken up in the discussion section of the two-sample median test (see Section 8.3.7) and will therefore not be repeated here.

The test given here is only one of several median tests outlined in Mood's book (referenced in the previous chapter) for cases analogous to those dealt with by analysis-of-variance techniques. The additional tests are for situations where both a row variable and a column variable are being investigated, introducing the possibility of interaction effects, and where there may be more than one observation per cell. The median tests for these situations are analogous to the one given here, but they are not always distribution-free.

REFERENCES

1. Danziger, L., and S. A. Davis, "Tables of Distribution-Free Tolerance Limits," *Annals of Mathematical Statistics*, **35** (1964), 1361–1365.
2. Epstein, B., "Tables for the Distribution of the Number of Exceedances," *Annals of Mathematical Statistics*, **25** (1954), 762–768.
3. Gart, J. J., "A Median Test with Sequential Application," *Biometrika*, **50** (1963), 55–62.
4. Gumbel, E. J., and H. von Schelling, "The Distribution of the Number of Exceedances," *Annals of Mathematical Statistics*, **21** (1950), 247–262.
5. Haga, T., "A Two-Sample Rank Test on Location," *Annals of the Institute of Statistical Mathematics*, **11** (1960), 211–219.
6. Hanson, D. L., and D. B. Owen, "Distribution-Free Tolerance Limits Elimination of the Requirement that Cumulative Distribution Functions be Continuous," *Technometrics*, **5** (1963), 518–522.
7. Kamat, A. R., "A Two-Sample Distribution-Free Test," *Biometrika*, **43** (1956), 377–387.
8. Mathisen, H. C., "A Method of Testing the Hypothesis that Two Samples are from the Same Population," *Annals of Mathematical Statistics*, **14** (1943), 188–194.

9. Moses, L. E., "A Two-Sample Test," *Psychometrika*, **17** (1952), 239–247.

10. Murphy, R. B., "Non-Parametric Tolerance Limits," *Annals of Mathematical Statistics*, **19** (1948), 581–589.

11. Nelson, L. S., "Tables for a Precedence Life Test," *Technometrics*, **5** (1963), 491–499.

12. Rosenbaum, S., "Tables for a Nonparametric Test of Dispersion," *Annals of Mathematical Statistics*, **24** (1953), 663–668.

13. Rosenbaum, S., "Tables for a Nonparametric Test of Location," *Annals of Mathematical Statistics*, **25** (1954), 146–150.

14. Scheffé, H., and J. W. Tukey, "Nonparametric Estimation. I. Validation of Order Statistics," *Annals of Mathematical Statistics*, **16** (1945), 187–192.

15. Somerville, P. N., "Tables for Obtaining Non-Parametric Tolerance Limits," *Annals of Mathematical Statistics*, **29** (1958), 599–601.

16. Thompson, W. R., "On Confidence Ranges for the Median and Other Expectation Distributions for Populations of Unknown Distribution Form," *Annals of Mathematical Statistics*, **7** (1936), 122–128.

17. Tukey, J. W., "A Quick, Compact, Two-Sample Test to Duckworth's Specifications," *Technometrics*, **1** (1959), 31–48.

18. Wilks, S. S., "Order Statistics," *Bulletin of the American Mathematical Society*, **54** (1948), 6–50.

19. Wilks, S. S., "Recurrence of Extreme Observations," *Journal of the Australian Mathematical Society*, **1** (1959–60), 106–112.

20. Wilks, S. S., "Statistical Prediction with Special Reference to the Problem of Tolerance Limits," *Annals of Mathematical Statistics*, **13** (1942), 400–409.

10

THE MULTINOMIAL DISTRIBUTION

The multinomial distribution is important in the derivation of a number of distribution-free tests, very few of which are dealt with in this book. It is presented here partly to provide the background for further reading but primarily to complete the set of probability formulas for the case of sampling from a population composed of units residing in mutually exclusive categories. The univariate and multivariate hypergeometric probability formulas were appropriate to the cases of sampling without replacements from finite populations consisting of two and more than two categories, respectively. The binomial and multinomial probability formulas are involved in respectively analogous situations except that the population is infinite or sampling is with replacements.

10.1 THE k-VARIATE MULTINOMIAL DISTRIBUTION

Let an event have $k + 1$ possible outcomes, designated by subscripts 1, 2, . . . , $k + 1$, and let these outcomes be mutually exclusive and independent and have probabilities $p_1, p_2, \ldots, p_{k+1}$ such that $\sum_{i=1}^{k+1} p_i = 1$. (Thus, we may imagine an infinite event population divided into $k + 1$ mutually exclusive and exhaustive categories, the ith category containing a proportion p_i of the population.) If the event is allowed to occur n times, the probability that the respective frequencies of occurrence of the various outcomes will be exactly $n_1, n_2, \ldots, n_{k+1}$ is

$$P(n_1, n_2, \ldots, n_{k+1}) = \frac{n!}{n_1!\,n_2!\ldots n_{k+1}!} p_1^{n_1} p_2^{n_2} \ldots p_{k+1}^{n_{k+1}}$$

$$= n! \prod_{i=1}^{k+1} \frac{p_i^{n_i}}{n_i!}$$

Proof: The probability that the outcomes will occur exactly $n_1, n_2, \ldots, n_{k+1}$ times respectively *and in a completely specified order* (for example, the order in which the first n_1 outcomes are those whose probability is p_1, the next n_2, those whose probability is p_2, etc.) is $p_1^{n_1} p_2^{n_2} \ldots p_{k+1}^{n_{k+1}}$. To obtain the probability for these frequencies, but in *any* order, the preceding product must be multiplied by the number of distinguishable orders. The n outcomes can be permuted in $n!$ ways. But in any one of these permutations, there are n_1 outcomes of the first category which are the same and which can be permuted among themselves in $n_1!$ ways without changing the appearance of the order. And for each of these $n_1!$ permutations, the outcomes of the second category can be permuted with one another in $n_2!$ ways without changing the appearance of the original order, etc. There are thus $n_1!\,n_2!\ldots n_{k+1}!$ ways in which each *distinguishable* order pattern can be permuted without creating a pattern distinguishable from it. Since $n!$ is the number of distinguishable patterns times $n_1!\,n_2!\ldots n_{k+1}!$, the number of distinguishable patterns of order is

$$\frac{n!}{n_1!\,n_2!\ldots n_{k+1}!}$$

and the probability that in n trials the $k + 1$ categories of outcomes will occur $n_1, n_2, \ldots n_{k+1}$ times respectively is

$$\frac{n!}{n_1!\,n_2!\ldots n_{k+1}!} p_1^{n_1} p_2^{n_2} \ldots p_{k+1}^{n_{k+1}}$$

The following assumptions are implicit in the above derivation.

(a) Since the same value, p_i, was taken as the probability for outcome i in each of its n_i occurrences, p_i must not vary from event to event. The outcome of each event must therefore be *independent* of the outcomes of all of the $n - 1$ other events.

(b) The $k + 1$ categories must be *mutually exclusive* and *exhaustive*. This is necessitated by the facts that we have taken $\sum p_i = 1$ and $\sum n_i = n$. Were the categories mutually exclusive but not exhaustive, $\sum p_i$ would be less than 1 and $\sum n_i$ might be less than n. Were the categories exhaustive but not mutually exclusive, $\sum p_i$ would exceed 1 and $\sum n_i$ might exceed n. The assumption of mutual exclusiveness is also implicit in taking the simple product of the p_i to represent a compound probability.

(c) Since p_1, p_2, etc. are strictly unconditional chance probabilities, sampling must be *random*, i.e., the n events or trials must be selected on a chance basis from the infinite number of potential events available. Specifically this means, among other things, that no bias shall have operated to exclude valid but "unfavorable" data from the test.

Use of the multinomial distribution in statistical tests requires that the probabilities for all of the possible outcomes be known exactly and be included in the formula

$$n! \prod_{i=1}^{k+1} \frac{p_i^{n_i}}{n_i!}$$

It is important, however, to recognize that the experimenter is free to define both the sample space in which he is interested and the categories which divide that sample space into $k + 1$ mutually exclusive parts. The experimenter must, in fact, be careful to do this in such a way as to define precisely that situation in which he is interested. If he fails he will obtain an exact probability for a situation in which he is not interested, and this probability will differ, perhaps considerably, from the exact probability for the situation in which his interest lies. For example, in coin-tossing, in addition to "heads" and "tails," the outcome category "on the rim" has finite probability which usually cannot be specified. Therefore, although heads and tails have equal probabilities, these probabilities are unknown since their sum is not 1. By defining his sample space as that including only those outcome categories in which the coin lands flat, the experimenter enables himself to specify as .50 the probability of heads and the probability of tails. The experimenter is no better off, however, unless his interest is confined to the sample space consisting only of heads and tails, i.e., is confined to the frequency of heads relative to "heads and tails" rather than tosses. Again, the experimenter may be interested in broader categories than those into which his data are fitted. In such cases he should use the categories in which he is interested rather than those in which the data are available. For example, in tossing two coins simultaneously the possible outcomes will be defined to be two heads ($P = \frac{1}{4}$), a head and a tail ($P = \frac{1}{2}$), and two tails ($P = \frac{1}{4}$). Suppose that the two coins have been simultaneously tossed n times and that the frequencies of the respective outcomes named above are n_1, n_2, and n_3. If the experimenter is interested in the point probability of the obtained frequencies for the outcomes stated, the proper formula is

$$\frac{n!}{n_1!\, n_2!\, n_3!}\left(\frac{1}{4}\right)^{n_1+n_3}\left(\frac{1}{2}\right)^{n_2}$$

On the other hand, if he is interested in the probability of the obtained frequencies for the recategorized outcomes, "coins have same side up" ($P = \frac{1}{2}$) and "coins have different sides up" ($P = \frac{1}{2}$), the obtained frequencies are $n_1 + n_3$ and n_2 respectively and the probability is

$$\frac{n!}{(n_1 + n_3)!\, n_2!}\left(\frac{1}{2}\right)^{n_1+n_3}\left(\frac{1}{2}\right)^{n_2}$$

The probabilities for the same data under the two different categorizations of outcome are not the same. The reason for the discrepancy between the two probabilities is that one states merely that $n_1 + n_3$ tosses result in either two heads or two tails without specifying precisely how many of these shall be two heads; the other probability *does* specify this further and much more restrictive information. The latter probability is, therefore, much smaller than the former.

Tables of multinomial probabilities are virtually nonexistent, except for the special case where $k = 1$, i.e., except for the case where the multinomial distribution becomes the binomial. When k exceeds 1 there are simply too many parameters to permit extensive tabling of multinomial probabilities in the general case (although they could be tabled in some interesting special cases such as all $p_i = p = 1/(k + 1)$ when k is very small). Furthermore, except when n is quite tiny, obtaining cumulative probabilities by calculating and summing point probabilities is prohibitively laborious. Consequently, even though an exact test would consist of referring the multinomial test statistic to its unavailable cumulative multinomial probability distribution, the common practice is to obtain imprecise probability levels by referring an alternative test statistic to a well-tabled distribution which only approximates its true distribution.

10.2 THE CHI-SQUARE ALTERNATIVE

Consider the point multinomial probability

$$P(n_1, n_2, \ldots, n_{k+1}) = \frac{n!}{n_1!\, n_2! \ldots n_{k+1}!} p_1^{n_1} p_2^{n_2} \ldots p_{k+1}^{n_{k+1}}$$

Under the assumption that all of the n_i are "infinitely" large, the factorials in the above formula can be replaced by their Stirling approximations and through some further mathematical manipulations described in [2, see also 5], the following approximate formula can be derived:

$$P(n_1, n_2, \ldots, n_{k+1}) \cong C \exp -\frac{1}{2} \sum_{i=1}^{k+1} \frac{(n_i - np_i)^2}{np_i}$$

where C is a constant for given values of n, k, and the various p_i. Thus the probability on the left is a monotonic decreasing function (approximately) of the summation. Now if the n_i are not too small, that same summation is distributed approximately as chi-square with k degrees of freedom. That is,

$$\chi_k^2 \cong \sum_{i=1}^{k+1} \frac{(n_i - np_i)^2}{np_i}$$

and the distribution of the summation coincides exactly with that of χ_k^2 when the smallest n_i becomes infinite. Thus the upper-tail cumulative probability of the summation in the well-tabled distribution of χ_k^2 gives us approximately the cumulation of the point multinomial probability over those sets of n_i causing

$$\sum_{i=1}^{k+1} \frac{(n_i - np_i)^2}{np_i}$$

to take on its greatest values.

Now, n_i is simply the observed, and np_i the expected, frequency of event-outcomes falling in the ith category. And

$$\sum_{i=1}^{k+1} \frac{(n_i - np_i)^2}{np_i}$$

is simply the test statistic for the familiar chi-square test of goodness of fit when no population parameters are estimated from the data and the only restriction is that $n = \sum n_i$, causing the loss of a single degree of freedom. That test is, of course, only one of many chi-square tests.

Chi-square tests as such will not be taken up in this book. They are thoroughly discussed in most textbooks on classical statistics with which they have much more in common, even though some of them are technically distribution-free. Like most classical tests, chi-square tests are inexact unless totally unrealistic assumptions (such as infinite n_i's) are met, their derivation is mathematically complex, and their application is hagridden with particularistic considerations in which the advocated procedure is often fallacious. Among the hazards associated with chi-square tests are the following. The outcome of the test of fit may be strongly affected by the choice of interval widths and endpoints [6]. Yates' correction is disastrous in certain situations, but even where it is deemed appropriate it may not correct, having the opposite effect instead [1, 3]. "Rules" about minimum expected frequencies ignore the facts that the test is inexact when any expected frequency is less than infinity and that for the *same* expected frequency np_i, the degree of inexactitude varies with at least three factors, n, p_i, and the probability level of the test. Finally, the prohibition against small expected frequencies has led to the widely accepted practice of pooling categories in order to bring the expected frequencies for the combined categories up to the required size. Such pooling, however, involves an arbitrary decision which (as pointed out earlier) changes the character of the test (i.e., of the tested hypothesis) and which, in the case of contingency tables, must usually be made subsequent to the collection of data. Such a posteriori manipulation of categories in effect violates the assumption of random sampling since outcomes are being influenced by a factor other than chance. These and other hazards are discussed in many of the references at the end of this chapter. The article by Lewis and Burke [7] strongly suggests that chi-square is one of the most misused classical statistics.

REFERENCES

1. Adler, F., "Yates' Correction and the Statisticians," *Journal of the American Statistical Association*, **46** (1951), 490–501.
2. Birnbaum, Z. W., *Introduction to Probability and Mathematical Statistics.* New York: Harper & Row, Publishers, Inc., 1962, pp. 243–270.
3. Cochran, W. G., "The χ^2 Test of Goodness of Fit," *Annals of Mathematical Statistics*, **23** (1952), 315–345.
4. Freeman, G. H., and J. H. Halton, "Note on an Exact Treatment of Contingency, Goodness of Fit, and Other Problems of Significance," *Biometrika*, **38** (1951) 141–149.
5. Fry, T. C., "The χ^2-Test of Significance," *Journal of the American Statistical Association*, **33** (1938), 513–525.
6. Gumbel, E. J., "On the Reliability of the Classical Chi-Square Test," *Annals of Mathematical Statistics*, **14** (1943), 253–263.
7. Lewis, D., and C. J. Burke, "The Use and Misuse of the Chi-Square Test," *Psychological Bulletin*, **46** (1949), 433–489 [see also **47** (1950), 331–355].
8. Williams, C. A., "On the Choice of the Number and Width of Classes for the Chi-Square Test of Goodness of Fit," *Journal of the American Statistical Association*, **45** (1950), 77–86 [see also **58** (1963), 678–689].

11

RUNS OF EVENTS HAVING CONSTANT PROBABILITY UNDER H_0

In a series of n elements of which n_1 are a's, the rest b's, the pattern in which the obtained a's and b's arrange themselves will be random if each position in the series had the same a priori probability of being occupied by an a, i.e., if the complementary events a and b had constant probabilities of p and $1 - p$ throughout the series. However the pattern will be nonrandom if the probability of an a varied from one position to the next. In that case, and especially if there was sequential dependency among elements of the same type, like events may tend to cluster, and this may be indicated by an unusually small number of runs, or clusters of like objects, in the pattern, or by runs of unexpected length. Thus the total number of runs, the length of the longest run, and various other run statistics can be used as the sample information with which to test for randomness of pattern of arrangement against the alternative of sequential dependency. By judicious definition of the two types of event, this test can be employed to test whether a trend exists in a sequentially sampled population, whether performance is improving, whether two sampled populations are identical, etc. Run tests are often rather weak and inefficient, depending upon the type of application contemplated. However, even when efficiency is very low, if data are plentiful enough to insure adequate power, the simplicity of run tests may prove an irresistibly attractive feature.

11.1 BASIC FORMULAS

11.1.1 General Case

A run is an unbroken sequence of similar events or like objects. For example, in the series *a a b a b b b a a* there are five runs: one run of *a*'s of length 1, two runs of *a*'s of length 2, one run of *b*'s of length 1, and one run of *b*'s of length 3. The following notation will be used in the derivation of run formulas when there are two kinds of objects [see 6]. Let r_{ij} be the number of runs of objects of type i whose length is j, and let r_i be the number of runs of objects of type i irrespective of length, i.e., of all lengths. Let n_i be the number of objects of type i and let n be the number of objects of both types. The two types of objects will be designated 1 and 2 respectively. The only things which can interrupt or terminate a run of like objects are a run of the other type objects or else termination of the entire series. Therefore r_1, the number of runs of 1's, can either be one greater than, equal to, or one less than r_2, the number of runs of 2's. When $r_1 = r_2 + 1$ the series can begin (and end) in only one way—with a run of 1's. Likewise when $r_2 = r_1 + 1$ the series must begin and end with runs of 2's. However, when $r_1 = r_2$ the series can either begin with a run of 1's and end with a run of 2's, or begin with a run of 2's and end with a run of 1's. Therefore it will be convenient to introduce the notation

$$\begin{aligned} F(r_1, r_2) &= 1 \qquad \text{if } r_1 \neq r_2 \\ &= 2 \qquad \text{if } r_1 = r_2 \end{aligned}$$

The r_1 runs of 1's of various lengths can be permuted in $r_1!$ ways. But a permutation that merely exchanges the positions of runs of 1's of the same length does not change the appearance of the series. The r_{1j} runs of 1's of length j can be permuted in $r_{1j}!$ ways without changing the appearance of the series. Therefore, the number of distinguishable permutations of the r_1 runs of 1's is

$$\frac{r_1!}{r_{11}!\, r_{12}! \ldots r_{1n_1}!}$$

For each of these distinguishable permutations of runs of 1's, there are

$$\frac{r_2!}{r_{21}!\, r_{22}! \ldots r_{2n_2}!}$$

distinguishable permutations of the runs of 2's. And since, if $r_1 = r_2$ the series can begin in two ways, the total number of distinguishable permutations given that there are r_1 runs of 1's and r_2 runs of 2's of specified lengths is

$$\left(\frac{r_1!}{r_{11}!\, r_{12}! \ldots r_{1n_1}!}\right)\left(\frac{r_2!}{r_{21}!\, r_{22}! \ldots r_{2n_2}!}\right)F(r_1, r_2)$$

Finally, since there are $n!/(n_1!\, n_2!)$ distinguishable permutations of n_1 1's and n_2 2's, the probability that there will be exactly r_{11} runs of 1's of length 1, r_{12} of length 2, etc., as well as exactly r_{21} runs of 2's of length 1, r_{22} of length 2, etc., *given that* there are n_1 1's and n_2 2's in the series is

$$P(r_{ij}) = \left(\frac{r_1!}{r_{11}!\, r_{12}! \ldots r_{1n_1}!}\right)\left(\frac{r_2!}{r_{21}!\, r_{22}! \ldots r_{2n_2}!}\right)\frac{F(r_1, r_2)}{n!/(n_1!\, n_2!)}$$

Suppose that we are interested in the breakdown of runs of 1's according to length, but that we are not interested in the corresponding breakdown of runs of 2's. Considering only the 1's, there are

$$\frac{r_1!}{r_{11}!\, r_{12}! \ldots r_{1n_1}!}$$

distinguishable permutations of the r_1 runs of 1's. Now imagine the n_2 2's arranged in a line. There are $n_2 - 1$ spaces between 2's, and the r_2 runs of 2's can be obtained by selecting $r_2 - 1$ of these $n_2 - 1$ spaces and "widening" them for occupation by runs of 1's. This can be done in $\binom{n_2-1}{r_2-1}$ ways. If $r_1 = r_2 - 1$, then any given permutation of runs of 1's can be fitted into a specified $r_2 - 1$ spaces between 2's in only one way, since the series must start and end with a run of 2's. If $r_1 = r_2 + 1$, in addition to the $r_2 - 1$ spaces between 2's the runs of 1's also occupy the space to the left of the leftmost 2 and to the right of the rightmost 2. The series starts and ends with runs of 1's, and the $r_2 - 1$ spaces between 2-runs are occupied by the second-to-$(r_1 - 1)$st 1-runs. Again, this can be accomplished in only one way. However, if $r_1 = r_2$, the first 1-run can be placed either to the left of the leftmost 2-run, or between the first and second 2-runs. Therefore the probability of exactly r_{11} runs of 1's of length 1, r_{12} of length 2, etc., and r_2 runs of 2's of any lengths given that there are n_1 1's and n_2 2's is

$$P(r_{1j}, r_2) = \frac{\dfrac{r_1!}{r_{11}!\, r_{12}! \ldots r_{1n_1}!}\dbinom{n_2 - 1}{r_2 - 1}F(r_1, r_2)}{\dfrac{n!}{n_1!\, n_2!}}$$

Suppose now that we are interested in neither the lengths nor the total number of runs of 2's. The r_1 runs of 1's can be inserted into any r_1 of the $n_2 + 1$ spaces before, between, and after the 2's, i.e., into any of the $n_2 - 1$ spaces between 2's as well as the space to the left of the leftmost 2 and the space to the right of the rightmost 2. This can be done in $\binom{n_2+1}{r_1}$ ways. Therefore (the rest of the derivation being analogous to that given earlier), the probability of exactly r_{11} runs of 1's of length 1, r_{12} of length 2, etc., given that there are n_1 1's and n_2 2's is

$$P(r_{1j}) = \frac{\dfrac{r_1!}{r_{11}!\, r_{12}! \ldots r_{1n_1}!}\dbinom{n_2 + 1}{r_1}}{\dfrac{n!}{n_1!\, n_2!}}$$

Since the number of runs of 2's is unspecified, it may be $r_1 - 1$, r_1, or $r_1 + 1$, and the term $F(r_1, r_2)$ is not required in the formula.

The preceding formulas give probabilities for the entire run pattern in the sense that the exact number of runs of each possible length is specified, at least for runs of one type. In order to obtain the more general probability for only certain specified r_{ij}, one fixes these r_{ij} as constants and sums the formula over all other values for which the relationships, $\sum_{j=1}^{n_i} jr_{ij} = n_i$ and $\sum_j r_{ij} = r_i$, are satisfied. For example, if $n_1 = 7$ and $n_2 = 9$ the probability of exactly one run of 1's of length 4 would be

$$P(r_{14} = 1) = \frac{\sum \dfrac{r_1!}{r_{11}!\, r_{12}!\, r_{13}!\, 1!}\dbinom{10}{r_1}}{\dfrac{16!}{7!\,9!}}$$

$$= \frac{\dfrac{4!}{3!\,1!}\dbinom{10}{4} + \dfrac{3!}{1!\,1!\,1!}\dbinom{10}{3} + \dfrac{2!}{1!\,1!}\dbinom{10}{2}}{\dfrac{16!}{7!\,9!}}$$

since a run of length 4 could be accompanied by three runs of length 1, one of length 1 and one of length 2, or by one of length 3, while still fulfilling the condition that $n_1 = 7$, and since the number of runs of 1's in these three cases is 4, 3, and 2 respectively.

In theory, point probabilities for patterns of arrangement having any specified set of run characteristics can be obtained by summations of the type just given. However, the process of selective summation is likely to prove tedious, cumbersome, and conducive to error. Much more compact and concise formulas are therefore derived in the sections to follow.

11.1.2 Number of Runs

Suppose that we are interested in number of runs only, and not in their lengths. Imagine the n_1 1's arranged in a line. There are $n_1 - 1$ spaces between 1's, and the 1's can be separated into r_1 runs by selecting and "widening" $r_1 - 1$ of these spaces, then filling them with runs of 2's. The $r_1 - 1$ spaces can be selected in $\binom{n_1-1}{r_1-1}$ ways. For each of these ways, the r_2 runs of 2's (which will eventually be interlaced with the 1's) can, by analogous reasoning, be selected in $\binom{n_2-1}{r_2-1}$ ways. Any given set of r_1 runs of 1's and r_2 runs of 2's can be fitted together in one way if $r_1 = r_2 \pm 1$ and in two ways if $r_1 = r_2$. The number of ways of obtaining r_1 runs of 1's and r_2 runs of 2's in a sequence of n_1 1's and n_2 2's is therefore $\binom{n_1-1}{r_1-1}\binom{n_2-1}{r_2-1} F(r_1, r_2)$. The number of distinguishable arrangements of n_1 1's and n_2 2's without restriction as to numbers of runs is $\binom{n}{n_1}$. Therefore the probability of exactly r_1 runs of 1's and r_2 runs of 2's given that there are n_1 1's and n_2 2's is

$$P(r_1, r_2) = \frac{\binom{n_1 - 1}{r_1 - 1}\binom{n_2 - 1}{r_2 - 1}F(r_1, r_2)}{\binom{n}{n_1}}$$

Let U stand for the total number of runs of both 1's and 2's. Since the number of runs of 1's can be one less than, equal to, or one greater than the number of runs of 2's, U can be an even number in only one way, but can be an odd number in two ways. Substituting into the above formula,

$$\text{if } r_1 = r_2 = r, \quad P(r_1, r_2) = \frac{2\binom{n_1 - 1}{r - 1}\binom{n_2 - 1}{r - 1}}{\binom{n}{n_1}}$$

$$\text{if } r_1 = r \text{ and } r_2 = r + 1, \quad P(r_1, r_2) = \frac{\binom{n_1 - 1}{r - 1}\binom{n_2 - 1}{r}}{\binom{n}{n_1}}$$

$$\text{and if } r_1 = r + 1 \text{ and } r_2 = r, \quad P(r_1, r_2) \frac{\binom{n_1 - 1}{r}\binom{n_2 - 1}{r - 1}}{\binom{n}{n_1}}$$

Therefore the probability that the total number of runs will be some even number $2r$ is

$$P(U = 2r) = \frac{2\binom{n_1 - 1}{r - 1}\binom{n_2 - 1}{r - 1}}{\binom{n}{n_1}}$$

and the probability that it will be some odd number $2r + 1$ is

$$P(U = 2r + 1) = \frac{\binom{n_1 - 1}{r - 1}\binom{n_2 - 1}{r} + \binom{n_1 - 1}{r}\binom{n_2 - 1}{r - 1}}{\binom{n}{n_1}}$$

If we are interested only in the number of runs of 1's and are indifferent to whether r_2 equals $r_1 - 1$, r_1, or $r_1 + 1$, we still select the r_1 runs of 1's by selecting $r_1 - 1$ of the $n_1 - 1$ spaces between 1's for widening. However, now the spaces before and after the 2's as well as the spaces between 2's are available for occupation by 1's because the number of runs of 2's is not fixed. Therefore there are $n_2 + 1$ spaces available for occupation by the r_1 runs, and they can be chosen in $\binom{n_2+1}{r_1}$ ways. The rest of the derivation is analogous to that described earlier. Therefore the probability that there will be exactly r_1 runs of 1's given that there are n_1 1's and n_2 2's is

$$P(r_1) = \frac{\binom{n_1 - 1}{r_1 - 1}\binom{n_2 + 1}{r_1}}{\binom{n}{n_1}}$$

The above formula can be written in the equivalent form

$$P(r_1) = \frac{\binom{n_1 - 1}{n_1 - r_1}\binom{n_2 + 1}{r_1}}{\binom{n_1 + n_2}{n_1}}$$

in which case it is easily seen to be a hypergeometric point probability (since the "upper-level" numbers visible in the denominator are the sum of the "upper-level" numbers visible in the numerator, and likewise for "lower-level" numbers). It is, in fact, the point probability that of the $n_2 + 1$ "cells" formed by the 2's $n_2 + 1 - r_1$ of them will be empty, i.e., will not contain a run of 1's. (See Section 8.7.1.)

11.1.3 Lengths of Runs

Consider now the probability [see 8] of obtaining an arrangement (i.e., sequence) containing at least one run of 1's of length S or greater, given that the arrangement contains n_1 1's and n_2 2's. The denominator of the probability fraction will be the number of distinguishable arrangements of n_1 1's and n_2 2's when no restrictions are imposed, and in all cases this will be $\binom{n}{n_1}$. (This may be thought of as the number of ways of choosing from a sequence of $n = n_1 + n_2$ elements a set of n_1 elements to be called 1's and a set of n_2 elements to be called 2's.) The numerator of the probability fraction will be the number of distinguishable arrangements of n_1 1's and n_2 2's that contain one or more runs of 1's of length S or longer.

The simplest case occurs when all of the 1's must belong to a single run of length S, i.e., when $n_1 = S$. The single run of 1's can be obtained from the n_1 1's in only one way and must be located in one of the $n_2 + 1$ intervals or cells created by the n_2 2's. Therefore, there are $n_2 + 1$ distinguishable arrangements in which a run of $S = n_1$ 1's occurs, and $n_2 + 1$ is therefore the numerator of the probability fraction.

The next simplest case occurs when $S > n_1/2$, so that no more than one run of 1's of length $\geqq S$ can occur in a single arrangement. As before, there are $n_2 + 1$ ways of choosing the cell to contain the run of 1's of length $\geqq S$. (Note that choosing one of the cells that will be formed by the n_2 2's does *not* involve identifying *which* of the n elements will be 2's; it merely involves recognizing that *whatever* sequential positions the 2's turn out to have, there will be $n_2 + 1$ intervals between, above, and below, the 2's). The remaining $n_1 - S$ 1's which do not belong to the first S 1's in the run of 1's of length $\geqq S$ can be chosen from the $n - S$ remaining elements in $\binom{n-S}{n_1-S}$ ways. So there are $(n_2 + 1)\binom{n-S}{n_1-S}$ ways of obtaining a run of 1's of length $\geqq S$, and this is the numerator of the probability fraction. This may be easier to grasp under the following alternative explanation. We require that there be S consecutive 1's in the sequence of n elements, but we do not yet specify its location. Whatever be its location, the n_2 2's (and the $n_1 - S$ remaining 1's) must be chosen from the $n - S$ elements remaining, and this can be done in $\binom{n-S}{n_2} = \binom{n-S}{n_1-S}$ ways. Now that they are chosen, the n_2 2's form $n_2 + 1$ cells, which constitute the only meaningful specification for the location of the run of 1's of length *S or more*. (The possible presence of 1's in a cell *before* the insertion of the run of length S is immaterial to the "location" of the run for the simple reason that they become part of it.) Thus there are $n_2 + 1$ ways of picking the location of the run of length $\geqq S$. So the number of ways of obtaining a run of 1's of length $\geqq S$ when $S > n_1/2$ is $(n_2 + 1)\binom{n-S}{n_2}$, and this is also the number of *arrangements* containing at least one run of S or more 1's and is therefore the numerator of the probability fraction.

Finally, consider the general case in which no restriction is placed upon the size of S relative to n_1. In analogy with the preceding derivation, we may derive the number of ways W_i of obtaining i runs of 1's, each of length $\geqq S$. There are $\binom{n_2+1}{i}$ ways of selecting from the $n_2 + 1$ cells the i cells to be occupied by the i runs of length $\geqq S$. And there are $\binom{n-iS}{n_2}$ ways of selecting the n_2 2's (and the $n_1 - iS$ remaining 1's) from the $n - iS$ remaining elements. So $W_i = \binom{n_2+1}{i}\binom{n-iS}{n_2}$. But the term $\binom{n-iS}{n_2}$ permits all possible dispositions of the remaining $n_1 - iS$ 1's, and some of these dispositions include additional runs of 1's of length $\geqq S$. Thus if $S = 4$, the single arrangement 1 1 1 1 2 2 1 1 1 1 1 2 2 2 1 1 1 1 2 1 1 2 2 will contribute $\binom{3}{2} = 3$ "ways" to W_2, one when the first two cells are considered (i.e., are identified by one of the combinations of $n_2 + 1$ things taken 2 at a time), one when the first and third cells are considered, and one when the second and third cells are considered to contain the two specified runs of four or more 1's. In short, when $iS \leqq n_1 - S$ the the number of "ways" of obtaining i runs of 1's of length $\geqq S$ exceeds the total number of arrangements containing exactly i runs of 1's of length $\geqq S$. Let A_i stand for the latter. Then not only does A_{i+1} contribute $\binom{i+1}{i}$ ways to W_i, as illustrated above, but A_{i+2} must contribute $\binom{i+2}{i}$ additional "ways" to W_i, etc. Thus it becomes clear that

$$W_i = \sum_{k=0}^{(n_1-iS)/S} \binom{i+k}{i} A_{i+k}$$

where k takes only integral values, and the integral part of $(n_1 - iS)/S$ is the maximum number of additional (beyond i) runs of 1's of length $\geqq S$ that a single arrangement can contain. Using this formula we obtain the expression for each W_i, and for even values of i we simply multiply each side of the equation by minus one, obtaining the following equations and finally their sum:

$$
\begin{aligned}
W_1 &= A_1 + \tbinom{2}{1} A_2 + \tbinom{3}{1} A_3 + \tbinom{4}{1} A_4 + \tbinom{5}{1} A_5 + \ldots \\
-W_2 &= \qquad - A_2 - \tbinom{3}{2} A_3 - \tbinom{4}{2} A_4 - \tbinom{5}{2} A_5 - \ldots \\
W_3 &= \qquad A_3 + \tbinom{4}{3} A_4 + \tbinom{5}{3} A_5 + \ldots \\
-W_4 &= \qquad - A_4 - \tbinom{5}{4} A_5 - \ldots \\
W_5 &= \qquad A_5 + \ldots \\
&\ \vdots
\end{aligned}
$$

$$\sum_{i=1}^{n_1/S} (-1)^{i+1} W_i = A_1 + A_2 + A_3 + A_4 + A_5 + \ldots$$

Since A_i is the number of arrangements containing *exactly* i runs of 1's of length $\geqq S$, the expression on the right is simply the total number of arrangements containing one or more runs of 1's of length S or greater which is the numerator we seek. And the expression on the left gives it to us in terms of W_i for which we already have the equation,

$$W_i = \binom{n_2 + 1}{i}\binom{n - iS}{n_2}$$

Therefore, the general formula for the probability of at least one run of 1's of length S or greater, given that there are n_1 1's and n_2 2's, is

$$P(r_{1, (\geqq S)} \geqq 1) = \sum_{i=1}^{n_1/S} \frac{(-1)^{i+1}\binom{n_2 + 1}{i}\binom{n - iS}{n_2}}{\binom{n}{n_1}}$$

and this formula subsumes as special cases all previously treated cases in this section.

We next turn our attention to the probability of obtaining at least one run of length S or greater among *both* the 1's and the 2's. The r_1 runs of 1's of all lengths could be obtained by selecting "for widening" $r_1 - 1$ of the $n_1 - 1$ spaces between 1's if there were no restrictions as to run-length. But, if we specify that i of the runs are of length at least S, then we must select the $r_1 - 1$ spaces from the $(n_1 - 1) - i(S - 1)$ spaces remaining under the restriction that i sets of $S - 1$ consecutive spaces are unavailable for widening because they are "within" one of the i runs of length $\geqq S$. Therefore there are $\binom{(n_1-1)-i(S-1)}{r_1-1}$ ways of creating the r_1 runs of 1's given that an as yet unlocated i of them are of length $\geqq S$. For each of these ways, there are $\binom{r_1}{i}$ ways of selecting from among the r_1 identified locations of 1-runs the i

locations for the i runs of length $\geqq S$. But some of the nonspecified locations may contain additional runs of length $\geqq S$, so for reasons analogous to those given in the preceding derivation, the number of ways of obtaining an entire arrangement of 1's containing one or more runs of 1's of length S or greater is

$$\sum_{i=1}^{(n_1-r_1)/(S-1)} (-1)^{i+1}\binom{r_1}{i}\binom{(n_1-1)-i(S-1)}{r_1-1}$$

(The upper limit of summation is explained by the fact that each of the r_1 runs must consist of at least one 1. A run of length S is obtained by adding $S - 1$ more 1's, which must be obtained from the remaining $n_1 - r_1$ 1's. So the maximum number of runs of length S or greater is $(n_1 - r_1)/(S - 1)$ unless the latter exceeds r_1, in which case the upper limit for the maximum is automatically provided by the fact that i cannot exceed r_1 in $\binom{r_1}{i}$.) Now if there are r_1 runs of 1's, the number of runs of 2's can be $r_1 - 1$, r_1 or $r_1 + 1$, and for each of these cases we can derive a completely analogous expression for the number of ways of obtaining an entire arrangement of 2's containing one or more runs of 2's of length S or greater. We need only replace n_1 by n_2 and r_1 by $r_1 - 1$, r_1, or $r_1 + 1$, as the case may be, in the expression already obtained for the 1's. The runs of 2's can be interlaced with the runs of 1's in only one way when the number of runs of 2's is $r_1 - 1$ or when it is $r_1 + 1$, but in two ways when it is r_1. Therefore, the number of arrangements containing at least one run of length $\geqq S$ among *both* the 1's and the 2's is $F(r_1, r_2)$ times the product of the 1-expression and the 2-expression. The sum of this product over the values $r_1 - 1$, r_1, and $r_1 + 1$, for r_2, and over all values of r_1, for r_1, is the desired numerator, and the probability fraction therefore becomes

$$P(r_{1,(\geqq S)} \geqq 1 \textit{ and } r_{2,(\geqq S)} \geqq 1)$$

$$= \frac{1}{\binom{n}{n_1}} \sum_{r_1=1}^{n_1-S+1} \left\{ \sum_{i=1}^{(n_1-r_1)/(S-1)} (-1)^{i+1}\binom{r_1}{i}\binom{(n_1-1)-i(S-1)}{r_1-1} \right\}$$

$$\left\{\left[\sum_{i=1}^{(n_2-r_1+1)/(S-1)} (-1)^{i+1}\binom{r_1-1}{i}\binom{(n_2-1)-i(S-1)}{r_1-2}\right]\right.$$

$$+ 2\left[\sum_{i=1}^{(n_2-r_1)/(S-1)} (-1)^{i+1}\binom{r_1}{i}\binom{(n_2-1)-i(S-1)}{r_1-1}\right]$$

$$\left. + \left[\sum_{i=1}^{(n_2-r_1-1)/(S-1)} (-1)^{i+1}\binom{r_1+1}{i}\binom{(n_2-1)-i(S-1)}{r_1}\right]\right\}$$

It is now easy to obtain the probability of obtaining at least one run of length $\geqq S$ among *either* the 1's or the 2's. We have already derived the probability $P(r_{1,(\geqq S)} \geqq 1)$ that there will be at least one run of length $\geqq S$ among the 1's, and the events covered by this probability include those cases where there is also such a run among the 2's. Likewise, the probability that there will be at least one run of length $\geqq S$ among the 2's (i.e.,

$$P(r_{2,(\geqq S)} \geqq 1) = \sum_{i=1}^{n_2/S} \frac{(-1)^{i+1}\binom{n_1+1}{i}\binom{n-iS}{n_1}}{\binom{n}{n_1}}$$

obtained by direct analogy from the formula for runs among the 1's) covers events which include the simultaneous occurrence of at least one run of length $\geqq S$ among the 1's. Thus the sum of $P(r_{1,(\geqq S)} \geqq 1)$ and $P(r_{2,(\geqq S)} \geqq 1)$ includes $P(r_{1,(\geqq S)} \geqq 1$ *and* $r_{2,(\geqq S)} \geqq 1)$ twice. Therefore the probability of at least one run of length $\geqq S$ among *either* the 1's or the 2's, it being unspecified which type of element is to contain the run, is

$$P(r_{?,(\geqq S)} \geqq 1) = P(r_{1,(\geqq S)} \geqq 1) + P(r_{2,(\geqq S)} \geqq 1) - P(r_{1,(\geqq S)} \geqq 1 \text{ and } r_{2,(\geqq S)} \geqq 1)$$

This may be regarded as the sum of the probabilities for three mutually exclusive classes of event: at least one run of length $\geqq S$ occurs among the 1's but not among the 2's; at least one such run occurs among the 2's but not among the 1's; at least one such run occurs among both the 1's and the 2's. Thus when we refer to $P(r_{?,(\geqq S)} \geqq 1)$ as the probability for at least one run of length $\geqq S$ among either the 1's or the 2's, it is to be understood that "either" includes the possibility of "both."

11.1.4 Extensions to the Case Where n_1 and n_2 Are Not Fixed

All of the run formulas heretofore listed take n_1 and n_2 as given. They give probabilities conditional upon the existence of exactly n_1 1's and n_2 2's in the obtained sample. If one is interested in the pattern of arrangement of 1's and 2's but not in the probability of obtaining a 1 or a 2, the foregoing formulas are generally the appropriate ones. However if "1" and "2" are mutually exclusive outcomes of a binomial event, with probabilities p and q respectively of occurrence on a single trial, the experimenter may be interested in the compound probability that there will be n_1 1's and n_2 2's and that their arrangement will contain a configuration of runs having a specified characteristic. This compound probability is obtained [see 6] by taking the product of the binomial probability

$$\binom{n}{n_1} p^{n_1} q^{n_2}$$

and whichever one of the probability formulas listed earlier gives the appropriate conditional probability for the specified characteristic. Thus if $P(C|n_1, n)$ is the probability of some specified run characteristic *given* n_1 and $n_2 = n - n_1$, the product $[P(C|n_1, n)][\binom{n}{n_1}p^{n_1}q^{n_2}]$ is the probability of "obtaining" *both C and* n_1 *given n* and p. And the sum of that product over all possible values of n_1 is the probability of obtaining C, completely *irrespective* of n_1, given only n and p.

11.2 NUMBER OF RUNS AS A TEST OF RANDOMNESS

11.2.1 Rationale

The derivations in Section 11.1.2 for number-of-run probabilities regarded every distinguishable arrangement as equally likely, since every such arrangement was, in effect, given a weight of 1. But if every arrangement has the same a priori probability, then by definition an actually obtained arrangement is a random one (or, perhaps better, the "process" by which it was "generated" is a random one). Thus the probability formulas derived in Section 11.1.2 permit us to use number of runs as the test statistic for tests of randomness. We could use any r_{ij}, or combination thereof, as the test statistic; however, we shall concentrate upon U, the total number of runs, irrespective of length, among both types of element.

The randomness for which we test is randomness in *order* of appearance (i.e., permutational arrangement) along a single dimension, but that dimension may be either a temporal or a spatial one. Appropriate examples from the literature on runs are the randomness of appearance of wet versus dry days in a sequence of days, or the randomness of arrangement of occupied versus vacant seats at a lunch counter. Furthermore, the original observations to which we apply the test may be the members of a natural, i.e., intrinsic, dichotomy such as occupied or vacant seats, or they may have a multiplicity of possible values which are dichotomized by some more or less arbitrary definition, such as above or below the sample median for observations on a continuously distributed variate.

11.2.2 Null Hypothesis

Given that the obtained sequence contains n_1 elements of one kind (arbitrarily designated 1's) and n_2 elements of the other kind (arbitrarily labeled 2's), each of the $\binom{n_1+n_2}{n_1}$ distinguishable arrangements of 1's and 2's was equally likely to have been obtained prior to sampling. This will be the case if the process determining relative ordinal position of 1's and 2's in the sequence is a random one and all assumptions are met.

H_0 implies that at every position in the sequence the probability that the element will be a 1 is the same, i.e., that the probability of a 1 remains constant throughout the sequence, and likewise, of course, for the complementary probability that the element will be a 2. Otherwise, every distinguishable arrangement would not be equally likely. For example, if the probability of a

1 decreased throughout the sequence, arrangements with the preponderance of 1's at the beginning would be more likely than the mirror-image arrangements with the preponderance of 1's at the end.

11.2.3 Assumptions

Sampling is *random* (unless sampling is considered to be part of the process whose randomness is being tested).

All observations can be unmistakably classified into one of two *mutually exclusive* and unconfusable categories.

When the two kinds of elements are values of a variate X falling above or below some dichotomizing value D, cases of $X = D$ create ambiguity which should be dealt with by method A of Section 3.3.1 (considering such cases as "tied ranks"). Such ambiguity can be avoided by using the dichotomy $X \geqq D$ versus $X < D$ or the dichotomy $X > D$ versus $X \leqq D$.

11.2.4 Efficiency

The total-number-of-runs test is appropriate and useful when the two kinds of elements constitute a natural dichotomy, i.e., when the variate whose randomness is to be tested can assume only two "values" on a nominal or categorical scale. However, when it can assume a multiplicity of values on an ordinal or higher scale, distribution-free tests based upon ranks, rather than dichotomous categories, will exploit more of the available information and would be expected to be more efficient, on that basis. Of course, if data are plentiful the number-of-runs test may have quite adequate power, and it has the undeniable advantage of being extremely simple and easy to perform.

Power functions were obtained by Bateman [2] for the total number of runs and also for the length of longest run as tests of randomness against the alternative of a simple Markov chain, i.e., the alternative that each event is dependent upon the preceding event but no other. For this case the total-number-of-runs test was found to be more powerful than the test based on length-of-longest-run. See also [1, 4].

11.2.5 Tables

Table IX of the Appendix gives critical values of U for a one-tailed test which rejects for too few runs. It tables the largest value of U' for which $P(U \leqq U') \leqq \alpha$ for $n_1 = 2\,(1)\,n_2$ and $n_2 = 4\,(1)\,20$ and for α's of .05 and .01. Since it is arbitrary which type of element is designated "1" and which is labeled "2," this table yields critical values for all cases where n_1 and n_2 are each $\leqq 20$.

Swed and Eisenhart [11] give $P(U \leq U')$ to seven decimal places for all possible values of U' (except the maximum possible value for which the probability is always 1) for $n_1 = 2\,(1)\,n_2$ and $n_2 = 2\,(1)\,20$. Thus they give, in effect, the complete cumulative probability distribution of U for all cases where n_1 and n_2 are each ≤ 20. They also give tables listing critical values of U for "too few" runs and critical values of $U - 1$ for "too many" runs for tests at the one-tailed α levels of .005, .01, .025, and .05. This is done for all values of n_1 and n_2 for which $2 \leq n_1 \leq n_2 \leq 20$, and, using an approximation rather than exact formulas, it is also done for all cases of $n_1 = n_2 = k$ where $k = 10\,(1)\,100$. Swed and Eisenhart's tables can be used for two-tailed tests, whereas Table IX cannot.

The mean and variance of U are

$$\frac{2n_1n_2}{n_1 + n_2} + 1 \qquad \text{and} \qquad \frac{2n_1n_2(2n_1n_2 - n_1 - n_2)}{(n_1 + n_2)^2(n_1 + n_2 - 1)}$$

respectively, and U is asymptotically normally distributed if the ratio of n_1 to n_2 remains constant while both approach infinity [12]. Therefore, when samples are too large for the tables to apply, approximate probabilities can be obtained by treating U as a normally distributed variate and referring the critical ratio

$$\frac{U - \dfrac{2n_1n_2}{n_1 + n_2} - 1}{\sqrt{\dfrac{2n_1n_2(2n_1n_2 - n_1 - n_2)}{(n_1 + n_2)^2(n_1 + n_2 - 1)}}}$$

to normal tables. (To correct for continuity, reduce the absolute value of the numerator by $\frac{1}{2}$.)

11.2.6 Example

Eighteen primary-school children, of whom $n_1 = 8$ are boys (B) and $n_2 = 10$ are girls (G) are in a queue before a drinking fountain. The queue has the following pattern:

$$B\,B\,B\,G\,G\,G\,G\,B\,B\;\;G\,G\,G\,G\,G\,B\,B\,B\,G$$

containing a total of 6 runs. A sociologist wishes to test, at the .05 level, the hypothesis that groupings of small children are random with regard to sex against the alternative hypothesis that small children of like sex tend to congregate. Entering Table IX with $n_1 = 8$ and $n_2 = 10$, $U = 6$ is found to be the critical value for "too few" runs at $\alpha = .05$. The hypothesis of randomness is therefore rejected in favor of the stated alternative.

11.2.7 Discussion

A particularly interesting application [8] of run tests is as follows. Let X be a continuously distributed variate the temporal randomness of whose generating process is suspect. A sample of n observations is drawn one at a time in temporal sequence from the X-population and arranged in a line in the same sequence in which they were drawn. Now let the n_1 observations having the n_1 lowest values be replaced by 1's (i.e., be regarded as elements of one type) and the $n_2 = n - n_1$ observations having the n_2 highest values be replaced by 2's (i.e., be regarded as elements of the other type). A run test applied to the resulting arrangement of 1's and 2's tests whether or not the temporal sequence in which the n_1 lowest observations were generated is a random one. Or, if X_{n_1} is the n_1th observation in order of increasing size, it tests roughly whether or not $P(X \leq X_{n_1})$ remains constant over the time period during which observations were drawn. It is to be noted that the test of randomness is with reference to the dichotomized elements and therefore is qualified by the criterion chosen for dichotomization, i.e., by n_1, which should, of course, be designated prior to sampling.

So far we have concentrated attention upon the total number of runs U as the test statistic for "number of runs" tests. Since the number of runs of 1's, r_1, and the number of runs of 2's, r_2, cannot differ by more than 1, and since $U = r_1 + r_2$, it is clear that r_1, r_2, and U contain or imply very nearly the same sets of information and in that sense are virtually redundant test statistics. There is one respect, however, in which r_i has a considerable practical advantage. The distribution of r_i is hypergeometric (see Section 11.1.2) and therefore much more extensive tables of probabilities exist for r_i than for U.

11.3 THE WALD-WOLFOWITZ TOTAL-NUMBER-OF-RUNS TEST FOR IDENTICAL POPULATIONS

This test is appropriate when n_1 observations have been drawn randomly and independently from a continuously distributed X-population and n_2 from a Y-population. The $n_1 + n_2$ observations are arranged in ascending order of size, irrespective of the population from which they came. Each observation is then replaced by an X or a Y depending upon the sample to which it originally belonged. The total number of runs U of like elements, i.e., X's or Y's, is then counted and used as the test statistic. If the populations differ, elements of one type would be expected to cluster, tending to make U "too small," whereas if the populations are identical the arrangement of X's and Y's should be a random one and U should have the same null distribution as in the U test for randomness.

This is one of the least powerful distribution-free tests. It has A.R.E. of zero both as a test for equal means relative to Student's t and as a test for equal variances relative to the F ratio, under conditions meeting all the assumptions of the comparison tests. Furthermore, it typically ranks last in empirical power comparisons among several distribution-free tests applied to the same data. The test is consistent if the ratio n_1/n_2 of sample sizes remains constant as n_1 and n_2 approach infinity and if certain other very mild conditions are met. If the ratio n_1/n_2 does not remain constant, but approaches zero or infinity, the test tends to be inconsistent. That is to say, if one sample is of much greater size than the other, observations from the sample of smaller size are almost certain to be separated from each other by observations from the larger sample (provided, of course, that the two populations are continuously distributed and occupy the same interval of abscissa values); thus the number of runs will tend to be maximum regardless of whether the null hypothesis of identical populations is true or false.

11.4 LENGTH OF LONGEST RUN AS A TEST OF RANDOMNESS

11.4.1 Rationale

For reasons pointed out in Section 11.2.1, we could use any r_{ij}, or combination thereof, as the test statistic for a test of randomness. Therefore, we may use as test statistic the length of longest run of *one* predesignated type of element, of *either* type of element, i.e., the type to contain the run being left unspecified, or of *each* type, i.e., the test statistic is the longest length of run which can be found both among the 1's and among the 2's.

11.4.2 Null Hypothesis

Same as Section 11.2.2.

11.4.3 Assumptions

Same as Section 11.2.3.

11.4.4 Efficiency

See Section 11.2.4.

11.4.5 Tables

Tables derived from exact formulas and covering all situations in a sufficiently general case, e.g., applying to all combinations of $n_1 \leq 20$ and $n_2 \leq 20$, are apparently not to be found. Olmstead [8] gives complete tables of exact probabilities for a run of length S or greater among elements of one predesignated type, among elements of either type, and among elements of each type for the special cases $n_1 + n_2 = 10$, $n_1 + n_2 = 20$, and $n_1 = n_2 = k$ for $k = 5, 10, 20, 30, 50$, and 100. For the special case $n_1 = n_2 = n/2$ he also gives tables of critical values of n, derived from approximate formulas, for $S = 1\ (1)\ 20$, $P(S$ or greater$) = .01, .10, .50, .90$, and $.99$, for a run of length S or greater among elements of one type, either type, and each type. A fragment of Olmstead's tables has been recapitulated and slightly supplemented by Mosteller [7]. Bateman [2], in effect, tables the exact point probability for a run of either kind of element of length *exactly* S for $n = 10$ (1) 15 (5) 20 and $n_1 = 1\ (1)\ n - 1$. Burr and Cane [3] examine the probability that the length of the longest run among n_1 elements of a single predesignated type will be of length S or *smaller*. They table the critical values of n for lower-tailed tests at the .05 and .01 levels for $S = 1\ (1)\ 3$ and $n_1 = 4\ (2)\ 60$ [also for $n_1 = 60\ (2)\ 80$ for the case $S = 3$].

11.4.6 Example

A psychologist hypothesizes that success feeds upon itself, e.g., that winning a trophy one year increases an athletic team's motivation to win it the next year. (He does not yet wish to make the hypothesis a compound one by including the postulate that failure feeds upon failure.) By chance he obtains the following list of wins (W) and losses (L) of an annual trophy, in sequence, for a certain team:

$$L\,L\,L\,W\,W\,L\,L\,L\,L\,W\,W\,W\,W\,L\,L\,W\,W\,W\,L\,L\,L\,L\,L\,L$$

Does this sequence support the hypothesis? There are nine W's and fifteen L's and the longest run of W's contains four of them, so $n_1 = 9$, $n_2 = 15$, $n = 24$, and $S = 4$. Substituting these values into the formula for the probability of one or more runs of 1's of length S or greater we have

$$P(r_{1,(\geqq S)} \geqq 1) = \frac{\sum_{i=1}^{n_1/S} (-1)^{i+1}\binom{n_2+1}{i}\binom{n-iS}{n_2}}{\binom{n}{n_1}}$$

$$= \frac{\sum_{i=1}^{9/4} (-1)^{i+1}\binom{16}{i}\binom{24-4i}{15}}{\binom{24}{9}}$$

$$= \frac{+\left(\frac{16!}{1!\,15!}\right)\left(\frac{20!}{15!\,5!}\right) - \left(\frac{16!}{2!\,14!}\right)\left(\frac{16!}{15!\,1!}\right)}{\frac{24!}{9!\,15!}}$$

$$= \frac{(16)(15{,}504) - (120)(16)}{1{,}307{,}504} = \frac{246{,}144}{1{,}307{,}504}$$

$$= .188$$

Thus the chance probability that a sequence of 9 W's and 15 L's would contain a run of four or more consecutive W's is about .19, so the null hypothesis of randomness cannot be rejected at a standard significance level, and the data provide no significant support for the hypothesis proposed by the psychologist.

11.4.7 DISCUSSION

As the above example shows, so long as n_1/S is small and one is interested only in the length of longest run of elements of a single predesignated type, cumulative probabilities can be obtained with relative ease directly from the formula, without resort to any tables. When n is moderately large, however, the task may be greatly facilitated by tables of binomial coefficients, such as [10], which gives the numerical value of $\binom{n}{r}$ for $r = 2\,(1)\,n/2$ and $n = 2\,(1)\,200$ as well as for a variety of additional cases. Actually, testing by formula is easier than it might appear, since when n is moderately small all S's that are significantly large exceed $n_1/2$ and therefore involve a "summation" over the single value $i = 1$. For example, in no case where $n_1 + n_2 = 20$ is an S value which is $\leqq n_1/2$ significant at the .05 level.

11.5 LENGTH OF LONGEST RUN IN BINOMIAL TRIALS AS A TEST FOR IMPROVEMENT

11.5.1 RATIONALE

We have already established that the "null" probability of obtaining a run of 1's of length S or greater, *given that* there are n_1 1's and $n_2 = n - n_1$ 2's is

$$P(r_{1,(\geqq S)} \geqq 1 \mid n_1, n) = \frac{\sum_{i=1}^{n_1/S} (-1)^{i+1} \binom{n_2 + 1}{i} \binom{n - iS}{n_2}}{\binom{n}{n_1}}$$

Now if the 1's and 2's are the outcomes of n binomial trials where n_1 and n_2 are free to vary and p is the constant probability of a 1 on a single trial, then the null probability of obtaining exactly n_1 1's is

$$P(n_1 \mid n, p) = \binom{n}{n_1} p^{n_1} (1 - p)^{n_2}$$

Therefore (see Section 11.1.4) the product

$$[P(r_{1,(\geqq S)} \geqq 1 \mid n_1, n)][P(n_1 \mid n, p)]$$

summed over all possible values of n_1, is the probability of obtaining a run of S or more 1's in n binomial trials, given that the probability of a 1 on a single trial is p, but completely irrespective of n_1, i.e., it is an a priori probability unqualified by n_1 or n_2 (but qualified by their sum n). Thus

$$P(r_{1,(\geqq S)} \geqq 1 \mid n, p) = \sum_{n_1=S}^{n} \sum_{i=1}^{n_1/S} (-1)^{i+1} \binom{n_2+1}{i} \binom{n-iS}{n_2} p^{n_1}(1-p)^{n_2}$$

is the probability of a run of S or more 1's in n binomial trials in which the probability of a 1 on a single trial is p.

Now let 1's be "successes" and suppose that an experimenter knows the probability of success to be $p = p_0$ at the beginning of the n trials and wishes to test the null hypothesis that p remains constant at p_0 throughout the n trials, using a test that is especially likely to reject when p increases monotonically. If H_0 is true, the probability of a run of S or more consecutive successes is obtained by substituting p_0 for p and solving the above formula for $P(r_{1,(\geqq S)} \geqq 1 \mid n, p)$, so that formula gives the (upper-tail cumulative) null distribution of S. On the other hand, if p increases monotonically then (a) n_1 will tend to be greater than in the null case; (b) the local density of successes will tend to increase (i.e., the n_1 successes will tend to bunch up increasingly) as one moves from the beginning to the end of the series. *Both* (a) and (b) increase the probability of obtaining a run of S or more successes: making length of longest run of successes a test statistic that is particularly sensitive to monotonic increase in p. The rejection region, of course, would consist of the upper tail of the null distribution representing longest runs that were "too long" and had small upper-tail cumulative probabilities. (Note that by taking as rejection region longest runs that are "too short," e.g., by rejecting when $P(r_{1,(\geqq S+1)} \geqq 1 \mid n, p) > .95$, where S is the length of the longest obtained run of 1's, we would have a test especially sensitive to the condition that p *decreases* monotonically throughout the n trials. Thus if 1's are "successes" we would have a test for degradation. However, if 1's are "failures," such as defectives in quality control, we would again have a test for improvement and this test might be preferable to the original test for improvement if defectives are rare.)

11.5.2 Null Hypothesis

The probability that in n trials there will be exactly n_1 1's is

$$\binom{n}{n_1} p_0^{n_1}(1-p_0)^{n-n_1}$$

where p_0 is the known probability of success on the first trial, and for any obtained value of n_1 each of the $\binom{n}{n_1}$ distinguishable arrangements of 1's and 2's is equally probable.

This will be the case if the probability of a 1 is p_0 on each of the n trials and all assumptions are met.

11.5.3 Assumptions

Sampling is *random*; 1 and 2 are *mutually exclusive* outcomes of a binomial event.

11.5.4 Efficiency

A sequentially monotonic increase in p over p_0 increases the probability of long runs of 1's in two separate and distinct ways, only one of which would be involved in the conventional length-of-longest-run-of-1's test. Therefore the present test should be considerably more efficient than the latter when sensitivity against increasing p is desired and p_0 is known. However, no explicit information appears to be available.

11.5.5 Tables

Grant [5] has tabled $P(r_{1,(\geqq S)} \geqq 1 \mid n, p_0)$ to four decimal places for p_0's of $\frac{1}{2}$, $\frac{1}{3}$, $\frac{1}{4}$, and $\frac{1}{5}$ for $n = 16\ (4)\ 20\ (5)\ 50$ and apparently for all values of S yielding probabilities that are $\leqq .1000$ and $\geqq .0001$. For the same values of p_0 and S, he also gives the largest value of n for which $P(r_{1,(\geqq S)} \geqq 1 \mid n, p_0) \leqq \alpha$ for α's of .001, .005, .01, .05, and .10.

11.5.6 Example

An experimenter wishes to test whether or not a monkey can learn to associate a red light with food. The monkey's food is always hidden in one of five boxes, and the "reward" box is always illuminated by a red light. The probability of "success" on a single trial is therefore $\frac{1}{5}$ if the null hypothesis of no learning is true. Consulting Grant's tables [5], the experimenter finds that when $p = \frac{1}{5}$, a run of 4 or more successes in 40 trials is significant at the .05 level. Therefore he decides to run not more than 40 trials and to reject the null hypothesis whenever the number of consecutive successes reaches 4. The monkey's successes (S) and failures (F) to go first to the red-illuminated box are

$$F\,F\,F\,S\,F\,F\,F\,F\,F\,S\,S\,F\,S\,S\,S\,S$$

so only 16 of the maximum of 40 trials had to be run. The significance level, however, is not reduced, but remains .05, since it had originally been intended to run as many as 40 trials if necessary.

11.5.7 Discussion

The question arises as to which type of test is appropriate, that which treats n_1 and n_2 as given or that which treats them as variable. Obviously a test [7, 8] which defines a 1 as a run above the sample median or as one of the n_1 lowest variate values in the sample preordains the value of n_1 as soon as n is decided upon (at least this is the case if the assumption of continuity is satisfied). Clearly it would be improper to treat "above the sample median" as a binomial event with probability $\frac{1}{2}$, since in n trials of such an event, n_1

should be able to assume any value from zero to n, which is obviously impossible if n_1 is the number of the n observations above the median of the same n observations.

On the other hand, if (under the null hypothesis) n_1 is a binomially distributed variate, either type of test would be valid. If one does not know the value of p_0, he will be obliged to use the more elementary test which takes n_1 and n_2 as given. (For example, in testing the randomness of a seating arrangement, one would take the observed numbers of occupied and vacant seats as given if he did not know the "null" probability of occupancy of a single seat.) If one does know the value of p_0, he may use the present, more sophisticated, test, and that test will be especially appropriate (and, presumably, efficient) if the condition against which sensitivity is required implies strong, more or less monotonic, sequential change in p.

11.6 EXTENSIONS OF RUN THEORY

Runs discussed so far have involved only two kinds of elements arranged in a linear sequence. However, various probability formulas have also been derived for runs of like elements when there are more than two kinds of elements [6] and for runs where adjacency among elements can occur along two or more dimensions. Such multiple-category and polydimensional runs are generally analyzed on the basis of large sample theory, using critical ratios, rather than exact probabilities, since the exact distribution of such runs rapidly becomes difficult to tabulate as sample size increases.

REFERENCES

1. Barton, D. E., and F. N. David, "Non-Randomness in a Sequence of Two Alternatives II. Runs Test," *Biometrika*, **45** (1958), 253–256.
2. Bateman, G., "On the Power Function of the Longest Run as a Test for Randomness in a Sequence of Alternatives," *Biometrika*, **35** (1948), 97–112.
3. Burr, E. J., and Gwenda Cane, "Longest Run of Consecutive Observations Having a Specified Attribute," *Biometrika*, **48** (1961), 461–465.
4. David, F. N., "A Power Function for Tests of Randomness in a Sequence of Alternatives," *Biometrika*, **34** (1947), 335–339.
5. Grant, D. A., "Additional Tables of the Probability of 'Runs' of Correct Responses in Learning and Problem-Solving," *Psychological Bulletin*, **44** (1947), 276–279.
6. Mood, A. M., "The Distribution Theory of Runs," *Annals of Mathematical Statistics*, **11** (1940), 367–392.

7. Mosteller, F., "Note on an Application of Runs to Quality Control Charts," *Annals of Mathematical Statistics*, **12** (1941), 228–232.

8. Olmstead, P. S., "Runs Determined in a Sample by an Arbitrary Cut," *Bell System Technical Journal*, **37** (1958), 55–82.

9. Prairie, R. R., W. J. Zimmer, and J. K. Brookhouse, "Some Acceptance Sampling Plans Based on the Theory of Runs," *Technometrics*, **4** (1962), 177–185.

10. Royal Society Mathematical Tables, Vol. 3. *Table of Binomial Coefficients*. London and New York: Cambridge University Press, 1954.

11. Swed, Frieda S., and C. Eisenhart, "Tables for Testing Randomness of Grouping in a Sequence of Alternatives," *Annals of Mathematical Statistics*, **14** (1943), 66–87.

12. Wald, A., and J. Wolfowitz, "On a Test Whether Two Samples are from the Same Population," *Annals of Mathematical Statistics*, **11** (1940), 147–162.

12

RUNS UP AND DOWN

A type of run test for randomness can be obtained by defining a run as an unbroken sequence of increasing or decreasing observations. In this case, the two kinds of events, "greater than the preceding observation" and "smaller than the preceding observation," are neither fixed in number nor of constant probability (since their probabilities depend on how "extreme" was the preceding observation) under the null hypothesis of randomness. Thus the formulas developed in the preceding chapter are inappropriate. By investigating the probability for a given pattern of observation magnitudes, rather than a given pattern of dichotomized "events," the necessary formulas are obtained. Run tests of this type using the total number of runs or the length of the longest run as the test statistic, are appropriate for testing the randomness of the process generating the values of a continuously distributed variate.

12.1 BASIC FORMULAS

Suppose that n observations have been taken on a continuously distributed variable and arranged in the order in which recorded. A steadily ascending sequence of observations will be defined as a run "up" and a monotonically decreasing sequence will be called a run "down." Now suppose that each observation is subtracted from the succeeding observation. There will be $n - 1$ algebraic signs to replace the n original observations. A run "up" will now be more definitively indicated by a sequence of $+$ signs, and a run "down" will be unambiguously identified by a run of $-$ signs. If the series is random, the farther an observation is from the median of the series, the less likely it will be that the succeeding observation will depart from the median still farther. Therefore "plus" and "minus" are not constant probability events under the null hypothesis of randomness, and probability formulas for runs up and down must be derived in the light of that fact.

12.1.1 Total Number of Runs

Probabilities for the total number of runs in an arrangement can be obtained from a recursion formula [see 1]. Let $P(i \mid j)$ be the a priori probability that in a random linear arrangement of j unequal numbers the total number of runs up and down will be exactly i. Then

$$P(r \mid n) = \frac{rP(r \mid n-1) + 2P(r-1 \mid n-1) + (n-r)P(r-2 \mid n-1)}{n}$$

Proof: Consider a single linear arrangement (i.e., a single permutation) of $n - 1$ unequal numbers containing a total of r runs of which r_+ are runs up and r_- are runs down. No generality is lost, and considerable simplification of conceptualization is gained, if these $n - 1$ unequal numbers are replaced by their ranks, so consider the numbers to be the integers from 1 to $n - 1$. These $n - 1$ integers may be regarded as the partitions between n cells, i.e., there are n spaces between, before, and after the integers. Now if we drop the integer n into any one of these n cells we obtain an arrangement, i.e., a permutation, of n integers. Thus each distinguishable permutation of the first $n - 1$ integers yields n distinguishable permutations of the first n integers. And the total number of distinguishable permutations of the first n integers may be regarded as having been generated by inserting the integer n into one of the n cells formed by a distinguishable permutation of the first $n - 1$ integers, and doing so for each of the n cells and each of the $(n - 1)!$ distinguishable permutations.

Dropping the integer n into a cell increases the number of integers by 1 and has one of three possible effects upon the total number of runs: it may leave the total unchanged, increase it by 1, or increase it by 2. First consider the situations in which the total number of runs is unchanged. There are three:

(a) the n is inserted at the end of a run up;
(b) the n is inserted at the next to the end position of a run up that is followed by a run down (in which case the n becomes the last integer of the run up, and the previous last integer of the run up becomes part of the following run down);
(c) the n is inserted at the beginning of a permutation that began with a run down.

Now if the original permutation starts and ends with a run up (so that $r_- = r_+ - 1$), r_+ cells yield situation (a), $r_+ - 1 = r_-$ cells yield situation (b), and no cells yield (c). So there are $r_+ + r_- = r$ ways of inserting the n without changing the number of runs. If the original permutation starts and ends with a run down (so that $r_+ = r_- - 1$), r_+ cells yield (a), $r_+ = r_- - 1$ cells yield (b), and 1 cell yields (c). So the number of ways is $r_+ + r_- - 1 + 1 = r$. If the permutation starts with a run up and ends with a run down (so that $r_+ = r_-$), r_+ cells yield (a), $r_+ = r_-$ yield (b), and none yields (c). So the number of ways is $r_+ + r_- = r$. If the permutation starts with a run down and ends with a run up (so that $r_+ = r_-$), $r_+ = r_-$ yield (a), $r_+ - 1$ yield (b), and 1 yields (c). So the number of ways is $r_- + r_+ - 1 + 1 = r$. (Finally, if the entire permutation consists of a single run, one cell will yield (a), none (b), and none (c) if it is a run up, and none will yield (a), none (b), and one (c) if it is a run down; in either case the number of ways is $1 = r$.) The above represent all possible cases, i.e., all relevant types of permutation, and in every case there are r cells into which the n can be inserted without changing the total number of runs.

Next consider the situations in which the total number of runs is increased by 1. There are four:

(a) the first run is a run up and the n is placed at the beginning of it, i.e., in the leftmost cell, thereby adding a run down (consisting of a single minus) at the beginning of the sequence;
(b) the first run is a run down and the n is inserted between the first two numbers, i.e., in the next-to-leftmost cell, thereby creating a run up (consisting of a single plus) at the beginning of the sequence;
(c) the last run is a run up and the n is inserted between the last two numbers, thereby creating a final run down consisting of one minus;
(d) the last run is a run down and the n is inserted at the end of it, thereby adding a run up consisting of one plus.

There is only one way of meeting the requirements of each of the above situations, given that the initial or terminal run is of the specified type. Furthermore, the initial run must be either up or down but cannot be both, and likewise for the terminal run, i.e., the specified runs for (a) and (b) are mutually exclusive and exhaustive, and likewise for (c) and (d). Therefore for any permutation, there are two ways of inserting the n so as to produce one more run, one way corresponding to (a) or (b) and one to (c) or (d).

The situation in which the number of runs is increased by 2 is exemplified by the case where the n is inserted in the middle of a long run of either type, thereby splitting the long run into two parts separated by a single sign of the opposite type. The number of runs cannot be increased by more than 2. Therefore, since for any permutation inserting the n into r of the n cells produces no increase in runs and insertion into 2 cells increases runs by 1, insertion into $n - r - 2$ of the cells must increase the number of runs by 2.

Now consider the totality of all $(n - 1)!$ distinguishable permutations of the $n - 1$ original integers and let $A_{r,n-1}$, $A_{r-1,n-1}$, and $A_{r-2,n-1}$ be the number of these permutations or arrangements containing r, $r - 1$, and $r - 2$ runs respectively, and let $A_{r,n}$ be the total number of arrangements of n integers that contain r runs. Then, we have shown, in effect, that

$$\begin{aligned} A_{r,n} &= rA_{r,n-1} + 2A_{r-1,n-1} + [n - (r - 2) - 2]A_{r-2,n-1} \\ &= rA_{r,n-1} + 2A_{r-1,n-1} + (n - r)A_{r-2,n-1} \end{aligned}$$

But $P(r \mid n) = A_{r,n}/n!$ and $P(r \mid n - 1) = A_{r,n-1}/(n - 1)!$, etc., so substituting for the A's in the above equation we have

$$\begin{aligned} n!\,P(r \mid n) = (n - 1)!\,rP(r \mid n - 1) &+ (n - 1)!\,2P(r - 1 \mid n - 1) \\ &+ (n - 1)!\,(n - r)P(r - 2 \mid n - 1) \end{aligned}$$

$$P(r \mid n) = \frac{rP(r \mid n - 1) + 2P(r - 1 \mid n - 1) + (n - r)P(r - 2 \mid n - 1)}{n}$$

12.1.2 Length of Longest Run

The expected number of runs of like *difference signs*, of length S or longer resulting from a random linear arrangement of n unequal *numbers* is

$$E(r_{\cdot,(\geqq S)}) = \frac{2 + 2(n - S)(S + 1)}{(S + 2)!}$$

when the type of difference sign is unspecified, i.e., if the runs of length $\geqq S$ can consist either of S or more plusses or of S or more minuses, and is

$$E(r_{+,(\geqq S)}) = E(r_{-,(\geqq S)}) = \frac{1 + (n - S)(S + 1)}{(S + 2)!}$$

for runs consisting of a single specified algebraic sign.

When $S \geqq n/2$, the expected number of runs of length $\geqq S$ of the designated type is also the probability of obtaining an arrangement yielding a run of length $\geqq S$ of that type, i.e.,

$$\begin{aligned} E(r_{\cdot,(\geqq S)}) &= P(r_{\cdot,(\geqq S)} \geqq 1) \\ E(r_{+,(\geqq S)}) &= P(r_{+,(\geqq S)} \geqq 1) \\ E(r_{-,(\geqq S)}) &= P(r_{-,(\geqq S)} \geqq 1) \end{aligned}$$

When S is less than $n/2$, the expected number exceeds the corresponding probability. However, when the latter is $\leqq .05$, it tends to be well approximated by the former. So if one uses the expected number as an estimate of the probability when conducting an upper-tail test (i.e., rejecting for "too long" a longest run) at a standard significance level, (a) the decision to accept or reject will tend to be the correct one (i.e., when H_0 is true the test will reject very nearly α of the time and accept very nearly $1 - \alpha$ of the time); (b) when the test rejects the expected number will very nearly equal the true probability level of the test; (c) when the test accepts the expected number may badly overestimate the true probability level, especially when the latter is much larger than .05.

Proof: Consider the probability that the ith observation obtained (i.e., the ith number in the sequence of n unequal numbers) initiates a run up of $S + 1$ or more observations so that the difference-sign obtained by subtracting the ith from the $(i + 1)$th observation is the first $+$ in a sequence of S or more plusses. Consider first the probability that the series begins with a run up of $S + 1$ or more observations. If the series is random, each of the $(S + 1)!$ permutations of the first $S + 1$ observations is equally probable. But only one of these permutations results in an unbroken run up of length $S + 1$. So the probability that the series begins with a run up of S or more plusses is $1/(S + 1)!$. (This is also the unqualified probability of obtaining a run up of S or more plusses when $n = S + 1$, i.e., when the entire series contains only $S + 1$ observations.) Now consider the general case where $n \geqq S + 2$ and $2 \leqq i \leqq n - S$. Let the $(i - 1)$th to the $(i + S)$th observations be ranked from 1 to $S + 2$ in order of increasing magnitude. If the series is random, each of the $(S + 2)!$ distinguishable permutations of these $S + 2$ observations is equally probable. But in order for the ith observation in the total sequence to initiate a run up of $S + 1$ observations, the $S + 2$ ranks must be arranged so that: (a) rank 1, i.e., the lowest among the $S + 2$ observations occupies position i in the sequence; (b) any one of the remaining $S + 1$ ranks occupies position $i - 1$; (c) the remaining S ranks are arranged in increasing order of size, occupying positions $i + 1$ to $i + S$. Of these three requirements, (a) can be fulfilled in only one way, (b) can be accomplished in $S + 1$ ways, and (c) can then take place in only one way. So $(1)(S + 1)(1) = S + 1$ of the $(S + 2)!$ equally likely permutations produce the desired effect whose probability is therefore $(S + 1)/(S + 2)!$. Notice that by requiring the $(i - 1)$th observation to be higher than the ith, we have insured that the run up beginning with the ith observation is not a continuation of a previous run up, but rather initiates the run.

We have seen that when $n \geqq S + 2$, the probability that a run of $S + 1$ or more ascending observations begins with the ith observation is $1/(S + 1)!$ when $i = 1$ and is $(S + 1)/(S + 2)!$ when i is any one of the $n - S - 1$ positions from 2nd to $(n - S)$th. These are probabilities that the ith observation *initiates* a run up of specified minimum length, i.e., each probability is conditional upon the $(i-1)$th observation, if there is one, *not* being a continuation of the run. Otherwise viewed, each probability is conditional upon the ith observation not being a continuation of any run up that began at some point earlier in the series. Therefore, since the probabilities do not refer to overlapping events, they can be summed over all relevant values of i to obtain the expected number of runs of the specified type. Thus, when $n \geqq S+2$, the expected number of runs of ascending observations of length $S+1$ or more, or of plus difference-signs of length S or more, is

$$\begin{aligned} E(r_{+,(\geqq S)}) &= \frac{1}{(S+1)!} + \frac{(n-S-1)(S+1)}{(S+2)!} \\ &= \frac{S+2+(n-S)(S+1)-(S+1)}{(S+2)!} \\ &= \frac{1+(n-S)(S+1)}{(S+2)!} \end{aligned}$$

(It should be noted that this derivation is based upon the n observations being in a random order, *not* upon each difference-sign of a given type being equally likely, which is not the case.)

By completely analogous reasoning, one obtains exactly the same value for the expected number of runs of *minuses* of length greater than or equal to S. A run up and a run down commencing with the ith observation are mutually exclusive events, and they have equal probabilities. Therefore to obtain the expected number of runs up *or* down of length $\geqq S$, the expected number of runs up of length $\geqq S$ should be doubled.

Consider the n observations ranked from 1 to n in order of increasing magnitude. There are $n!$ permutations of these ranks, and the *expected number* of runs of a specified type is simply the total number of such *runs* which can be found in these $n!$ permutations divided by the number of permutations $n!$. On the other hand, the *probability* of at least one run of the type specified is the total number of *permutations* in which such a run can be found divided by the number of permutations $n!$. Therefore, the probability and expected number do not coincide when it is possible for more than one run of the specified type to be found in a single permutation. However, when $S > (n-1)/2$ (i.e., when $S \geqq n/2$) the formulas already presented for the expected number of runs of a given variety also give the exact probability of occurrence for such runs. This appears to be the only situation, when dealing with runs up or down, in which an exact probability can be calculated without resort to a recursion formula.

Olmstead [5] has published tables of probabilities for runs of like difference-signs of length $\geqq S$. The probabilities are obtained from an exact recursion formula for $n \leqq 14$ and from approximate formulas for higher n's. We shall now compare the *expected number* of runs of like difference-signs of length $\geqq S$, calculated from the formula given at the beginning of this section, with Olmstead's corresponding *probabilities*, in cases where $S < n/2$ and the probability is the largest probability $\leqq .10$ (thus applying to $\alpha = .10$ for runs of either type or to $\alpha = .05$ for runs in a single direction). We shall do likewise for the largest probability $\leqq .02$. It is not until $n = 9$ that we can find an $S < n/2$ yielding an upper-tail cumulative probability $\leqq .10$,

n	S	*Expected number*	*Probability*	S	*Expected number*	*Probability*
9	4	.072222	.071836	5	.009921	.009921
11	4	.100000	.098362	5	.014683	.014670
40	5	.08373	.0810	6	.01186	.0118
1000	7	.04379	.0428	8	.00492	.0049

so we start there. The accompanying table shows close approximation between expected numbers and probabilities when the latter are $\leqq .10$. On the other hand, for $n = 14$ and $S = 3$, the expected number and the exact probability of runs that long or longer are .750000 and .583315, respectively. It seems clear from the above, therefore, the when S is so long as to fall in an upper-tail rejection region corresponding to $\alpha \leqq .05$, the expected number formula will yield a close approximation to the true cumulative probability. When $S \geqq n/2$, the "approximation" will be perfect, yielding the exact true probability; when $S < n/2$, the approximation will overestimate and therefore err in the "conservative" direction of acceptance of H_0.

12.2 NUMBER OF RUNS UP AND DOWN AS A TEST OF RANDOMNESS

12.2.1 RATIONALE

If the process generating n unequal numbers (i.e., generating the objects or events of which the numbers are measures) is a random one, then each of the $n!$ possible distinguishable arrangements of the n numbers had equal a priori probability of being the sequence actually observed, and the formulas derived in Section 12.1 under this supposition give the chance probability of the observed number of runs. However, if the generating process is nonrandom, then certain sequences are more probable than others and the observed sequence is excessively likely to be one of them. Consequently, the observed number of runs should tend to fall at one of the far tails of the chance distribution of the total number of runs. Thus we may use the total number of runs r to test a null hypothesis of randomness. If we wish to guard against an alternative of "insufficient" fluctuation, we shall reject for "too few" runs. Against an alternative of overly frequent fluctuation, we reject for "too many" runs. And against general alternatives, we use a two-tailed rejection region.

12.2.2 Null Hypothesis

Each of the $n!$ possible distinguishable arrangements of the n unequal numbers was equally likely, a priori, to have become the sequence of numbers actually observed.

This will be the case if the process generating the numbers is a random one and if all assumptions are met.

12.2.3 Assumptions

Sampling is *random* (unless sampling is considered to be part of the process whose randomness is being tested).

Each of the n observations occupies a *unique position* in sequence, i.e., no two observations are "tied for position."

Each of the n observations has a *unique value*, i.e., there are no tied observations. (For treatment of ties, see Method A of Section 3.3.1.)

12.2.4 Efficiency

This test was compared with one parametric and eight distribution-free tests as a test for randomness against normal regression alternatives. It had A.R.E. of zero with respect to all of them.

Because of its more sophisticated use of "information," one would expect this test to be superior to a number-of-runs test (see 11.2) that artificially dichotomized the values of a continuously distributed variate and took the number of elements falling in each member of the dichotomy as given (unless, of course, some special virtue derived from these constraints, e.g., if one were interested only in the randomness of the "direction" of an observation's value with respect to the point of dichotomization).

When data are plentiful, so that power is large even when efficiency is small, this test may prove highly desirable because of its ease and simplicity of application.

12.2.5 Tables

Table X of the Appendix gives the probability that the total number of runs of both types (i.e., the number of runs up plus the number of runs down, runs of all lengths being counted) will be *r or fewer* for $n = 2\,(1)\,25$ and $r = 1\,(1)\,n - 1$. Since the *number* of runs of monotonically changing (i.e., ascending or descending) observations is the same as the number of runs of like difference-signs, one may count either in determining r.

The mean and variance of the total number of runs r are $(2n - 1)/3$ and $(16n - 29)/90$, respectively, and r is asymptotically normally distributed. So, when n exceeds 25, an approximation to the probability of r may be obtained by treating r as a normal deviate and referring the critical ratio

$$\frac{r - [(2n - 1)/3]}{\sqrt{\frac{(16n - 29)}{90}}}$$

to normal tables. By reducing the absolute value of the numerator by $\frac{1}{2}$, the critical ratio can be corrected for continuity.

12.2.6 Example

A quality control engineer wishes to test, at the .05 level, the hypothesis that the process producing a certain manufactured article is random, with regard to sequence of size variations, against the alternative that there is a tendency toward "too few" runs up and down. Twenty-five articles produced at different points in time are randomly selected and measured. Variations occurred only in the last three digits of the measurements so they alone are reported. The measurements on the 25 articles in the same sequence in which the articles were produced are: 644, 640, 633, 626, 627, 644, 646, 654, 650, 650, 651, 655, 651, 655, 674, 686, 694, 695, 700, 700, 706, 714, 716, 717, 715. Subtracting each number from the number immediately following and recording only the sign of the difference (or recording a zero when the two numbers are equal) the following series of 24 difference-signs is obtained:

$$- - - + + + + - 0 + + - + + + + + + 0 + + + + -$$

The first zero lies between a minus and a plus, so regardless of what algebraic sign it truly represents the number of runs is the same, i.e., it is a "noncritical tie." The second zero represents a critical tie. If it is replaced by a + there are 7 runs of like difference-signs, and if it is replaced by a − there are 9. Entering Table X with $n = 25$, $r = 9$, the probability of 9 or fewer runs up and down is found to be .0003, and for $r = 7$ it is .0000. So the hypothesis of randomness is rejected in favor of the alternative of "too few" runs in either case and the ties present no problem.

12.2.7 Discussion

The context of the above example is fictitious, but the sequence of numbers is not. They are successive closing prices of a "growth stock" on the New York Stock Exchange, the first two digits representing the integral part of the price and the last digit representing eighths, rather than tenths, of a point. The reason for the substitution of context is that tests based on runs up and down are more appropriate in situations where the experimenter is interested in *what caused nonrandomness* than where he is interested in *extrapolating a trend.* In the latter case the experimenter can draw practically useful inferences from the direction of the last run or from the direction of overall trend based on the entire series, but neither the total number of runs nor the length of longest run (even of a specified type) necessarily contains, or even implies, this information. For example, consider the following series: 5, 3, 1, 2, 4, 7, 9, 10, 8, 6. It contains "too few" runs and "too long" a longest run, both significant at about the .01 level, so the evidence favors the conclusion of nonrandomness. But what future trend does this nonrandomness signify? The overall trendline for the entire series is essentially horizontal. The longest run is upward and significant. But the latest run is downward and nonsignificant. (One could just as easily have constructed a series in which both the longest run and the final run are significantly long but are opposite in direction.) The ambiguity stems from the fact that while too few runs and too long a run tend to identify a form of *nonrandomness*, they do not unequivocally identify a form of either overall or current *trend.* A significant longest run that occurs early in the sequence, for example, is too deeply imbedded in past history, and therefore too far removed from the present, to be a very efficient predictor of the future.

The quality control engineer, on the other hand, is not seeking to predict the nonrandomnesses of the future but rather to correct for those of the past. Too few runs, too long a run at any point in the series, and indeed any form of "significant" nonrandomness that is explicable in terms of a plausible defect in the manufacturing process, provide him with valuable clues in seeking an "assignable cause" which may be corrected, thereby restoring the process to randomness (with perhaps a smaller variance). Furthermore, it happens that runs up and down are natural indices of a fairly common type of defect in manufacturing processes. Thus, in the past, quality control seems to have been the major area of application for tests based on runs up and down.

12.3 LENGTH OF LONGEST RUN AS A TEST OF RANDOMNESS

This test makes the same assumptions and is used to test the same null hypothesis as the preceding test. However it tests that H_0 against the different, alternative hypothesis that the longest run up, run down, or run of either type, is overly long. With this modification, the rationale and discussion of the preceding test apply equally to the present test.

The a priori probability that a random arrangement of n unequal numbers will yield a run of either positive or negative difference-signs of length S or longer has been tabled by Olmstead [5] to at least four decimal places for $n = 2\ (1)\ 15\ (5)\ 20\ (20)\ 100$ and n's of 200, 500, 1000, and 5000 and in every case for all values of S yielding nonzero, rounded table entries. (For $n \leqq 14$, use Olmstead's Table 2, which gives exact probabilities to eight decimal places; for higher n's, use Table 5 when $S \leqq 6$ and Table 3 when $S \geqq 7$. The latter two tables give approximate probabilities.) The formulas for exact expected values for runs of like difference-signs of length $\geqq S$ given in Section 12.1.2 yield exact probabilities for such runs when $S \geqq n/2$ and approximate probabilities otherwise, the approximation being fairly good when the probability is $\leqq .05$.

As an example of application, consider only the first 20 observations (and therefore only the first 19 difference-signs) of the series given in Section 12.2.6. The longest run of like difference-signs contains 6 plusses if the second zero is counted as a minus. Entering Olmstead's Table 5 with $n = 20$, $S = 6$, we obtain .0049 as the probability for a run of 6 or more signs of the same, but unspecified, type. Alternatively, substituting into the formula given in Section 12.1.2, we obtain

$$\begin{aligned} P(r_{\cdot,(\geqq S)} \geqq 1) \cong E(r_{\cdot,(\geqq S)}) &= \frac{2 + 2(n - S)(S + 1)}{(S + 2)!} \\ &= \frac{2 + 2(20 - 6)(6 + 1)}{(6 + 2)!} \\ &= \frac{198}{40{,}320} \\ &= .00491 \end{aligned}$$

So the approximate probability of a run of 6 or more signs of the same but unspecified type is .00491. And if we had specified prior to sampling that we would base the test on the length of the longest run of plusses, the probability level for the test would be .00491/2, or .002455.

REFERENCES

1. André, D., "Sur le nombre de permutations de n éléments qui présentent S séquences," *Comptes Rendus (Paris)*, **97** (1883), 1356–1358.

2. Edgington, E. S., "Probability Table for Number of Runs of Signs of First Differences in Ordered Series," *Journal of the American Statistical Association*, **56** (1961), 156–159.

3. Kermack, W. O., and A. G. McKendrick, "Tests for Randomness in a Series of Numerical Observations," *Proceedings of the Royal Society of Edinburgh*, **57** (1936–37), 228–240.

4. Levene, H., and J. Wolfowitz, "The Covariance Matrix of Runs Up and Down," *Annals of Mathematical Statistics*, **15** (1944), 58–69.
5. Olmstead, P. S., "Distribution of Sample Arrangements for Runs Up and Down," *Annals of Mathematical Statistics*, **17** (1946), 24–33.
6. Wallis, W. A., and G. H. Moore, "A Significance Test for Time Series Analysis," *Journal of the American Statistical Association*, **36** (1941), 401–409.

13

MISCELLANEOUS TESTS

Up until now, the tests described in detail in this book have belonged to families of tests having mathematically, or at least conceptually, similar derivations, each such family being dealt with in a separate chapter. Once the concept and general type of derivation that define a family have been mastered, little additional effort is required to understand one more test belonging to that family. For this reason, some of the tests dealt with in the preceding chapters have been included more on account of their family membership than because of their individual importance. Likewise, certain moderately important tests have been excluded from this book largely because of their conceptual or mathematical isolation. Some tests, however, which are not members of a family, or which are members of a family not dealt with as such, are too important, too well known, or simply too appealing to be omitted entirely, and are therefore presented in this chapter. Kendall's test is a member of a rather large family of tests, and the various tests authored by Kolmogorov, Smirnov, and Tsao are all members of the same very large family. These tests are of primary importance, being practical, powerful, and well known, and they are described in detail. David's empty-cell test for goodness of fit is also described in detail, and a few other isolated tests are simply mentioned.

13.1 KENDALL'S TEST FOR CORRELATION

13.1.1 RATIONALE

Suppose that an X-measurement and a Y-measurement have been taken on each of n units and that tied X's and tied Y's are both impossible. Correlation can be tested by arranging the n units in increasing order on the X-variable and testing the resulting order of the Y-variable for randomness. If the two variates are independent (so that there is no correlation), the resulting sequence of Y-observations is equally likely to be any of the $n!$ possible permutations of the $n\,Y$'s. However, if the two variables are linearly (or even "monotonically") correlated, the Y-observations should tend to form an increasing or decreasing sequence, and any statistic that reflects this increase or decrease can be used to test for correlation.

Let the n units be arranged in increasing order of X-observation, and let I be the number of times a Y-observation is followed, in the resulting sequence of Y's, by a smaller Y-observation. The statistic I will be called the number of inversions, since it is the number of pairs of numbers in the sequence whose rank order is the inverse of the "natural" ascending order. Thus in the sequence of integers 3, 5, 1, 4, 2, 6, there are six inversions: 3 is followed by two smaller numbers, 1 and 2; 5 is followed by three smaller numbers, 1, 4, and 2; and 4 is followed by the smaller number 2. Actually, I could serve perfectly well as the test statistic. So also could the number of times a Y-observation, in the sequence of n Y's, is followed by a *larger* Y-observation, which number we shall call T. However, the test statistic we shall use is S, which equals $T - I$.

When there are no tied scores, the choice of test statistic is of little consequence, since in that case I, T, and S are mathematically equivalent test statistics. The maximum value I can take is simply the number of pairs of Y's that can be formed for comparison of its two members with each other, and this is $\binom{n}{2}$ or $[n(n-1)]/2$. Since T is the number of times a Y is followed by a *larger* Y, it is the complement of I and therefore equals $\binom{n}{2} - I$. And therefore $S = T - I = \binom{n}{2} - 2I$. So the variable part of all three statistics is simply I.

The test does not require it, but it is easier to conceptualize if the Y's (and also the X's) are replaced by their ranks from 1 to n. If the variates are independent, each of the $n!$ possible permutations of the Y's had equal a priori probability of becoming the actually obtained sequence. Therefore, if there is independence, the a priori probability that S will take the value V is simply the number of different permutations of the Y's for which S has the value V, divided by the number of different possible permutations $n!$. And the null distribution of S is simply the relative frequency distribution of S over this set of $n!$ permutations of the Y's.

That distribution is symmetric about a mean of zero, extending from $-[n(n-1)]/2$ to $+[n(n-1)]/2$; its exact probabilities (or, more specifically, their numerators) can be obtained by means of a recursion formula given by Kendall [15, page 67], from which tables have been prepared. If the variates are independent, $S = T - I$ will tend to fall near zero; if there is negative "monotonic" correlation, S will tend to assume extreme negative values; and if the X- and Y-variates have a positive rank-order correlation, S will tend to be large and positive.

13.1.2 Null Hypothesis

Prior to sampling, each of the $n!$ possible permutations of the Y's (or their ranks) was equally likely to become the sequence of Y's actually obtained when the n units are arranged in order of increasing X-values. This will be the case if the X- and Y-variates are independent and all assumptions are met.

13.1.3 Assumptions

Sampling is *random*, and *tied values cannot occur* within the n observations upon the X-variate nor among the n observations upon the Y-variate.

13.1.4 Efficiency

This test has A.R.E. of $9/(\pi^2)$ or .912 relative to the parametric t test based on Pearson's r when both tests are applied under common conditions meeting all the assumptions of the latter. Under the same conditions it has A.R.E. of 1.00 relative to Hotelling and Pabst's distribution-free test for rank-order correlation (see Section 5.3).

13.1.5 Tables

Table XI of the Appendix gives, for $n = 4\ (1)\ 40$, critical values of S for upper-tail tests at α's of .005, .01, .025, .05, and .10. Because of the symmetry of the S-distribution about zero, the negative of these cell entries are the critical values of S for lower-tail tests at the same α levels.

In effect, Kaarsemaker and van Wijngaarden [14] give the entire cumulative probability distribution of S for $n = 4\ (1)\ 40$. (Actually, they give upper-tail cumulative probabilities to three decimal places, for all $S \geq 0$.) Kendall

[15] does the same thing, but always for at least two significant figures, and only for $n = 4\,(1)\,10$.

The null distribution of S has a mean of zero, a variance of $[n(n-1)\cdot(2n+5)]/18$, and a shape that becomes increasingly normal as n increases and exactly normal when n becomes infinite. Therefore, when n is too large for the exact tables to apply, approximate probabilities for S can be obtained by referring the critical ratio

$$\frac{S}{\sqrt{\frac{n(n-1)(2n+5)}{18}}}$$

to standard normal tables. The approximation can be improved by a correction for continuity that consists of reducing by one unit the absolute value of the numerator S of the critical ratio. Further corrections "for ties" are given by Kendall [15].

13.1.6 Example

An experimenter wishes to test at the .05 level the hypothesis that an X-variate and a Y-variate are independent against the alternative that they have a positive rank-order correlation. He draws five units from the bivariate population and obtains the following data, parentheses enclosing the two measurements on a single unit, the first of the two always being the X-measurement: (39, 11), (60, 20), (29, 5), (40, 8), (32, 6). In the following table, each column represents a single unit. The first row contains the X's arranged in order of ascending magnitude (with their size ranks in parentheses beside them, simply because they are easier to work with). The second row gives the Y's corresponding respectively to the X's immediately above them (and beside each Y, in parentheses, its size rank). Each cell in the third row gives the number T_i of larger Y's that follow the Y in the cell immediately above it. Finally, each cell in the fourth row gives the number I_i of smaller Y's that follow the Y in the same column of the second row. Entering Table XI of

X	29 (1)	32 (2)	39 (3)	40 (4)	60 (5)	
Y	5 (1)	6 (2)	11 (4)	8 (3)	20 (5)	
T_i	4	3	1	1	0	$T = \sum T_i = 9$
I_i	0	0	1	0	0	$I = \sum I_i = 1$

$$S = T - I = \sum T_i - \sum I_i = 9 - 1 = 8$$

the Appendix with $n = 5$, the critical value of S for an upper-tailed test (the proper tail for a test against the alternative of positive correlation) at $\alpha = .05$ is found to be 8. So the null hypothesis of independence is rejected in favor of the conclusion that the X- and Y-variates have a positive rank-order correlation. These same data were used in a completely analogous way in the example for Hotelling and Pabst's test for rank-order correlation, where they also gave the test statistic its critical value at the one-tailed .05 level for a test against the alternative of positive correlation.

13.1.7 Discussion

Kendall's rank-order correlation test is one of the most important distribution-free tests. It is equalled in efficiency (under normal-theory conditions) and excelled in speed of computation by Hotelling and Pabst's test based on Spearman's rank-order correlation coefficient r_s. However, in most other respects it seems to be a better test. For a detailed comparison of the two tests, see the discussion section for Hotelling and Pabst's test (Section 5.3.7).

As in the case of Hotelling and Pabst's test, a coefficient of correlation can be calculated from the data used in Kendall's test. The maximum value S can attain is simply the number of pairs whose two members can be compared, $\binom{n}{2}$ or $[n(n-1)]/2$. Kendall therefore defines

$$\frac{S}{\frac{n}{2}(n-1)}$$

to be his coefficient of rank correlation, which he calls t. Its value ranges from -1 for perfect negative correlation to $+1$ for perfect positive correlation. It is indirectly related to r_s in ways that are discussed by Kendall [15].

Kendall's test has many interesting features. A number of other distribution-free tests are either special cases or generalizations of it. There are alternative (sometimes simpler) ways of calculating its test statistic, and the test can be modified and adapted to special situations, for example, to take account of ties. The test statistic and its associated correlation coefficient have a variety of meaningful and interesting interpretations. In short, the test, the coefficient, and the method all have great versatility and general utility. There is not sufficient space to do justice to this subject here. However, a large bibliography covering such matters can be found in Kendall's book [15], and many of these topics are succinctly discussed in the book itself.

13.2 THE MANN-KENDALL TEST FOR TREND

Suppose that, in Kendall's test for correlation, the X-variable were the time at which a unit "appeared," or was generated, and the Y-variable were some quantitative measure upon the unit itself. Kendall's test would then test whether or not the size order of the Y-measurement is randomly related to (i.e., is independent of) the temporal order in which the units were generated.

And it would be particularly likely (in an appropriately-tailed test) to reject the hypothesis of randomness if there were a monotonic trend in the generating process.

In such an application, of course, if the Y-measurements are recorded in temporal order of appearance of their respective units, there is no need to bother with recording the X's. Indeed, a test for trend is usually conceptualized as a univariate test on the temporal randomness of the Y's rather than as a bivariate test on the independence between the Y's and a second variable, time.

When both tests are used as tests of randomness against normal regression alternatives, this test has A.R.E. of $(3/\pi)^{1/3}$ or .98 relative to the parametric test based on the regression coefficient b. The analogous use of Hotelling and Pabst's test as a test for trend (see Section 5.4) has the same A.R.E. relative to the same parametric test under the same conditions. So under these conditions the two distribution-free tests have A.R.E. of 1.00 with respect to each other. They are among the most powerful distribution-free tests for trend.

The test is consistent and unbiased under general conditions stated by Mann [18].

The test can be performed simply as a particular application of Kendall's test for correlation. However, Mann [18] who proposed it, took T rather than S as the test statistic and provided a table of lower-tail cumulative probabilities for T in the cases $n = 3\,(1)\,10$. Since the S tables are more extensive, however, it is probably better to employ the Kendall version of the test.

13.3 SMIRNOV'S MAXIMUM DEVIATION TESTS FOR IDENTICAL POPULATIONS

Many distribution-free tests are in fact tests for identical populations, but most such tests are especially sensitive to some particular *form* of nonidentity, such as unequal locations or unequal "scales," and are used with the primary purpose of detecting that *special form* of nonidentity. Indeed, such tests may be very *in*sensitive to other forms of nonidentity, and may therefore be very poor tests against alternatives including all forms of nonidentity. The present test is appropriate when the experimenter wishes to test the H_0 of identity against general alternatives of nonidentity and does not wish the sensitivity of the test to be concentrated upon one aspect of nonidentity at the expense of most others.

13.3.1 RATIONALE

Suppose that an experimenter has made n random observations upon a continuously distributed X-variate and m observations upon a continuously distributed Y-variate and wishes to test, against very general alternatives, the hypothesis that the X- and Y-variates have identical distributions (or, roughly, that the X- and Y-populations are identical). Let the $n + m$ observations in the combined sample of X's and Y's be arranged in increasing order of size and called Z's, subscripts indicating rank, so that Z_i is the ith smallest Z. Now let r_i be the number of Z's $\leq Z_i$ that were originally X's, and let s_i be the number that were originally Y's. For each of the $n + m$ values of i, and therefore of Z_i, calculate

$$d_i = \frac{r_i}{n} - \frac{s_i}{m}$$

Finally, let D^+ be the largest positive value, and let D be the largest absolute value, among the $n + m$ values of d_i. We shall use $D^+ = \max d_i$ and $D = \max |d_i|$ as test statistics, the choice depending upon whether we wish a "one-sided" or "two-sided" test.

Now if the H_0 of identical populations is true, we may regard the $n + m$ Z's as a single sample from a single population. And we may regard the X- and Y-samples as randomly determined subsets of n and m observations to which the arbitrary labels "X" and "Y" are then applied. There are $\binom{n+m}{n}$ distinguishably different ways of separating the $n + m$ Z's into n observations to be labeled X's and m observations to be labeled Y's. And under H_0 each of these $\binom{n+m}{n}$ pairs of X- and Y-samples were equally likely to become the X- and Y-samples actually obtained. Therefore, the null distribution of D (or of D^+) is simply the relative frequency distribution of D (or of D^+) over these $\binom{n+m}{n}$ pairs of samples. If we were to replace the Z's by their ranks, D would have the same value for the obtained sample and would have the same null distribution, and a similar statement holds for D^+. Therefore, the test is clearly independent of the form of the common distribution of the sampled variates, under H_0. Thus the null probability that $D \geq D'$ is simply $f/\binom{n+m}{n}$, where $\binom{n+m}{n}$ is the total number of distinguishably different pairs of "X" and "Y" samples of sizes n and m into which the integers from 1 to $n + m$ can be separated, and f is the number of these pairs which yield a D value greater than or equal to D'. Clearly, f can be obtained by a simple, although perhaps extensive, counting process, and we may imagine that this has been done by a giant electronic computer. Actually, however, in obtaining tables of probabilities for D, the numerator f of the probability fraction (or the entire fraction) usually is mathematically calculated, e.g., from random walk formulas [see 33].

The test statistics D^+ and D have a highly meaningful graphic interpretation. Let $F(X)$ and $G(Y)$ be the cumulative probability distribution functions of the X- and Y-variates, respectively, so that $F(K)$ is the proportion of the X-population lying at or below $X = K$ and $G(K)$ is the proportion of the Y-population lying at or below $Y = K$. Then, in effect, the null hypothesis states that $F(K) = G(K)$ for all values of K from $-\infty$ to $+\infty$. Or, in other words, H_0 is that if $F(X)$ and $G(Y)$ are plotted on the same graph, to the same scales, their two curves will coincide exactly, lying one on top of the other. The extent to which the $F(X)$ and $G(Y)$ curves depart from one another would be a meaningful index of the "degree of falsity" of H_0. We do not have $F(X)$ and $G(Y)$, of course, but we do have sample estimates of them. In analogy with $F(X)$, let $S_n(X)$ be the *empirical* cumulative probability distribution function of the n *obtained* observations comprising the X-sample. Thus $S_n(K)$ is the proportion of the observations in the X-sample whose X-values are $\leq K$, and if k of the X's are $\leq K$, then $S_n(K) = k/n$. Clearly, for any value of K from $-\infty$ to $+\infty$, $S_n(K)$ must be an integral multiple of $1/n$. So if we plot $S_n(X)$ as ordinate against X as abscissa, the resulting "curve" will be a step function in which as X goes from $-\infty$ to $+\infty$, $S_n(X)$ rises from 0 to 1.00 in ordinate jumps of $1/n$, the jumps occurring at abscissas corresponding to the X-values of the n observations in the X-sample. Likewise, in analogy with $G(Y)$, let $T_m(Y)$ be the empirical cumulative probability distribution function of the m obtained observations comprising the Y-sample. Thus $T_m(K)$ is the proportion of Y's in the Y-sample that are $\leq K$. And if we plot $T_m(Y)$ as ordinate against Y as abscissa, the resulting curve will be a step function whose ordinates rise from 0 to 1.00 in jumps of $1/m$ as Y goes from $-\infty$ to $+\infty$, the jumps occurring at abscissas corresponding to obtained Y's.

Now suppose we plot $S_n(X)$ and $T_m(Y)$ on the same graph and, therefore, against common ordinate and abscissa scales. Since the abscissa scale is common, this amounts to plotting $S_n(Z)$ and $T_m(Z)$ as ordinates against Z as abscissa. As before, let $Z_1, Z_2, \ldots, Z_i, \ldots, Z_{n+m}$ be, in order of increasing size, the values of the $n + m$ actually obtained X- and Y-observations. Then, the only ordinate changes in either $S_n(Z)$ or $T_m(Z)$, i.e., *all* of the n ordinate changes in $S_n(Z)$ and the m ordinate changes in $T_m(Z)$, occur at these $n + m$ Z_i-values, and $S_n(Z)$ and $T_m(Z)$ are horizontal lines (and therefore parallel with each other) in between. Therefore, if $O = S_n(Z) - T_m(Z)$ is the ordinatewise difference between the two step functions at a common abscissa of Z, and if $O_i = S_n(Z_i) - T_m(Z_i)$ is the ordinatewise difference at the common abscissa Z_i (and which will generally have two values at that point), we may be sure that the set of values assumed by O is the same as the set of values assumed by O_i, so that the maximum value of O_i is also the maximum value of O and $\max|O_i| = \max|O|$. (In visualizing this, it helps if we remember that, since there are no ties, only one of the two step functions experiences an ordinate jump at any given value of Z_i.) Now $S_n(Z_i)$ and $T_m(Z_i)$ are, by definition, the *proportions* of observations in the X- and Y-samples, respectively, that have values $\leq Z_i$. And since we have already defined r_i to be the *number* of X's $\leq Z_i$ and s_i to be the number of Y's $\leq Z_i$, it follows that

$$O_i = S_n(Z_i) - T_m(Z_i) = \frac{r_i}{n} - \frac{s_i}{m} = d_i$$

and likewise that $|O_i| = |d_i|$. Finally, since the set of O values is the same as the set of O_i values, over the $n + m$ different i's, we have the result that the test statistic

$$\begin{aligned} D^+ &= \max d_i = \max O_i = \max O = \max [S_n(Z) - T_m(Z)] \\ &= \text{maximum ordinatewise excess of } S_n(X) \text{ over } T_m(Y) \\ &\quad \text{at a common abscissa, i.e., given that } X = Y \end{aligned}$$

and, likewise, we have the result that the test statistic

$$\begin{aligned} D &= \max |d_i| = \max |O_i| = \max |O| = \max |S_n(Z) - T_m(Z)| \\ &= \text{maximum ordinatewise difference between } S_n(X) \text{ and } T_m(Y) \\ &\quad \text{at a common abscissa} \end{aligned}$$

13.3.2 Null Hypothesis

Each of the $\binom{n+m}{n}$ pairs of distinguishably different samples of sizes n and m that can be formed from the $n + m$ observations in the combined sample was equally likely, prior to sampling, to have become the actually obtained pair of X- and Y-samples. This will be the case if the X- and Y-variates have identical distributions and all assumptions are met.

13.3.3 Assumptions

Sampling is *random*, the sampled populations are of *infinite* size, and *tied observations cannot occur*. (The last two assumptions will be met if the populations are continuous and measurement is precise.)

13.3.4 Efficiency

The test statistic D does not have all the characteristics required for exact calculation of conventional A.R.E.'s. However, Capon [5] has obtained lower and upper bounds for an index that is essentially similar to the conventional A.R.E. His index, which we shall call "A.R.E.," assumes that, as $n + m$ approaches infinity, the ratio n/m does not become zero or infinity. The lower bound of the "A.R.E." of the D test relative to the likelihood ratio test when the two sampled populations differ only in location is .637 for normal populations, .811 for Cauchy populations, and 1.00 for double-exponential populations; but when the two populations differ only in scale, the "A.R.E." lower bounds become .117 for normal populations, .203 for Cauchy populations, and .541 for exponential populations. In all six cases the upper bound is 1.00, and the true "A.R.E." lies somewhere between the two bounds. The "A.R.E." of D relative to Student's t test when the two populations have the same unspecified type of distribution, differing only in location, has a lower bound of $\frac{1}{3}$, but can become "arbitrarily large" for certain populations.

Various power comparisons indicate that when applied to normal populations differing only in location, the D test is more powerful than the Westenberg-Mood median test, the Wald-Wolfowitz total-number-of-runs test, or the exceedances test, but is less powerful than the Wilcoxon rank-sum test, van der Waerden's inverse normal scores test, or Student's t test [10, 11, 32]. When applied to large samples against the nonparametric alternatives investigated by Lehmann [17], the D test appears to be much more powerful than the total-number-of-runs test. For power functions see [10, 11, 17].

The D test is consistent against all alternatives $F(X) \neq G(Y)$, as the smaller of n and m approaches infinity while $0 < n/m < \infty$; however, the test is biased at finite sample sizes [20].

13.3.5 Tables

The values that the test statistics can take are always a ratio of two integers. When sample sizes are equal so that $n = m$, that ratio becomes simply k/n where k is an integer $\leq n$, and it becomes more convenient to table critical values of k than to table critical values of the test statistic. For $n = m = 3$ (1) 40 and for α's of .05, .025, .01, .005, .001, and .0005, Table XII of the Appendix gives the smallest value of k for which $P(D^+ \geq k/n) \leq \alpha$, and also gives the exact value of $P(D^+ \geq k/n)$ for the listed value of k. The table is therefore appropriate for an upper-tail test whose rejection region consists of those cases where at some common abscissa the ordinate of $S_n(X)$ too greatly *exceeds* that of $T_m(Y)$. (There is no need to provide tables for a lower-tail test. If $S_n(X)$ has fallen too far below $T_m(Y)$, then $T_m(Y)$ has too greatly exceeded $S_n(X)$, and this can be tested simply be relabeling X's as Y's and Y's as X's and performing the above upper-tail test.)

A test based on the statistic $D = \max |d_i|$ is *not* a two-tailed version of a test based on the statistic $D^+ = \max d_i$; that is, it is not true that $P(D \geq D') = 2P(D^+ \geq D')$. In order to appreciate this, we need only reflect that in order for D to be ≤ 0 (actually, to equal zero, since D cannot be < 0) the two step functions $S_n(X)$ and $T_m(Y)$ must coincide exactly, which is impossible if $n \neq m$ and requires fantastically improbable ties if $n = m$; but in order for D^+ to be ≤ 0 it is only necessary that the ordinates of $S_n(X)$ never exceed those of $T_m(Y)$ at a common abscissa, i.e., that $S_n(X)$ lie entirely at or below $T_m(Y)$, which has quite an appreciable probability at small sample sizes. However, whereas it is *not* true *in general* that $P(D \geq k/n) = 2P(D^+ \geq k/n)$, it *is* true, to the number of decimal places used, of all but the leftmost column of Table XII of the Appendix. So for $\alpha \leq .025$, the k values listed are *also* the smallest values of k for which $P(D \geq k/n) \leq 2\alpha$, and the probabilities listed beside them are exactly half the exact value of $P(D \geq k/n)$.

The complete cumulative probability distribution of D has been given by Massey [21] to five decimal places for all values $n \leq 10$, $m \leq 10$. For the cases of equal sample sizes $n = m = 1\,(1)\,40$, the entire cumulative probability distribution of D can be obtained, almost always to six and never to less than four decimal places, from tables published by Birnbaum and Hall [2]. (Massey [22] has done the same thing, but somewhat less completely at the upper tail of the distribution.)

The complete cumulative probability distribution of D^+ for the cases $n = m = 1\,(1)\,40$ can be obtained to at least five and usually six decimal places from tables published by Birnbaum and Hall [2].

When sample sizes are quite large, approximate probabilities can be obtained from limiting distributions of functions of the test statistics. Smirnov [30] has published the limiting cumulative distribution of

$$z = D\sqrt{\frac{nm}{n+m}}$$

giving cumulative probabilities $L(z)$ to at least six decimal places, for the cases $z = .28\,(.01)\,2.50\,(.05)\,3.00$. When both n and m are large, $P(D < D') \cong L(z')$ and $P(D \geq D') \cong 1 - L(z')$, where

$$z' = D'\sqrt{\frac{nm}{n+m}}$$

And to roughly the number of decimal places involved in Smirnov's table, we may expect that so long as $1 - L(z') \leq .05$ the limiting

$$P(D^+ \geq D') \cong [1 - L(z')]/2$$

Or, more generally, to obtain probability levels for D^+ we may make use of the fact that $[4\,(D^+)^2\,nm]/(n+m)$ is, for large sample sizes, approximately distributed as chi-square with 2 degrees of freedom. Using any of the above procedures, one can obtain from the limiting distributions of the test statistics the critical values corresponding to standard significance levels. Let

$$P\left(D \geq \lambda\sqrt{\frac{n+m}{nm}}\right) = \alpha_\lambda$$

and let

$$P\left(D^+ \geq \theta\sqrt{\frac{n+m}{nm}}\right) = \alpha_\theta$$

Then we have the following table, for the limiting case:

$\alpha_\lambda =$	.10	.05	.02	.01	.002	.001
$\lambda =$	1.224	1.358	1.517	1.628	1.858	1.950
$\alpha_\theta =$	.05		.01		.001	
$\theta =$	1.224		1.517		1.858	

Of course, λ is simply the appropriate value of z, and θ is

$$\sqrt{\frac{\chi_2^2}{4}}$$

where χ_2^2 is the appropriate value of chi-square with 2 degrees of freedom.

13.3.6 EXAMPLE

Twelve psychotics of a certain type, all admitted to a mental institution on the same day, are randomly divided into two groups of six patients each. The X-group receives psychoanalytic treatment and the Y-group receives none. The measured variable in both cases is the time from admission to discharge from the hospital. The two groups belonged to the same population before treatment (of the X-group), so the question is whether or not they still belong to the same population after treatment, i.e., whether psychoanalysis had any effect, either good or bad, upon the distribution of remission times. Thus the tested null hypothesis will be that $F(X) = G(Y)$, and the alternative hypothesis will be that $F(X) \neq G(Y)$, so $D = \max|d_i| = \max |S_n(X) - T_m(Y)|$ will be the appropriate test statistic. The accompanying table gives the data and computational procedures for performing the test at a level of significance taken to be $\alpha = .05$. The Z_i's are the 12 observations in the combined sample. The Z_i's that are boldface are the 6 observations from the experimental X-group, the lightface Z_i's being the 6 observations from the control Y-group. In both cases, the figure given is the number of days spent in the hospital. Entering Table XII of the Appendix with $n = 6$ and $k = 5$,

Z_i	$\frac{r_i}{n}$	$\frac{s_i}{m}$	$d_i = \frac{r_i}{n} - \frac{s_i}{m}$
47	1/6	0	1/6
72	1/6	1/6	0
91	2/6	1/6	1/6
115	3/6	1/6	2/6
122	4/6	1/6	3/6
129	5/6	1/6	4/6
166	6/6	1/6	5/6
201	6/6	2/6	4/6
255	6/6	3/6	3/6
284	6/6	4/6	2/6
407	6/6	5/6	1/6
962	6/6	6/6	0

$D = \max|d_i| = \frac{5}{6}$

we find that $P(D^+ \geq k/n = \frac{5}{6})$ is .01299. Multiplying this by 2 we get $P(D \geq k/n = \frac{5}{6}) \cong .02598$, which compares with the exact probability level of .025975 obtained for D from Birnbaum and Hall's tables [2]. The null hypothesis of identical X- and Y-populations is therefore rejected at the predesignated .05 level in favor of the alternative hypothesis that they are not identical.

13.3.7 Discussion

In testing the H_0 that $F(X) = G(Y)$ against one-sided alternatives, it is well to bear in mind that if the cumulative distribution function $F(X)$ for the X's lies entirely *above* that for the Y's $G(Y)$, this means that the X's tend to be *smaller* than the Y's.

The Smirnov tests are only the best known of a large variety of "maximum deviation" tests for identical populations. One such test, which has considerable intuitive appeal, is the two-sample Cramer–von Mises test based upon the test statistic

$$\frac{nm}{(n+m)^2} \sum d_i^2$$

Another is based upon the test statistic

$$\frac{nm}{(n+m)^2} \sum (d_i - \bar{d})^2$$

See [4] for a small table of probabilities. A major advantage of Smirnov's tests over their maximum-deviation-test rivals is that the former are far more extensively tabled than the latter.

The Smirnov tests can be generalized from the two-sample to the multi-sample case, and this has been done by a number of authors. A fairly extensive table of probabilities, for the case of three equal-sized samples, has been provided by Birnbaum and Hall [2], the test statistic being the maximum absolute ordinatewise difference between any two of the three sample cumulative distributions.

The literature on maximum deviation tests is quite extensive. A concise expository article on the subject, containing a large bibliography, has been published by Darling [8].

13.4 TSAO'S TRUNCATED SMIRNOV TEST

In certain types of experiments, observations may become available in increasing order of size. This is the case, for example, when the observations are the time required for some event to happen (such as failure of a manufactured product on life test, or discharge of a mental patient, as in the example for the Smirnov test) after some common temporal starting point. But time durations are often positively skewed so that an inordinately long wait may be required to obtain the last few data points. In such cases, it may be advantageous to modify the Smirnov test, using as test statistic the maximum absolute ordinatewise difference between $S_n(X)$ and $T_m(Y)$ *at or below* the rth smallest X, or at or below the larger of the rth smallest X and the rth smallest Y.

Let

$$D_r = \max_{r_i \leqslant r} \left| \frac{r_i}{n} - \frac{s_i}{m} \right| = \max_{Z \leqslant X_r} | S_n(Z) - T_m(Z) |$$

and

$$D_r^* = \max_{\min(r_i, s_i) \leqslant r} \left| \frac{r_i}{n} - \frac{s_i}{m} \right| = \max_{Z \leqslant \max(X_r, Y_r)} | S_n(Z) - T_m(Z) |$$

Tsao [31] has tabled both $P(D_r \leq c/n)$ and $P(D_r^* \leq c/n)$ for $n = m = 3\,(1)\,10\,(5)\,20\,(10)\,40$, $r = 2\,(1) \min\,(n - 1, 10)$, $c = 1\,(1) \min\,(n, 12)$.

The probability that $D_r \geq D_r'$ is simply $K/\binom{n+m}{n}$, where the denominator is the number of different ways of assigning n X-labels and m Y-labels to the $n + m$ observations in the combined sample (or to the sequence of integers from 1 to $n + m$), and K is the number of these "ways" or assignments that yield a value of $D_r \geq D_r'$.

The test appears to be more powerful than the exceedances test or the total-number-of-runs test, but less powerful than the Wilcoxon rank-sum test, when the two populations are normal with equal variances but different locations [11].

13.5 THE KOLMOGOROV-SMIRNOV MAXIMUM DEVIATION TESTS OF AN HYPOTHESIZED POPULATION DISTRIBUTION

Most one-sample tests are concerned with testing an hypothesis about a particular feature of the sampled population, such as its location, whose specification falls far short of completely identifying the population. However, in some cases the experimenter may wish to specify completely an entire distribution and then test whether or not his sampled population has exactly that distribution in every detail. The Kolmogorov-Smirnov tests are appropriate in this situation, if the entire specification can be made prior to sampling.

13.5.1 RATIONALE

Let $F(X)$ be the cumulative probability distribution function of a continuously distributed variate X upon which n random observations have been made, and let $S_n(X)$ be the *empirical* cumulative probability distribution function of the *n obtained* observations on X. Thus $F(C) = P(X \leq C)$ is the proportion of the X-population at or below $X = C$, and if r of the n obtained X's are $\leq C$, then $S_n(C) = r/n$ is the proportion of observations in the X-sample at or below $X = C$. Now, let $F(X)$ and $S_n(X)$ be plotted on the same graph, to the same ordinate and abscissa scales. So, with increasing values of X, $F(X)$ is a monotonically increasing curve and $S_n(X)$ is a step function which rises in steps of $1/n$ or multiples thereof. Finally, let $S_n(X) - F(X)$ be understood to be the ordinatewise difference between the two cumulatives at a common abscissa value, i.e., at the common value of X designated by the value enclosed within parentheses. Then the distributions of the statistics,

$$K^+ = \max\,[S_n(X) - F(X)]$$
$$K^- = \min\,[S_n(X) - F(X)] = -\max\,[F(X) - S_n(X)]$$
$$K = \max\,|S_n(X) - F(X)|$$

are known and are independent of $F(X)$, i.e., of the location, variance, shape, and all other features of the X-population except continuity. But, of course, they are not independent of whether or not the n sample observations were actually drawn from the population identified by $F(X)$. Therefore, to test the null hypothesis H_0 that $F(X) = F_0(X)$ (where $F_0(X)$ is a *completely specified* cumulative distribution hypothesized to be that of the sampled variate), one need only substitute $F_0(X)$ for $F(X)$ in one of the above formulas and refer the resulting value of the test statistic to the distribution the test statistic is known to have when $F_0(X) = F(X)$, i.e., when $F_0(X)$ is actually the cumulative distribution function of the sampled variate. The Kolmogorov statistic K would be used to guard against general alternatives $F(X) \neq F_0(X)$, whereas Smirnov's modification K^+ would be used if one wished sensitivity only against the one-sided alternative that $F(X) > F_0(X)$ for some values of X, and K^- would be used to guard against the opposite one-sided alternative that $F(X) < F_0(X)$ for some X.

Mathematically efficient proofs that the three statistics are distribution-free and derivations of their probabilities in the general case tend to require sophisticated mathematical methods (see [33]). However, the necessary insights are provided by the following method, which is mathematically simpler and theoretically adequate, but which would be a very inefficient way to obtain tables for the null distribution of one of the test statistics. The method will deal with a single one of the three statistics, since its application to the other two is entirely analogous.

Let the true distribution (i.e., the density function) of the X-variate be divided into n vertical strips each containing $(1/n)$th of the area of the distribution, and let $X^{(i)}$ be the upper endpoint of the abscissa range, i.e., of the X-interval, corresponding to the ith lowest strip. Then

$$F(X^{(1)}) = P(X \leq X^{(1)}) = \frac{1}{n}$$

$$F(X^{(2)}) = P(X \leq X^{(2)}) = \frac{2}{n}$$

$$\vdots$$

$$F(X^{(i)}) = P(X \leq X^{(i)}) = \frac{i}{n}$$

so from one interval endpoint to the next $F(X)$ rises by exactly $1/n$.

Now let a random sample of n observations be drawn from the X-population, and let n_i be the number of these observations whose X-values fall in the ith lowest interval. Then the a priori point probability for the obtained set of n_i values corresponding to all n values of i is given by the multinomial probability

$$P(n_1, n_2, \ldots, n_i, \ldots, n_n)$$

$$= \frac{n!}{n_1!\, n_2! \ldots n_i! \ldots n_n!}\left(\frac{1}{n}\right)^{n_1}\left(\frac{1}{n}\right)^{n_2} \cdots \left(\frac{1}{n}\right)^{n_i} \cdots \left(\frac{1}{n}\right)^{n_n}$$

$$= \frac{n!}{n_1!\, n_2! \ldots n_i! \ldots n_n!}\left(\frac{1}{n}\right)^{n}$$

This is the probability of a specified pattern of interval occupancy, $n_1, n_2, \ldots, n_i, \ldots, n_n$. But for a given pattern of interval occupancy, the n values of $S_n(X)$ at the interval endpoints are determined, since

$$S_n(X^{(1)}) = \frac{n_1}{n}$$

$$S_n(X^{(2)}) = \frac{n_1 + n_2}{n}, \qquad \text{etc.}$$

And since $F(X^{(i)}) = i/n$, which is a known constant, the n values of $S_n(X^{(i)}) - F(X^{(i)})$ are also determined. Therefore, the above multinomial probability for a specified pattern of interval occupancy is also the a priori probability for the corresponding set of n values of $S_n(X^{(i)}) - F(X^{(i)})$ where $i = 1, 2, \ldots n$.

Now we may imagine that a giant electronic computer obtains the probability distribution of max $[S_n(X^{(i)}) - F(X^{(i)})]$ by forming each of the different possible patterns of interval occupancy (i.e., each of the distinguishable partitions of n into the sequence $n_1, n_2, \ldots n_n$, where $\sum_{i=1}^{n} n_i = n$), determining

$$\max\,[S_n(X^{(i)}) - F(X^{(i)})] = \max\left[\frac{\sum_{j=1}^{i} n_j}{n} - \frac{i}{n}\right]$$

for each different pattern and summing the multinomial probabilities for patterns of interval occupancy yielding the same value of max $[S_n(X^{(i)}) - F(X^{(i)})]$. Alternatively, we may imagine that the computer distributes or assigns the n observations to the n intervals or "cells" in each of the n^n different possible ways. (The first observation drawn can be assigned to one of the n cells in n ways, and for each of these ways there are n ways of placing the second observation in a cell, etc., so that there are n^n ways of assigning the entire set of n observations.) The computer then counts the number of these ways for which

$$\max [S_n(X^{(i)}) - F(X^{(i)})] = \max \left[\frac{\left(\sum_{j=1}^{i} n_j \right) - i}{n} \right]$$

has a given value k/n, and then divides that number by n^n to obtain the point probability that max $[S_n(X^{(i)}) - F(X^{(i)})]$ equals k/n. Obviously k must be an integer (since $\sum_{j=1}^{i} n_j$ and i are both integers), so max $[S_n(X^{(i)}) - F(X^{(i)})]$ can only take on the $n + 1$ different values of k/n corresponding to k's of $0, 1, 2, \ldots, n$.

Using the above method, we can obtain $P\{\max [S_n(X^{(i)}) - F(X^{(i)})] \geq k/n\}$. But, provided only that k is an integer, the above probability equals $P\{\max [S_n(X) - F(X)] \geq k/n\}$, which is the probability we seek. This can be seen as follows. Let $d = S_n(X) - F(X)$ and let $d_{\max}$ occur in the ith interval at the abscissa value X'. Since $S_n(X)$ is always an integral multiple of $1/n$ and $F(X)$ is an integral multiple of $1/n$ only at interval endpoints (here we are assuming, merely for the sake of reducing explanation, that $F(X)$ constantly rises) it follows that $d = S_n(X) - F(X)$ can be an integral multiple of $1/n$ only at interval endpoints. Therefore let $d_{\max}$ lie between k/n and $(k + 1)/n$, where k is an integer. Now as X increases from X' to $X^{(i)}$, the upper endpoint of the interval, $F(X)$ rises by less than $1/n$ (since $1/n$ is the total rise for the entire interval) and $S_n(X)$ does not rise at all. For if it did, it would have to rise by at least the size of a jump $1/n$, which being greater than the corresponding rise in $F(X)$ would cause a "new" $d_{\max}$. The rise in $F(X)$ from X' to $X^{(i)}$ reduces d from $d_{\max}$ to k/n, since (a) $d_{\max}$ lies between k/n and $(k + 1)/n$; (b) the rise in $F(X)$ is less than $1/n$; (c) d must be an integral multiple of $1/n$ at interval endpoints. Thus if $d_{\max}$ lies between k/n and $(k + 1)/n$, then at the upper endpoint of the interval containing $d_{\max}$, d will have the value k/n. And k/n will have to be max $[S_n(X^{(i)}) - F(X^{(i)})]$, since the latter statistic obviously cannot exceed $d_{\max}$ and must be a multiple of $1/n$. So the point probability that max $[S_n(X^{(i)}) - F(X^{(i)})] = k/n$ must also be the probability that $k/n \leq d_{\max} < (k + 1)/n$, and consequently

$$P\{\max [S_n(X^{(i)}) - F(X^{(i)})] \geq k/n\} = P(d_{\max} \geq k/n)$$

13.5.2 Null Hypothesis

The a priori probability that a sample observation on X(i.e., "drawn" from $f(X)$, the density function corresponding to the cumulative distribution $F(X)$) will fall into a given one of the n abscissa intervals (that divide into equal areas $f_0(X)$, the density function corresponding to the hypothesized cumulative distribution $F_0(X)$) is the same for each of the n intervals, always being $1/n$. This will be the case if the actual cumulative probability distribution function $F(X)$ is identical to the hypothesized cumulative probability distribution function $F_0(X)$ and all assumptions are true.

13.5.3 Assumptions

Sampling is *random*, the sampled population is *continuously distributed*, and precision of measurement is such that there are *no tied observations*. Also, the hypothesized distribution must be capable of being divided into n nonoverlapping vertical strips each containing $(1/n)$th of the distribution and must be continuous, since H_0 equates it with the sampled population. Finally, the hypothesized distribution must be specified completely and without regard to any information contained in the sample.

13.5.4 Efficiency

The present test may be regarded as a special case of Smirnov's two-sample test in which one sample size is infinite, thereby making it, in effect, the "population," in the null case. It is not surprising, therefore, that the test should have properties similar to those of the two-sample Smirnov test. The test is consistent against any alternative $F(X) \neq F_0(X)$, provided only that X is continuously distributed; at finite sample sizes, however, the test is biased [20].

Since the sensitivity of the K test is not concentrated upon a particular type or class of alternatives, it is, in effect, a test of "goodness-of-fit," and the most appropriate classical test against which to compare it is the chi-square test of goodness-of-fit. For the situation in which $F(X)$ and $F_0(X)$ are separated by a constant "distance" Δ, Massey [23] determined and compared the smallest values of Δ detectable with probability .50 by the K test and by the chi-square test for α's of .05 and .01 and n's ranging from 200 to 2000. The K test was found to be superior to chi-square in all of the 46 cases examined. Upper and lower bounds for the power of the K^+ test are given for various values of Δ and n in [6].

13.5.5 Tables

Table XIII of the Appendix gives critical values of K^+ for $n = 1\ (1)\ 100$ and (upper-tail) α's of .10, .05, .025, .01, and .005. These values are calculated from exact formulas for $n \leq 20$, but for $n > 20$ they are based upon a good approximate formula. A test based upon the test statistic K is *not* a two-tailed version of a test based upon K^+. However, as long as $\alpha \leq .05$, the upper-tail critical values given for K^+ at significance level α may be regarded as upper-tail critical values for K at significance level 2α. At $n = 20$ the difference between the critical value for K^+ at level α and the true critical value for K at level 2α appears to be less than one unit at the fifth decimal place for $.005 \leq \alpha \leq .05$; so this use of the table would seem to afford a sufficiently close approximation to the desired critical values of K. The tabled critical values of K^+ for an upper-tail test at level α are, of course, also the critical values of K^- for a lower-tail test at level α, except that the negative algebraic sign is omitted and must be added. That this must be the case can be seen by taking the graph upon which $S_n(X)$ and $F(X)$ are plotted and turning it upside down. We now have the sample and population cumulatives with the cumulation taking place from largest to smallest values, and the "new" F^+ is the "old" F^- if we measure in the "new" direction. Therefore, since the derivation of the distribution of F^+ is independent of the direction of cumulation, the probability distributions of F^- and F^+ must be mirror-images of each other about an axis through zero.

Birnbaum [1] has provided tables to five decimal places, of $P(K < C/n)$ for $n = 1\ (1)\ 100$ and, in effect, $C = 1\ (1)\ 15$. A three-decimal-place table of critical values for K has been published by Massey [23] for $n = 1\ (1)\ 20\ (5)\ 35\ (\infty)\ \infty$ and upper-tailed α's of .20, .15, .10, .05 and .01. Smirnov [30] has published the limiting cumulative distribution of $z = K\sqrt{n}$, giving cumulative probabilities $L(z)$ to at least six decimal places for the cases $z = .28\ (.01)\ 2.50\ (.05)\ 3.00$. Let $P(K \geq \lambda/\sqrt{n}) = \alpha_\lambda$ and let $P(K^+ \geq \theta/\sqrt{n}) = \alpha_\theta$. Then critical values of λ and θ corresponding to standard values of α are given in the small table of Section 13.3.5 on the two-sample Smirnov test.

13.5.6 Example

An experimenter hypothesizes that (the variate underlying) a certain population is normally distributed with mean of 3 and a standard deviation of 2. He decides to test this hypothesis at the .05 level of significance using as test statistic K based upon 5 observations. He randomly draws the 5 observations from the population in question, and quickly obtains $S_n(X)$ for each of them. For each of the 5 values of X he then obtains $F_0(X)$, the cumulative probability of X in the hypothesized normal distribution with mean of 3 and standard deviation of 2. This is accomplished by looking up the cumulative probability of $Z = (X - 3)/2$ in a table of cumulative probabilities for a standardized normal variate Z, i.e., a normally distributed variate with mean of zero and standard deviation of 1.00. For each of the 5 values of X he obtains the difference $|S_n(X) - F_0(X)|$, the largest one of which is K, the test statistic. The accompanying table incorporates these data and shows the steps in the procedure. For these data, the maximum absolute deviation is $K = .5791$, which exceeds the critical value of .56328 found in Table XIII of the Appendix for an upper-tail test at the .05 level of significance. Therefore the experimenter rejects the hypothesis that the distribution of the sampled variate is normal with mean of 3 and standard deviation of 2.

X	$S_n(X)$	$F_0(X)$	$\lvert S_n(X) - F_0(X) \rvert$
−.311	.2	.0489	.1511
−.078	.4	.0619	.3381
.555	.6	.1108	.4892
1.462	.8	.2209	.5791
5.711	1.0	.9124	.0876

13.5.7 Discussion

The relative merits of the chi-square and maximum absolute deviation tests of goodness of fit have been discussed by a number of authors [1, 23]. The K test family is superior to chi-square in the following ways. The K test requires only the relatively modest assumptions that sampling is random and that the sampled population is continuous, whereas chi-square assumes, among other things, conditions that can be fulfilled only when sample size is infinite. Not surprisingly, therefore, the exact distribution of K is known and tabled for small sample sizes, whereas the exact distribution of the chi-square test statistic is known and tabled only for infinite-sized samples. Consequently, the chi-square test is only an approximate test, at all sample sizes, and the degree of approximation is hard to assess, whereas the K test is exact at small sample sizes and its degree of approximation at large sample sizes is more readily assessable. The K^+ and K^- test statistics were designed to test for deviations *in a given direction*, and do so easily, while the chi-square test must be specially modified and conducted in unconventional fashion in order to do so at all. The K test uses ungrouped data, every observation representing a point at which "goodness of fit" is examined; chi-square loses this information (if the hypothesized distribution is continuous) by requiring that data be grouped into cells. Furthermore, by using ungrouped data, the K test avoids the hazards and pitfalls associated with the choice of interval width and selection of starting point in chi-square tests of fit, and no correction for continuity is required by the K test. The K test can be applied to data which become available sequentially from smallest to largest, computations being continued only up to the point at which rejection occurs; it thus has an "efficiency" aspect not present in chi-square. Confidence bands can be easily established on the basis of the distribution of K, while chi-square has no such analogous property. More seems to be known about the power of the K test than of chi-square, although the information presently available as to which test is the more powerful is somewhat ambiguous.

Chi-square, on the other hand, is superior to K in the following ways. Chi-square does not require that the hypothesized population be completely specified in advance of sampling. Certain population parameters can be estimated from the sample and the resulting degree of "artificial" fit between obtained sample and hypothesized population can be taken account of, and prevented from unduly biasing the probability of significance, by making the appropriate reduction in degrees of freedom. No such adjustment is possible with the K test, which requires that the hypothesized population be completely known and specified a priori. Chi-square values can be meaningfully added, a useful property which the K statistic does not possess. Finally, chi-square can be applied to discrete populations. The K test, however, is not incapable of such applications (see ref. [16]; cf. [13]). When the assumption of continuity is not met, and when the probability level of K is obtained from its limiting distribution, the true probability P that $K \geq K_\alpha$ or that $K^+ \geq K_\alpha^+$ (where the subscripted value is a critical value corresponding to a nominal significance level of α) will not exceed α, i.e., $P(K \geq K_\alpha) \leq \alpha$. Thus, in tests of significance the true probability of rejection may be smaller, but not greater, than the nominal probability α. And in setting confidence bands about $S_n(X)$, the true confidence level for the statement that $F(X)$ is enclosed within the bands may be greater, but not smaller, than the nominal confidence level $1 - \alpha$. In both cases, the probability error is a "conservative" one.

The Kolmogorov-Smirnov tests are only the best known of many "maximum deviation" tests for an hypothesized population cumulative distribution. The basic information can be exploited in a variety of ways, each of which has its unique advantages and disadvantages. For example, the test statistic could be based upon the sum of n squared deviations between $S_n(X)$ and $F(X)$, the n deviations being measured at the n actually obtained values of X. And this would allow the test to be influenced by a variety of deviations rather than simply by one deviation (the maximum) as is the case with the Kolmogorov-Smirnov tests. The one-sample Cramer-von Mises test uses an approach analogous to this, except that the squared deviation is integrated over all values of X(see reference [8]). A major advantage of the Kolmogorov-Smirnov tests over other tests based on the deviations between $S_n(X)$ and $F(X)$ is that the former are far more extensively tabled than the latter.

The literature on maximum deviation tests is quite extensive. A large bibliography can be found in [8].

13.6 CONFIDENCE BANDS FOR THE POPULATION CUMULATIVE DISTRIBUTION

The derived and tabled null distribution of K is the distribution of max $|S_n(X) - F(X)|$, since, in the null case, $F_0(X) = F(X)$. Let K_α be the critical value of K corresponding exactly to a significance level of α. Then $P(\max |S_n(X) - F(X)| \geq K_\alpha) = \alpha$, or contrariwise, $P(\max |S_n(X) - F(X)| < K_\alpha) = 1 - \alpha$. This means that the a priori probability that $S_n(X)$ will lie entirely within a band of values $F(X) \pm K_\alpha$, i.e., within $\pm K_\alpha$ of $F(X)$, over all values of X is $1 - \alpha$. Or, in other words, it means that there is an a priori probability of $1 - \alpha$ that when the sample is drawn it will be the case that over the entire range of X-values $[F(X) - K_\alpha] < S_n(X) < [F(X) + K_\alpha]$, i.e., the step function will always stay between the two curves whose ordinates are the bracketed values.

But if $S_n(X)$ stays within $\pm K_\alpha$ of $F(X)$, it is also true that $F(X)$ stays within $\pm K_\alpha$ of $S_n(X)$. We do not wish to speak of any probability that it will do so, since it is $S_n(X)$ that varies from one sample to the next and $F(X)$ that remains fixed, rather than vice versa. However, we can attach a confidence level of $1 - \alpha$ to the statement that $F(X)$ lies within $\pm K_\alpha$ of our obtained $S_n(X)$, i.e., to the statement that $S_n(X) - K_\alpha < F(X) < S_n(X) + K_\alpha$. Since there is a probability of $1 - \alpha$ that $S_n(X)$ will lie within $\pm K_\alpha$ of $F(X)$, if we took an infinite number of samples $F(X)$ would lie within $\pm K_\alpha$ of $S_n(X)$ in a proportion of exactly $1 - \alpha$ of them. And if, for every one of the infinite number of samples, we alleged that $F(X)$ lies within $\pm K_\alpha$ of $S_n(X)$, we would be right $1 - \alpha$ of the time; so the proportion of correct allegations or the expectation of being right is what the confidence level refers to. Thus we may plot two step functions, $S_n(X) - K_\alpha$ and $S_n(X) + K_\alpha$, enclosing an area always $2K_\alpha$ in vertical width, and state with confidence of level $1 - \alpha$ that the unknown curve representing $F(X)$ lies entirely within the enclosed area. The two step functions are confidence bands for $F(X)$, and the enclosed area is a confidence belt.

Let K_α^+ be the critical value of K^+ corresponding exactly to a significance level of α, so that $P(K^+ \geq K_\alpha^+) = \alpha$ and $P(K^+ < K_\alpha^+) = 1 - \alpha$. Then, by similar reasoning, we may state with confidence of level $1 - \alpha$ that $F(X) + K_\alpha^+$ lies entirely above $S_n(X)$ and therefore that $F(X)$ lies entirely above $S_n(X) - K_\alpha^+$. Analogously, we may state at confidence level $1 - \alpha$ that $F(X)$ lies entirely below $S_n(X) + |K_\alpha^-|$.

13.7 DAVID'S EMPTY-CELL TEST OF AN HYPOTHESIZED POPULATION DISTRIBUTION

The following test may be regarded as an alternative to the Kolmogorov-Smirnov tests. Like the latter, it is a goodness-of-fit test and it requires that the hypothesized distribution of the sampled population be completely specified prior to sampling. The early part of its derivation is also similar to the derivation presented for the Kolmogorov-Smirnov tests.

13.7.1 Rationale

Suppose that an experimenter wishes to test an hypothesis that completely specifies the distribution of a certain continuous population prior to the drawing of observations (so that no population parameters need be estimated from the sample), and wishes to base the test upon a random sample of n observations drawn from the population in question. Since the hypothesized distribution is known and continuous, it can, at least if it is extensively tabled, be divided into C vertical strips, each containing the same area $1/C$. The "base" of each strip will be a range or interval of abscissa values, and these intervals will be called cells. Since the C strips contain equal areas, the value of an observation drawn randomly from the hypothesized distribution is equally likely to fall into each of the C cells; and if the experimenter's hypothesis is correct, this is also true of an observation drawn randomly from the actual, sampled, population. Therefore if the experimenter's H_0 is true, i.e., if the sampled population has exactly the distribution used to create the cells, the n sample observations should be well dispersed among the C cells. On the other hand, if the H_0 is badly false, the cells, although equiprobable for the hypothesized distribution, should be very unequally probable for the sampled population, with the result that the n observations tend excessively to congregate or "bunch up" in certain cells, leaving an excessive number of cells empty. Thus the number E of empty cells is a reasonable index of the truth or falsity of H_0.

The null point probability of E is easily obtained. It is the same as the null point probability of the number O of occupied cells, since $O = C - E$. So, since O is easier to work with, we shall derive the desired probability in terms of occupied, rather than empty, cells. Let P_i be the null probability that the n sample observations all fall within a *specified* set of i cells. Since all C cells are equiprobable, the null probability that a single random observation will fall within the predesignated set of i cells (i.e., will fall into one of the i cells constituting the set) is i/C, and the null probability that all n observations will fall into some of the i cells is $(i/C)^n$, which equals P_i. In defining P_i, we have required only that all n observations fall into a set of i cells, without specifying how they shall be distributed; and, in particular, we have not required that every one of the i cells contain observations. Let P_{i} be the null probability that the n sample observations all fall within a *specified* set of i cells in such a way that *all* of the i cells contain at least one of the n observations, so that none of the i cells remains empty. Obviously, $P_{\mathbf{1}}$ must equal P_1, which we know to be $(1/C)^n$. When $i = 2$, the n observations can all fall into the set of two cells in such a way that both cells are occupied, the probability for which is $P_{\mathbf{2}}$, or they can all fall into one of the two cells. The probability that they will all fall into a specified one of the two cells is $P_{\mathbf{1}} =$

P_1, and the number of possible ways of specifying the one occupied cell within the set of two cells is $\binom{2}{1}$, so

$$P_2 = P_{\mathbf{2}} + \binom{2}{1}P_1 \quad \text{or} \quad P_{\mathbf{2}} = P_2 - \binom{2}{1}P_1$$

When $i = 3$, the n observations can all fall into the set of three cells in such a way that all three cells are occupied, the probability for which is $P_{\mathbf{3}}$. Or they can all fall into two cells occupying both of them, the probability for which is $\binom{3}{2}P_{\mathbf{2}}$, since there are $\binom{3}{2}$ ways of specifying two of three cells, and since $P_{\mathbf{2}}$ is the probability that all observations fall into both members of a specified pair of cells. Alternatively, they can all fall into a single one of the three cells, the probability for which is $\binom{3}{1}P_1$, since there are $\binom{3}{1}$ ways of specifying the cell and since P_1 is the probability that all n observations fall into the cell specified. Therefore

$$P_3 = P_{\mathbf{3}} + \binom{3}{2}P_{\mathbf{2}} + \binom{3}{1}P_1$$

so

$$P_{\mathbf{3}} = P_3 - \binom{3}{2}P_{\mathbf{2}} - \binom{3}{1}P_1$$

And, in general,

$$P_{\mathbf{i}} = P_i - \binom{i}{i-1}P_{\mathbf{i-1}} - \binom{i}{i-2}P_{\mathbf{i-2}} \ldots - \binom{i}{2}P_{\mathbf{2}} - \binom{i}{1}P_1$$

We already know that $P_i = (i/C)^n$ and $P_1 = (1/C)^n$, so the above formula defines $P_{\mathbf{i}}$ in terms of known values and of P's with **boldface** subscripts *smaller* than i and *greater* than 1. Therefore, we can obtain each of the required P's with boldface subscripts by solving the formula first for $P_{\mathbf{i}} = P_{\mathbf{2}}$, then using the result to solve for $P_{\mathbf{i}} = P_{\mathbf{3}}$, etc., obtaining each successive $P_{\mathbf{i}}$ from the known values of its predecessors. Following this procedure, we can determine the value of $P_{\mathbf{i}} = P_{\mathbf{O}}$, i.e., we can solve the formula

$$P_{\mathbf{O}} = P_O - \binom{O}{O-1}P_{\mathbf{O-1}} - \binom{O}{O-2}P_{\mathbf{O-2}} \ldots - \binom{O}{2}P_{\mathbf{2}} - \binom{O}{1}P_1$$

Now, let $P(O)$ be the null point probability that exactly O of the C cells will be occupied. There are $\binom{C}{O}$ ways of specifying a set of O cells from among the C cells and the probability that all n observations will fall into the specified set, occupying all of them, is $P_{\mathbf{O}}$, so $P(O) = \binom{C}{O}P_{\mathbf{O}}$, and we have finally

$$P(O) = \binom{C}{O}\left[P_O - \binom{O}{O-1}P_{\mathbf{O-1}} - \binom{O}{O-2}P_{\mathbf{O-2}} \ldots - \binom{O}{2}P_{\mathbf{2}} - \binom{O}{1}P_1\right]$$

$$= \binom{C}{O}\left[P_O - \sum_{i=1}^{O-1}\binom{O}{i}P_{\mathbf{i}}\right]$$

$$= P(E)$$

The above formula is serviceable but a bit cumbersome. It was given because its derivation is easy to follow. An equivalent formula which is more difficult to derive but which is more efficient to use, giving the same answers more easily, is

$$P(E) = P(O) = \binom{C}{E} \sum_{i=0}^{C-E} (-1)^i \binom{C-E}{i} \left(\frac{C-E-i}{C}\right)^n$$

In using either formula it is important to remember that the formula gives point, not cumulative, probabilities.

13.7.2 Null Hypothesis

The a priori probability that a single observation will fall into a given cell is $1/C$, and this is true for each of the C cells and each of the n observations. (So every cell has the same a priori probability of "receiving" every observation.) This will be the case if the hypothesized distribution, i.e., the one divided into C vertical strips of equal area in order to create the C cells, is identical to the actual distribution of the sampled variate and all assumptions are true.

13.7.3 Assumptions

Sampling is *random*. The sampled variate is *continuously distributed.* And the hypothesized distribution is *completely specified* prior to sampling and is *continuous*. (Actually, the continuity assumptions are sufficient, rather than necessary. They can be replaced by the assumptions that (a) either the population is infinite or sampling is with individual replacements; (b) whenever adjacent cells have a common endpoint, there is zero probability that an observation will have exactly the value of the shared interval endpoint. However, it is seldom easy, and often impossible, to divide a discrete distribution into C exactly equal areas.)

13.7.4 Efficiency

The test is consistent and unbiased against a certain class of alternative hypotheses (see [7] for reference).

David [9] compared the power of the empty-cell test with that of the chi-square test to reject the H_0 that the sampled population is normal with mean of zero and standard deviation of 1 in favor of the alternative that it is normal with mean of zero and standard deviation of $\frac{4}{3}$. Sample size was 30 and α was .05 in both cases, but C was 25 for the empty-cell test and (in order to produce an expected frequency of 5) was 6 for the chi-square test. Under these circumstances the power of the empty-cell test was .183, whereas that of the chi-square test was .189.

There is reason to believe that the test should be most sensitive when C and n are not too unequal. If n greatly exceeds C, there is too much likelihood of no empty cells; whereas if C greatly exceeds n, there is too much likelihood that each occupied cell will contain only one observation, so that the number of occupied cells equals the number of observations and $E = C - n$. In either case, an overwhelming proportion of the null probability distribution of E would be concentrated over a single value of E, only a tiny remaining probability being divided among the other possible values of E. And under many alternatives to H_0, E would continue to be overwhelmingly likely to assume its modal value under H_0, thus depriving E of much of its sensitivity to the *falsity* of H_0. It would seem prudent therefore, to let C be $\geq n$ and $\leq 2n$ when this is not too inconvenient (a slight excess of C over n has the beneficial effect of spreading out the upper, i.e., testing, tail of the distribution of E over a greater range of values, thereby making the test more fine-grained and bringing the actual cumulative probability of the critical value closer to the nominal significance level of the test), and in any case not to let the ratio between C and n stray exceedingly far from 1.00, especially when n exceeds C. When n is very large, following this rule will entail creating a laboriously large number of cells, thus calling practical efficiency into question.

13.7.5 Tables

Table XIV of the Appendix gives critical values of E (and, in parentheses, the corresponding exact cumulative probability) for upper-tailed tests at levels $\alpha = .05$ or $\alpha = .01$ for the cases $n = 5\ (1)\ 50$ and $C = 2\ (1)\ 10$. It thus gives the smallest value of E' for which $P(E \geq E') \leq \alpha$ as well as the corresponding value of $P(E \geq E')$.

Nicholson [26] has published tables appropriate for either upper- or lower-tailed tests at one-tailed α's of .10, .05, .025, and .01 in the cases $C = 2\ (1)\ 20$, $O = 1\ (1)\ C - 1$, $n = 2\ (1)\ 30$. Let $P(\theta;\ C, n)$ stand for the point probability of θ occupied cells when the number of cells is C and the number of observations is n. For a lower-tail test on the number of occupied cells (equivalent to an upper-tail test on the number of empty cells), Nicholson tables the smallest value of n that satisfies the equation

$$\sum_{\theta=1}^{O} P(\theta;\ C, n) \leq \alpha$$

as well as the exact cumulative probability represented by the summation. For an upper-tail test, he tables the largest value of n which satisfies

$$\sum_{\theta=O}^{C} P(\theta;\ C, n) \leq \alpha$$

and also tables the exact cumulative probability. David [9] has tabled the entire point probability distribution of E, when $n = C$, for the cases $n = C = 3\ (1)\ 20$.

For reasons stated under *Efficiency* (Section 13.7.4), there is some question as to the desirability of this test in the large-sample case. However, a normal approximation suitable for obtaining approximate probabilities in the large-sample case is given in [7].

13.7.6 Example

An experimenter hypothesizes that a certain population has a normal distribution with a mean of 3 and a standard deviation of 2 and decides to test his hypothesis at $\alpha = .05$ by means of an upper-tail empty-cell test using $C = n = 5$. Since $C = 5$, each cell will have to contain $\frac{1}{5}$ or 20% of the hypothesized normal distribution. Consulting tables of cumulative probabilities for a standardized normal deviate Z (i.e., a normal variate with zero mean and unit variance), the experimenter looks up the twentieth, fortieth, sixtieth, and eightieth percentiles, i.e., the values of Z that divide its distribution into five equal areas, and finds that $P(Z \leq -.84162) = .20$, $P(Z \leq -.25335) = .40$, $P(Z \leq .25335) = .60$, and $P(Z \leq .84162) = .80$. Multiplying these Z values by 2 and adding 3 to the product, he obtains 1.31676, 2.49330, 3.50670, and 4.68324 as the twentieth, fortieth, sixtieth, and eightieth percentiles of a normally distributed variate with mean of 3 and standard deviation of 2. Therefore the five cells are the intervals (1) from minus infinity to 1.31676; (2) from 1.31676 to 2.49330; (3) from 2.49330 to 3.50670; (4) from 3.50670 to 4.68324; and (5) from 4.68324 to plus infinity; and any value of X for which $1.31676 < X < 2.49330$ places X in the second cell.

Now the experimenter draws a random sample of five observations from the population in question, obtaining the values $-.311$, $-.078$, .555, 1.462, and 5.711. The first three values listed fall into the first cell (i.e., the interval from $-\infty$ to 1.31676); the fourth value, 1.462, falls into the second cell (the interval from 1.31676 to 2.49330); and the fifth value, 5.711, falls into the fifth cell (the interval from 4.68324 to $+\infty$). So there are two empty cells. Entering Table XIV of the Appendix with $\alpha = .05$, $n = 5$, and $C = 5$, the critical value for the number of empty cells is found to be 4 and $P(E \geq 4) = .0016$. Since the number of empty cells, 2, is not equal to or greater than the critical value, 4, the hypothesis cannot be rejected.

The hypothesis and data used in the above example are the same as those used in the example for the Kolmogorov test, which rejected the hypothesis at the predesignated .05 level of significance. From David's tables we find that the actual upper-tail cumulative probability for two empty cells is $P(E \geq 2) = .5776$. In this particular case, therefore, the maximum absolute deviation test is far more powerful than the empty-cell test, and one would suspect that this is so in general.

13.7.7 Discussion

David [9] has proposed, and provided tables for, a second empty-cell test in which the hypothesized distribution is divided into $2n$ cells, only n of which are examined for emptiness. Thus the test statistic is the number of empty cells among a predesignated n "test cells." This feature gives the experimenter flexibility in choosing alternatives against which the test is to be especially sensitive. By taking the lower n cells as test cells, the test is made sensitive to alternatives in which the population's location is higher than hypothesized. By designating the lowest $n/2$ and highest $n/2$ cells as test cells, the test is sensitized to a smaller-than-hypothesized population variance, etc. The present test lacks this feature and, like the Kolmogorov test, is an omnibus test against general alternatives.

13.8 OTHER TESTS

The number of distribution-free tests is far too large for this book to include them all. And this is true even of those tests that are ingenious, exact, practical, and easy to derive. Most of the simple, exact, better-known tests that are members of families of tests having similar derivations or rationale have been included. However, some excellent isolated tests have been excluded, such as Olmstead and Tukey's corner test for association [27], Foster and Stuart's records test [12], and Mosteller's k-sample slippage test for an extreme population [25], as well as others. In the case of such unique tests, it is often true that all of the information about the test is concentrated in a single one or a pair of journal articles. So, once having obtained the reference, the reader can quickly learn all about the test from one or two primary sources.

This book has included very few multi-sample tests, i.e., distribution-free analogues of the analysis-of-variance tests. This has not been due to any paucity of such tests. Indeed, they are being developed at a frantic rate. Unfortunately, however, a great many of the multi-sample distribution-free tests lack the pristine simplicity of the one- and two-sample tests. And this is also true of many bivariate and multivariate analogues of simple distribution-free tests for the univariate case. Distribution-free tests for the multisample or multivariate case often require as much mathematical sophistication, as do parametric tests, in order to be understood; and this disqualifies them for inclusion in this book.

REFERENCES

1. Birnbaum, Z. W., "Numerical Tabulation of the Distribution of Kolmogorov's Statistic for Finite Sample Size," *Journal of the American Statistical Association*, **47** (1952), 425–441.

2. Birnbaum, Z. W., and R. A. Hall, "Small Sample Distributions for Multi-Sample Statistics of the Smirnov Type," *Annals of Mathematical Statistics*, **31** (1960), 710–720.

3. Birnbaum, Z. W. and F. H. Tingey, "One-Sided Confidence Contours for Probability Distribution Functions," *Annals of Mathematical Statistics*, **22** (1951), 592–596.

4. Burr, E. J., "Small Sample Distributions of the Two-Sample Cramér–von Mises' W^2 and Watson's U^2," *Annals of Mathematical Statistics*, **35** (1964), 1091–1098. [See also **34** (1963), 95–101, and **33** (1962), 1148–1159.]

5. Capon, J., "On the Asymptotic Efficiency of the Kolmogorov-Smirnov Test, *Journal of the American Statistical Association*, **60** (1965), 843–853.

6. Chapman, D. G., "A Comparative Study of Several One-Sided Goodness-of-Fit Tests," *Annals of Mathematical Statistics*, **29** (1958), 655–674.

7. Csorgo, M., and I. Guttman, "On the Empty Cell Test," *Technometrics*, **4** (1962), 235–247.

8. Darling, D. A., "The Kolmogorov-Smirnov, Cramér–von Mises Tests," *Annals of Mathematical Statistics*, **28** (1957), 823–838.

9. David, F. N., "Two Combinational Tests of Whether a Sample Has Come from a Given Population," *Biometrika*, **37** (1950), 97–110.

10. Dixon, W. J., "Power under Normality of Several Nonparametric Tests," *Annals of Mathematical Statistics*, **25** (1954), 610–614.

11. Epstein, B., "Comparison of Some Non-Parametric Tests against Normal Alternatives with an Application to Life Testing," *Journal of the American Statistical Association*, **50** (1955), 894–900.

12. Foster, F. G., and A. Stuart, "Distribution-Free Tests in Time-Series Based on the Breaking of Records," *Journal of the Royal Statistical Society, Series B*, **16** (1954), 1–22. [See also **14** (1952), 220–228, and **19** (1957), 149–153.]

13. Goodman, L. A., "Kolmogorov-Smirnov Tests for Psychological Research," *Psychological Bulletin*, **51** (1954), 160–168.

14. Kaarsemaker, L., and A. van Wijngaarden, *Tables for Use in Rank Correlation*, Report R73 of the Computation Department of the Mathematical Centre, Amsterdam. Also printed in *Statistica Neerlandica*, **7** (1953), 41–54.

15. Kendall, M. G., *Rank Correlation Methods*, 2nd Ed. New York: Hafner Publishing Co., 1955.

16. Kolmogorov, A., "Confidence Limits for an Unknown Distribution Function," *Annals of Mathematical Statistics*, **12** (1941), 461–463.

17. Lehmann, E. L., "The Power of Rank Tests," *Annals of Mathematical Statistics*, **24** (1953), 23–43.

18. Mann, H. B., "Nonparametric Tests against Trend," *Econometrica*, **13** (1945), 245–259.

19. Massey, F. J., "A Note on the Estimation of a Distribution Function by Confidence Limits," *Annals of Mathematical Statistics*, **21** (1950), 116–119.

20. Massey, F. J., "A Note on the Power of a Non-Parametric Test," *Annals of Mathematical Statistics*, **21** (1950), 440–443 [see also **23** (1952), 637–638].

21. Massey, F. J., "Distribution Table for the Deviation between Two Sample Cumulatives," *Annals of Mathematical Statistics*, **23** (1952), 435–441.

22. Massey, F. J., "The Distribution of the Maximum Deviation between Two Sample Cumulative Step Functions," *Annals of Mathematical Statistics*, **22** (1951), 125–128.

23. Massey, F. J., "The Kolmogorov-Smirnov Test for Goodness of Fit," *Journal of the American Statistical Association*, **46** (1951), 68–78.

24. Miller, L. H., "Table of Percentage Points of Kolmogorov Statistics," *Journal of the American Statistical Association*, **51** (1956), 111–121.

25. Mosteller, F., "A k-Sample Slippage Test for an Extreme Population," *Annals of Mathematical Statistics*, **19** (1948), 58–65 [see also **21** (1950), 120–123].

26. Nicholson, W. L., "Occupancy Probability Distribution Critical Points," *Biometrika*, **48** (1961), 175–180.

27. Olmstead, P. S., and J. W. Tukey, "A Corner Test for Association," *Annals of Mathematical Statistics*, **18** (1947), 495–513.

28. Owen, D. B., *Handbook of Statistical Tables*. Reading, Mass.: Addison-Wesley Publishing Co., Inc., 1962, pp. 396–399, 423–448, 454–457.

29. Schmid, P., "On the Kolmogorov and Smirnov Limit Theorems for Discontinuous Distribution Functions," *Annals of Mathematical Statistics*, **29** (1958), 1011–1027.

30. Smirnov, N., "Table for Estimating the Goodness of Fit of Empirical Distributions," *Annals of Mathematical Statistics*, **19** (1948), 279–281.

31. Tsao, C. K., "An Extension of Massey's Distribution of the Maximum Deviation between Two-Sample Cumulative Step Functions," *Annals of Mathematical Statistics*, **25** (1954), 587–592.

32. van der Waerden, B. L., "Order Tests for the Two-Sample Problem," *Proceedings Koninklijke Nederlandse Akademie van Wetenschappen, Series A*, **56** (1953), 303–310.

33. Wilks, S. S., *Mathematical Statistics*. New York: John Wiley & Sons, Inc., 1962, pp. 336–341, 433–438, 454–459.

APPENDIX

TABLES FOR THE TEST STATISTICS

In this appendix, tables are given for determining the significance of obtained values of the various test statistics. The tables are presented in the same order in which the corresponding tests were taken up in the body of the book. Each table is identified by a roman numeral followed, in parentheses, by the section number of the portion of the book in which the table is explained.

The tables contained in this appendix are as follows:

Table I* (5.3.5): *Critical Lower-Tail Values of D (Indicating Positive Correlation) for Hotelling and Pabst's Spearman Rank-Order Correlation Test*

	Largest value of D' for which $P(D \leq D') \leq \alpha$ Significance level, α					
n	.001	.005	.010	.025	.050	.100
4	—	—	—	—	0	0
5	—	—	0	0	2	4
6	—	0	2	4	6	12
7	0	4	6	12	16	24
8	4	10	14	24	32	42
9	10	20	26	36	48	62
10	20	34	42	58	72	90
11	32	52	64	84	102	126
12	50	76	92	118	142	170
13	74	108	128	160	188	224
14	104	146	170	210	244	288
15	140	192	222	268	310	362
16	184	248	282	338	388	448
17	236	312	354	418	478	548
18	298	388	436	510	580	662
19	370	474	530	616	694	788
20	452	572	636	736	824	932
21	544	684	756	868	970	1090
22	650	808	890	1018	1132	1268
23	770	948	1040	1182	1310	1462
24	902	1102	1206	1364	1508	1676
25	1048	1272	1388	1564	1724	1910
26	1210	1460	1588	1784	1958	2166
27	1388	1664	1806	2022	2214	2442
28	1584	1888	2044	2282	2492	2742
29	1798	2132	2304	2562	2794	3066
30	2030	2396	2582	2866	3118	3414

*Adapted from Table 2 in G. J. Glasser and R. F. Winter's "Critical Values of the Coefficient of Rank Correlation for Testing the Hypothesis of Independence," *Biometrika*, **48** (1961), 444–448, with permission of the authors and editor.

Table II* (5.5.5): *Critical and Quasi-Critical Lower-Tail Values of W_+ (and their Probability Levels) for Wilcoxon's Signed-Rank Test*

	W'_+ followed by $P(W_+ \leq W'_+)$							
n	$\alpha = .05$		$\alpha = .025$		$\alpha = .01$		$\alpha = .005$	
5	0	.0313						
	1	.0625						
6	2	.0469	0	.0156				
	3	.0781	1	.0313				
7	3	.0391	2	.0234	0	.0078		
	4	.0547	3	.0391	1	.0156		
8	5	.0391	3	.0195	1	.0078	0	.0039
	6	.0547	4	.0273	2	.0117	1	.0078
9	8	.0488	5	.0195	3	.0098	1	.0039
	9	.0645	6	.0273	4	.0137	2	.0059
10	10	.0420	8	.0244	5	.0098	3	.0049
	11	.0527	9	.0322	6	.0137	4	.0068
11	13	.0415	10	.0210	7	.0093	5	.0049
	14	.0508	11	.0269	8	.0122	6	.0068
12	17	.0461	13	.0212	9	.0081	7	.0046
	18	.0549	14	.0261	10	.0105	8	.0061
13	21	.0471	17	.0239	12	.0085	9	.0040
	22	.0549	18	.0287	13	.0107	10	.0052
14	25	.0453	21	.0247	15	.0083	12	.0043
	26	.0520	22	.0290	16	.0101	13	.0054
15	30	.0473	25	.0240	19	.0090	15	.0042
	31	.0535	26	.0277	20	.0108	16	.0051
16	35	.0467	29	.0222	23	.0091	19	.0046
	36	.0523	30	.0253	24	.0107	20	.0055
17	41	.0492	34	.0224	27	.0087	23	.0047
	42	.0544	35	.0253	28	.0101	24	.0055
18	47	.0494	40	.0241	32	.0091	27	.0045
	48	.0542	41	.0269	33	.0104	28	.0052
19	53	.0478	46	.0247	37	.0090	32	.0047
	54	.0521	47	.0273	38	.0102	33	.0054
20	60	.0487	52	.0242	43	.0096	37	.0047
	61	.0527	53	.0266	44	.0107	38	.0053

*Body of table is reproduced, with changes only in notation, from Table II in F. Wilcoxon, S.K.Katti, and Roberta A. Wilcox, *Critical Values and Probability Levels for the Wilcoxon Rank Sum Test and the Wilcoxon Signed Rank Test*, American Cyanamid Company (Lederle Laboratories Division, Pearl River, N.Y.) and The Florida State University (Department of Statistics, Tallahassee, Fla.), August 1963, with permission of the authors and publishers.

Table II (*Continued*)

n	$\alpha = .05$		$\alpha = .025$		$\alpha = .01$		$\alpha = .005$	
21	67	.0479	58	.0230	49	.0097	42	.0045
	68	.0516	59	.0251	50	.0108	43	.0051
22	75	.0492	65	.0231	55	.0095	48	.0046
	76	.0527	66	.0250	56	.0104	49	.0052
23	83	.0490	73	.0242	62	.0098	54	.0046
	84	.0523	74	.0261	63	.0107	55	.0051
24	91	.0475	81	.0245	69	.0097	61	.0048
	92	.0505	82	.0263	70	.0106	62	.0053
25	100	.0479	89	.0241	76	.0094	68	.0048
	101	.0507	90	.0258	77	.0101	69	.0053
26	110	.0497	98	.0247	84	.0095	75	.0047
	111	.0524	99	.0263	85	.0102	76	.0051
27	119	.0477	107	.0246	92	.0093	83	.0048
	120	.0502	108	.0260	93	.0100	84	.0052
28	130	.0496	116	.0239	101	.0096	91	.0048
	131	.0521	117	.0252	102	.0102	92	.0051
29	140	.0482	126	.0240	110	.0095	100	.0049
	141	.0504	127	.0253	111	.0101	101	.0053
30	151	.0481	137	.0249	120	.0098	109	.0050
	152	.0502	138	.0261	121	.0104	110	.0053
31	163	.0491	147	.0239	130	.0099	118	.0049
	164	.0512	148	.0251	131	.0105	119	.0052
32	175	.0492	159	.0249	140	.0097	128	.0050
	176	.0512	160	.0260	141	.0103	129	.0053
33	187	.0485	170	.0242	151	.0099	138	.0049
	188	.0503	171	.0253	152	.0104	139	.0052
34	200	.0488	182	.0242	162	.0098	148	.0048
	201	.0506	183	.0252	163	.0103	149	.0051
35	213	.0484	195	.0247	173	.0096	159	.0048
	214	.0501	196	.0257	174	.0100	160	.0051
36	227	.0489	208	.0248	185	.0096	171	.0050
	228	.0505	209	.0258	186	.0100	172	.0052
37	241	.0487	221	.0245	198	.0099	182	.0048
	242	.0503	222	.0254	199	.0103	183	.0050
38	256	.0493	235	.0247	211	.0099	194	.0048
	257	.0509	236	.0256	212	.0104	195	.0050
39	271	.0493	249	.0246	224	.0099	207	.0049
	272	.0507	250	.0254	225	.0103	208	.0051
40	286	.0486	264	.0249	238	.0100	220	.0049
	287	.0500	265	.0257	239	.0104	221	.0051

Table II (*Continued*)

n	$\alpha = .05$		$\alpha = .025$		$\alpha = .01$		$\alpha = .005$	
41	302	.0488	279	.0248	252	.0100	233	.0048
	303	.0501	280	.0256	253	.0103	234	.0050
42	319	.0496	294	.0245	266	.0098	247	.0049
	320	.0509	295	.0252	267	.0102	248	.0051
43	336	.0498	310	.0245	281	.0098	261	.0048
	337	.0511	311	.0252	282	.0102	262	.0050
44	353	.0495	327	.0250	296	.0097	276	.0049
	354	.0507	328	.0257	297	.0101	277	.0051
45	371	.0498	343	.0244	312	.0098	291	.0049
	372	.0510	344	.0251	313	.0101	292	.0051
46	389	.0497	361	.0249	328	.0098	307	.0050
	390	.0508	362	.0256	329	.0101	308	.0052
47	407	.0490	378	.0245	345	.0099	322	.0048
	408	.0501	379	.0251	346	.0102	323	.0050
48	426	.0490	396	.0244	362	.0099	339	.0050
	427	.0500	397	.0251	363	.0102	340	.0051
49	446	.0495	415	.0247	379	.0098	355	.0049
	447	.0505	416	.0253	380	.0100	356	.0050
50	466	.0495	434	.0247	397	.0098	373	.0050
	467	.0506	435	.0253	398	.0101	374	.0051

Table III* (5.8.5): *Critical Lower-Tail Values of W_n for Wilcoxon's Rank-Sum Test*

Largest value of W'_n for which $P(W_n \leq W'_n) \leq \alpha$ where α is given in boldface

	$n = 1$							$n = 2$							
m	**.001**	**.005**	**.01**	**.025**	**.05**	**.10**	$2\overline{W}$	**.001**	**.005**	**.01**	**.025**	**.05**	**.10**	$2\overline{W}$	m
2							4						—	10	2
3							5						3	12	3
4							6					—	3	14	4
5							7					3	4	16	5
6							8					3	4	18	6
7							9				—	3	4	20	7
8						—	10				3	4	5	22	8
9						1	11				3	4	5	24	9
10						1	12				3	4	6	26	10
11						1	13				3	4	6	28	11
12						1	14			—	4	5	7	30	12
13						1	15			3	4	5	7	32	13
14						1	16			3	4	6	8	34	14
15						1	17			3	4	6	8	36	15
16						1	18			3	4	6	8	38	16
17						1	19			3	5	6	9	40	17
18					—	1	20		—	4	5	7	9	42	18
19					1	2	21		3	4	5	7	10	44	19
20					1	2	22		3	4	5	7	10	46	20
21					1	2	23		3	4	6	8	11	48	21
22					1	2	24		3	4	6	8	11	50	22
23					1	2	25		3	4	6	8	12	52	23
24					1	2	26		3	4	6	9	12	54	24
25	—	—	—	—	1	2	27	—	3	4	6	9	12	56	25

*Body of table is reproduced, with changes only in notation, from Table 1 in L.R. Verdooren's "Extended Tables of Critical Values for Wilcoxon's Test Statistic," *Biometrika*, **50** (1963), 177–186, with permission of the author and editor.

Table III (*Continued*)

$n = 3$

m	.001	.005	.01	.025	.05	.10	$2\overline{W}$
3					6	7	21
4				—	6	7	24
5				6	7	8	27
6			—	7	8	9	30
7			6	7	8	10	33
8		—	6	8	9	11	36
9		6	7	8	10	11	39
10		6	7	9	10	12	42
11		6	7	9	11	13	45
12		7	8	10	11	14	48
13		7	8	10	12	15	51
14		7	8	11	13	16	54
15		8	9	11	13	16	57
16	—	8	9	12	14	17	60
17	6	8	10	12	15	18	63
18	6	8	10	13	15	19	66
19	6	9	10	13	16	20	69
20	6	9	11	14	17	21	72
21	7	9	11	14	17	21	75
22	7	10	12	15	18	22	78
23	7	10	12	15	19	23	81
24	7	10	12	16	19	24	84
25	7	11	13	16	20	25	87

$n = 4$

.001	.005	.01	.025	.05	.10	$2\overline{W}$	m
		—	10	11	13	36	4
	—	10	11	12	14	40	5
	10	11	12	13	15	44	6
	10	11	13	14	16	48	7
	11	12	14	15	17	52	8
—	11	13	14	16	19	56	9
10	12	13	15	17	20	60	10
10	12	14	16	18	21	64	11
10	13	15	17	19	22	68	12
11	13	15	18	20	23	72	13
11	14	16	19	21	25	76	14
11	15	17	20	22	26	80	15
12	15	17	21	24	27	84	16
12	16	18	21	25	28	88	17
13	16	19	22	26	30	92	18
13	17	19	23	27	31	96	19
13	18	20	24	28	32	100	20
14	18	21	25	29	33	104	21
14	19	21	26	30	35	108	22
14	19	22	27	31	36	112	23
15	20	23	27	32	38	116	24
15	20	23	28	33	38	120	25

$n = 5$

m	.001	.005	.01	.025	.05	.10	$2\overline{W}$
5		15	16	17	19	20	55
6		16	17	18	20	22	60
7	—	16	18	20	21	23	65
8	15	17	19	21	23	25	70
9	16	18	20	22	24	27	75
10	16	19	21	23	26	28	80
11	17	20	22	24	27	30	85
12	17	21	23	26	28	32	90
13	18	22	24	27	30	33	95
14	18	22	25	28	31	35	100
15	19	23	26	29	33	37	105
16	20	24	27	30	34	38	110
17	20	25	28	32	35	40	115
18	21	26	29	33	37	42	120
19	22	27	30	34	38	43	125
20	22	28	31	35	40	45	130
21	23	29	32	37	41	47	135
22	23	29	33	38	43	48	140
23	24	30	34	39	44	50	145
24	25	31	35	40	45	51	150
25	25	32	36	42	47	53	155

$n = 6$

.001	.005	.01	.025	.05	.10	$2\overline{W}$	m
—	23	24	26	28	30	78	6
21	24	25	27	29	32	84	7
22	25	27	29	31	34	90	8
23	26	28	31	33	36	96	9
24	27	29	32	35	38	102	10
25	28	30	34	37	40	108	11
25	30	32	35	38	42	114	12
26	31	33	37	40	44	120	13
27	32	34	38	42	46	126	14
28	33	36	40	44	48	132	15
29	34	37	42	46	50	138	16
30	36	39	43	47	52	144	17
31	37	40	45	49	55	150	18
32	38	41	46	51	57	156	19
33	39	43	48	53	59	162	20
33	40	44	50	55	61	168	21
34	42	45	51	57	63	174	22
35	43	47	53	58	65	180	23
36	44	48	54	60	67	186	24
37	45	50	56	62	69	192	25

Table III (*Continued*)

$n = 7$

m	.001	.005	.01	.025	.05	.10	$2\overline{W}$
7	29	32	34	36	39	41	105
8	30	34	35	38	41	44	112
9	31	35	37	40	43	46	119
10	33	37	39	42	45	49	126
11	34	38	40	44	47	51	133
12	35	40	42	46	49	54	140
13	36	41	44	48	52	56	147
14	37	43	45	50	54	59	154
15	38	44	47	52	56	61	161
16	39	46	49	54	58	64	168
17	41	47	51	56	61	66	175
18	42	49	52	58	63	69	182
19	43	50	54	60	65	71	189
20	44	52	56	62	67	74	196
21	46	53	58	64	69	76	203
22	47	55	59	66	72	79	210
23	48	57	61	68	74	81	217
24	49	58	63	70	76	84	224
25	50	60	64	72	78	86	231

$n = 8$

.001	.005	.01	.025	.05	.10	$2\overline{W}$	m
40	43	45	49	51	55	136	8
41	45	47	51	54	58	144	9
42	47	49	53	56	60	152	10
44	49	51	55	59	63	160	11
45	51	53	58	62	66	168	12
47	53	56	60	64	69	176	13
48	54	58	62	67	72	184	14
50	56	60	65	69	75	192	15
51	58	62	67	72	78	200	16
53	60	64	70	75	81	208	17
54	62	66	72	77	84	216	18
56	64	68	74	80	87	224	19
57	66	70	77	83	90	232	20
59	68	72	79	85	92	240	21
60	70	74	81	88	95	248	22
62	71	76	84	90	98	256	23
64	73	78	86	93	101	264	24
65	75	81	89	96	104	272	25

$n = 9$

m	.001	.005	.01	.025	.05	.10	$2\overline{W}$
9	52	56	59	62	66	70	171
10	53	58	61	65	69	73	180
11	55	61	63	68	72	76	189
12	57	63	66	71	75	80	198
13	59	65	68	73	78	83	207
14	60	67	71	76	81	86	216
15	62	69	73	79	84	90	225
16	64	72	76	82	87	93	234
17	66	74	78	84	90	97	243
18	68	76	81	87	93	100	252
19	70	78	83	90	96	103	261
20	71	81	85	93	99	107	270
21	73	83	88	95	102	110	279
22	75	85	90	98	105	113	288
23	77	88	93	101	108	117	297
24	79	90	95	104	111	120	306
25	81	92	98	107	114	123	315

$n = 10$

.001	·005	.01	.025	.05	.10	$2\overline{W}$	m
65	71	74	78	82	87	210	10
67	73	77	81	86	91	220	11
69	76	79	84	89	94	230	12
72	79	82	88	92	98	240	13
74	81	85	91	96	102	250	14
76	84	88	94	99	106	260	15
78	86	91	97	103	109	270	16
80	89	93	100	106	113	280	17
82	92	96	103	110	117	290	18
84	94	99	107	113	121	300	19
87	97	102	110	117	125	310	20
89	99	105	113	120	128	320	21
91	102	108	116	123	132	330	22
93	105	110	119	127	136	340	23
95	107	113	122	130	140	350	24
98	110	116	126	134	144	360	25

Table III (*Continued*)

$n = 11$

m	.001	.005	.01	.025	.05	.10	$2\overline{W}$
11	81	87	91	96	100	106	253
12	83	90	94	99	104	110	264
13	86	93	97	103	108	114	275
14	88	96	100	106	112	118	286
15	90	99	103	110	116	123	297
16	93	102	107	113	120	127	308
17	95	105	110	117	123	131	319
18	98	108	113	121	127	135	330
19	100	111	116	124	131	139	341
20	103	114	119	128	135	144	352
21	106	117	123	131	139	148	363
22	108	120	126	135	143	152	374
23	111	123	129	139	147	156	385
24	113	126	132	142	151	161	396
25	116	129	136	146	155	165	407

$n = 12$

.001	.005	.01	.025	.05	.10	$2\overline{W}$	m
98	105	109	115	120	127	300	12
101	109	113	119	125	131	312	13
103	112	116	123	129	136	324	14
106	115	120	127	133	141	336	15
109	119	124	131	138	145	348	16
112	122	127	135	142	150	360	17
115	125	131	139	146	155	372	18
118	129	134	143	150	159	384	19
120	132	138	147	155	164	396	20
123	136	142	151	159	169	408	21
126	139	145	155	163	173	420	22
129	142	149	159	168	178	432	23
132	146	153	163	172	183	444	24
135	149	156	167	176	187	456	25

$n = 13$

m	.001	.005	.01	.025	.05	.10	$2\overline{W}$
13	117	125	130	136	142	149	351
14	120	129	134	141	147	154	364
15	123	133	138	145	152	159	377
16	126	136	142	150	156	165	390
17	129	140	146	154	161	170	403
18	133	144	150	158	166	175	416
19	136	148	154	163	171	180	429
20	139	151	158	167	175	185	442
21	142	155	162	171	180	190	455
22	145	159	166	176	185	195	468
23	149	163	170	180	189	200	481
24	152	166	174	185	194	205	494
25	155	170	178	189	199	211	507

$n = 14$

.001	.005	.01	.025	.05	.10	$2\overline{W}$	m
137	147	152	160	166	174	406	14
141	151	156	164	171	179	420	15
144	155	161	169	176	185	434	16
148	159	165	174	182	190	448	17
151	163	170	179	187	196	462	18
155	168	174	183	192	202	476	19
159	172	178	188	197	207	490	20
162	176	183	193	202	213	504	21
166	180	187	198	207	218	518	22
169	184	192	203	212	224	532	23
173	188	196	207	218	229	546	24
177	192	200	212	223	235	560	25

$n = 15$

m	.001	.005	.01	.025	.05	.10	$2\overline{W}$
15	160	171	176	184	192	200	465
16	163	175	181	190	197	206	480
17	167	180	186	195	203	212	495
18	171	184	190	200	208	218	510
19	175	189	195	205	214	224	525
20	179	193	200	210	220	230	540
21	183	198	205	216	225	236	555
22	187	202	210	221	231	242	570
23	191	207	214	226	236	248	585
24	195	211	219	231	242	254	600
25	199	216	224	237	248	260	615

$n = 16$

.001	.005	.01	.025	.05	.10	$2\overline{W}$	m
184	196	202	211	219	229	528	16
188	201	207	217	225	235	544	17
192	206	212	222	231	242	560	18
196	210	218	228	237	248	576	19
201	215	223	234	243	255	592	20
205	220	228	239	249	261	608	21
209	225	233	245	255	267	624	22
214	230	238	251	261	274	640	23
218	235	244	256	267	280	656	24
222	240	249	262	273	287	672	25

Table III (*Continued*)

$n = 17$

m	.001	.005	.01	.025	.05	.10	$2\overline{W}$
17	210	223	230	240	249	259	595
18	214	228	235	246	255	266	612
19	219	234	241	252	262	273	629
20	223	239	246	258	268	280	646
21	228	244	252	264	274	287	663
22	233	249	258	270	281	294	680
23	238	255	263	276	287	300	697
24	242	260	269	282	294	307	714
25	247	265	275	288	300	314	731

$n = 18$

.001	.005	.01	.025	.05	.10	$2\overline{W}$	m
237	252	259	270	280	291	666	18
242	258	265	277	287	299	684	19
247	263	271	283	294	306	702	20
252	269	277	290	301	313	720	21
257	275	283	296	307	321	738	22
262	280	289	303	314	328	756	23
267	286	295	309	321	335	774	24
273	292	301	316	328	343	792	25

$n = 19$

m	.001	.005	.01	.025	.05	.10	$2\overline{W}$
19	267	283	291	303	313	325	741
20	272	289	297	309	320	333	760
21	277	295	303	316	328	341	779
22	283	301	310	323	335	349	798
23	288	307	316	330	342	357	817
24	294	313	323	337	350	364	836
25	299	319	329	344	357	372	855

$n = 20$

.001	.005	.01	.025	.05	.10	$2\overline{W}$	m
298	315	324	337	348	361	820	20
304	322	331	344	356	370	840	21
309	328	337	351	364	378	860	22
315	335	344	359	371	386	880	23
321	341	351	366	379	394	900	24
327	348	358	373	387	403	920	25

$n = 21$

m	.001	.005	.01	.025	.05	.10	$2\overline{W}$
21	331	349	359	373	385	399	903
22	337	356	366	381	393	408	924
23	343	363	373	388	401	417	945
24	349	370	381	396	410	425	966
25	356	377	388	404	418	434	987

$n = 22$

.001	.005	.01	.025	.05	.10	$2\overline{W}$	m
365	386	396	411	424	439	990	22
372	393	403	419	432	448	1012	23
379	400	411	427	441	457	1034	24
385	408	419	435	450	467	1056	25

$n = 23$

m	.001	.005	.01	.025	.05	.10	$2\overline{W}$
23	402	424	434	451	465	481	1081
24	409	431	443	459	474	491	1104
25	416	439	451	468	483	500	1127

$n = 24$

.001	.005	.01	.025	.05	.10	$2\overline{W}$	m
440	464	475	492	507	525	1176	24
448	472	484	501	517	535	1200	25

$n = 25$

m	.001	.005	.01	.025	.05	.10	$2\overline{W}$
25	480	505	517	536	552	570	1275

Table IV-A* (5.12.5): *Critical Values of S (and Their Probability Levels) for Friedman Test*

Smallest value of S' for which $P(S \geq S') \leq \alpha$, followed by $P(S \geq S')$

	Significance level, α			
R	.10	.05	.01	.001
		$C = 3$		
2	—	—	—	—
3	18 .028	18 .028	—	—
4	24 .069	26 .042	32 .0046	—
5	26 .093	32 .039	42 .0085	50 .00077
6	32 .072	42 .029	54 .0081	72 .00013
7	38 .085	50 .027	62 .0084	86 .00032
8	42 .079	50 .047	72 .0099	98 .00086
9	50 .069	56 .048	78 .010	114 .00072
10	50 .092	62 .046	96 .0073	122 .0010
11	54 .100	72 .043	104 .0066	146 .00067
12	62 .080	74 .050	114 .0080	150 .00087
13	62 .098	78 .050	122 .0076	168 .00081
14	72 .089	86 .049	126 .010	186 .00058
15	74 .096	96 .047	134 .010	194 .00095

*Adapted from Table 14.1 in D. B. Owen's *Handbook of Statistical Tables*, Reading, Massachusetts: Addison-Wesley Publishing Company, Inc., 1962, pp. 407–419, with permission of the author, the publisher, the Sandia Corporation, and the United States Atomic Energy Commission.

Table IV-A (*Continued*)

	Significance Level, α			
R	.10	.05	.01	.001
		$C = 4$		
2	20 .042	20 .042	—	—
3	33 .075	37 .026	45 .0017	—
4	42 .093	52 .036	64 .0056	74 .00094
5	53 .089	65 .049	83 .0092	105 .00040
6	64 .088	76 .043	102 .0097	128 .0010
7	75 .093	91 .040	121 .0091	161 .00084
8	84 .098	102 .049	138 .010	184 .00100

Table IV-B* (5.12.5): *"Critical Values" of S for Friedman Test, Based Upon z-Approximation*

Values of S' for which $P(S \geq S') \cong \alpha$

R	C = 3	4	5	6	7	Additional values for $C = 3$: R	S'
			Values at .05 level of significance				
3			64.4	103.9	157.3	9	54.0
4		49.5	88.4	143.3	217.0	12	71.9
5		62.6	112.3	182.4	276.2	14	83.8
6		75.7	136.1	221.4	335.2	16	95.8
8	48.1	101.7	183.7	299.0	453.1	18	107.7
10	60.0	127.8	231.2	376.7	571.0		
15	89.8	192.9	349.8	570.5	864.9		
20	119.7	258.0	468.5	764.4	1158.7		
			Values at .01 level of significance				
3			75.6	122.8	185.6	9	75.9
4		61.4	109.3	176.2	265.0	12	103.5
5		80.5	142.8	229.4	343.8	14	121.9
6		99.5	176.1	282.4	422.6	16	140.2
8	66.8	137.4	242.7	388.3	579.9	18	158.6
10	85.1	175.3	309.1	494.0	737.0		
15	131.0	269.8	475.2	758.2	1129.5		
20	177.0	364.2	641.2	1022.2	1521.9		

*Body of table is reproduced, with changes only in notation, from Table III in M. Friedman's "A Comparison of Alternative Tests of Significance for the Problem of *m* Rankings," *Annals of Mathematical Statistics*, **11**(1940), 86–92, with permission of the author and editor.

Table V* (6.1): *Expected Normal Scores*

Expected values, algebraic signs omitted, of the Rth smallest or of the Rth largest ($[N - R + 1]$th smallest) observations in a random sample of N observations drawn from a normal distribution with zero mean and unit variance. Omitted algebraic sign is minus for the smallest $N/2$ observations.

N	R	E_{NR}	N	R	E_{NR}	N	R	E_{NR}
2	1	.56418 95835	12	5	.31224 88787	17	5	.61945 76511
3	1	.84628 43753	12	6	.10258 96798	17	6	.45133 34467
4	1	1.02937 53730	13	1	1.66799 01770	17	7	.29518 64872
4	2	.29701 13823	13	2	1.16407 71937	17	8	.14598 74231
5	1	1.16296 44736	13	3	.84983 46324	18	1	1.82003 18790
5	2	.49501 89705	13	4	.60285 00882	18	2	1.35041 37134
6	1	1.26720 63606	13	5	.38832 71210	18	3	1.06572 81829
6	2	.64175 50388	13	6	.19052 36911	18	4	.84812 50190
6	3	.20154 68338	14	1	1.70338 15541	18	5	.66479 46127
7	1	1.35217 83756	14	2	1.20790 22754	18	6	.50158 15510
7	2	.75737 42706	14	3	.90112 67039	18	7	.35083 72382
7	3	.35270 69592	14	4	.66176 37035	18	8	.20773 53071
8	1	1.42360 03060	14	5	.45556 60500	18	9	.06880 25682
8	2	.85222 48625	14	6	.26729 70489	19	1	1.84448 15116
8	3	.47282 24949	14	7	.08815 92141	19	2	1.37993 84915
8	4	.15251 43995	15	1	1.73591 34449	19	3	1.09945 30994
9	1	1.48501 31622	15	2	1.24793 50823	19	4	.88586 19615
9	2	.93229 74567	15	3	.94768 90303	19	5	.70661 14847
9	3	.57197 07829	15	4	.71487 73983	19	6	.54770 73710
9	4	.27452 59191	15	5	.51570 10430	19	7	.40164 22742
10	1	1.53875 27308	15	6	.33529 60639	19	8	.26374 28909
10	2	1.00135 70446	15	7	.16529 85263	19	9	.13072 48795
10	3	.65605 91057	16	1	1.76599 13931	20	1	1.86747 50598
10	4	.37576 46970	16	2	1.28474 42232	20	2	1.40760 40959
10	5	.12266 77523	16	3	.99027 10960	20	3	1.13094 80522
11	1	1.58643 63519	16	4	.76316 67458	20	4	.92098 17004
11	2	1.06191 65201	16	5	.57000 93557	20	5	.74538 30058
11	3	.72883 94047	16	6	.39622 27551	20	6	.59029 69215
11	4	.46197 83072	16	7	.23375 15785	20	7	.44833 17532
11	5	.22489 08792	16	8	.07728 74593	20	8	.31493 32416
12	1	1.62922 76399	17	1	1.79394 19809	20	9	.18695 73647
12	2	1.11573 21843	17	2	1.31878 19878	20	10	.06199 62865
12	3	.79283 81991	17	3	1.02946 09889			
12	4	.53684 30214	17	4	.80738 49287			

*Body of table is reproduced, with changes only in notation, from Table 1 in D. Teichroew's "Tables of Expected Values of Order Statistics and Products of Order Statistics for Samples of Size Twenty and Less from the Normal Distribution," *Annals of Mathematical Statistics*, **27** (1956), 410–426, with permission of the author and editor.

Table VI* (6.2.5): *Critical or Quasi-Critical Upper-Tail Values of* $S = \sum_{i=1}^{n} E_{NR_i}$ *(and Their Probability Levels) for Terry-Hoeffding Expected Normal Scores Test*

		Value of S' for which $P(S \geq S') \cong \alpha$, followed by $P(S \geq S')$						
		Nominal levels						
N	n	.001	.005	.010	.025	.050	.075	.100
6	3					2.11051 .05000		1.70741 .10000
7	2					2.10955 .04762		1.70489 .09524
7	3				2.46226 .02857	2.10955 .05714	1.75685 .08571	1.70489 .11429
8	2				2.27583 .03571		1.89642 .07143	1.57611 .10714
8	3			2.74865 .01786	2.42834 .03571	2.12331 .05357	2.04894 .07143	1.74391 .10714
8	4			2.90116 .01429	2.59613 .02857	2.27583 .05714	1.95552 .07143	1.89642 .10000
9	2				2.41731 .02778	2.05698 .05556	1.75954 .08333	1.50427 .11111
9	3			2.98928 .01190	2.69184 .02381	2.33151 .04762	2.05698 .07143	1.78246 .09524
9	4		3.26381 .00794	2.98928 .01587	2.71476 .02381	2.41731 .04762	2.11987 .07143	1.78246 .10317
10	2				2.54011 .02222	2.19481 .04444	1.66142 .08889	1.65742 .11111
10	3		3.19617 .00833	2.91587 .01667	2.66278 .02500	2.31748 .05000	2.03719 .07500	1.81905 .10000
10	4		3.57193 .00476	3.31884 .00952	2.82040 .02381	2.44791 .05238	2.19481 .07619	1.94171 .09524
10	5		3.69460 .00397	3.44927 .00794	2.91587 .02778	2.57058 .04762	2.16435 .07540	2.03318 .10317
11	2			2.64835 .01818	2.31528 .03636	2.04841 .05455	1.81133 .07273	1.79076 .09091
11	3		3.37719 .00606	3.11033 .01212	2.77725 .02424	2.31528 .04848	2.09038 .07273	1.85330 .09697
11	4	3.83917 .00303	3.60208 .00606	3.37719 .00909	2.91521 .02424	2.54017 .04848	2.25273 .07576	2.02784 .10000
11	5	4.06406 .00216	3.83917 .00433	3.60208 .00866	3.00214 .02597	2.60638 .04978	2.31528 .07359	2.07819 .09740
12	2			2.74496 .01515	2.42207 .03030	2.16607 .04545	1.90857 .07576	1.65258 .10606

*Body of table is reproduced, with changes only in notation, from Table 1 in J.H. Klotz's "On the Normal Scores Two-Sample Rank Test," *Journal of the American Statistical Association*, **59** (1964), 652–664, with permission of the author and editor.

Table VI (*Continued*)

N	n	.001	.005	.010	.025	.050	.075	.100
		Nominal levels						
12	3		3.53780 .00455	3.28180 .00909	2.84755 .02273	2.43271 .05000	2.10982 .07273	1.88522 .10000
12	4	4.07464 .00202	3.85005 .00404	3.43521 .01010	3.00096 .02424	2.54800 .05051	2.31071 .07475	2.05471 .09899
12	5	4.38689 .00126	3.95264 .00505	3.69664 .00884	3.10354 .02525	2.65507 .05051	2.41778 .07449	2.15543 .09848
12	6	4.48948 .00108	3.86498 .00541	3.64039 .01082	3.17921 .02597	2.74496 .05195	2.42207 .07576	2.16607 .10173
13	2			2.83207 .01282	2.51782 .02564	2.05632 .05128	1.85851 .07692	1.66799 .10256
13	3	3.68190 .00350	3.43492 .00699	3.22039 .01049	2.83207 .02448	2.44374 .04895	2.15525 .07343	1.88251 .10140
13	4	4.28475 .00140	3.82324 .00559	3.50900 .00979	3.04659 .02517	2.61754 .05035	2.30330 .07552	2.07304 .09790
13	5	4.67308 .00078	4.07023 .00466	3.69953 .00932	3.23711 .02409	2.77561 .04973	2.44374 .07537	2.20444 .10023
13	6	4.67308 .00117	4.08695 .00524	3.82324 .01049	3.29357 .02506	2.83207 .05128	2.50020 .07517	2.24684 .10023
14	2			2.91128 .01099	2.60451 .02198	2.10903 .05495	1.86967 .07692	1.66347 .09890
14	3	3.81241 .00275	3.57305 .00549	3.26627 .01099	2.82312 .02473	2.45330 .04945	2.14894 .07418	1.94274 .09890
14	4	4.47417 .00100	3.90057 .00500	3.63415 .00999	3.11748 .02498	2.67192 .04995	2.36515 .07692	2.15895 .09790
14	5	4.74147 .00100	4.16787 .00500	3.81241 .01049	3.26869 .02498	2.82312 .04995	2.50437 .07493	2.24952 .09940
14	6	4.84158 .00100	4.28591 .00466	3.91252 .00999	3.39478 .02498	2.91128 .05162	2.59253 .07493	2.31523 .09990
14	7	4.92974 .00117	4.35614 .00495	3.99155 .00991	3.44541 .02477	2.94835 .05012	2.61693 .07517	2.33721 .10023
15	2			2.98385 .00952	2.45079 .02857	2.19562 .04762	1.90121 .07619	1.73591 .09524
15	3	3.93154 .00220	3.69873 .00440	3.31914 .01099	2.91050 .02418	2.46815 .05055	2.17827 .07473	1.96281 .09890
15	4	4.64641 .00073	3.93154 .00513	3.66485 .01026	3.18302 .02491	2.69622 .04982	2.41691 .07473	2.15054 .09963
15	5	4.81171 .00100	4.24948 .00500	3.91418 .00999	3.38196 .02498	2.88136 .04995	2.54133 .07493	2.28549 .10057
15	6	4.99682 .00100	4.43213 .00500	4.06766 .00999	3.51832 .02498	2.98502 .04975	2.64855 .07493	2.36566 .09990
15	7	5.16212 .00093	4.47642 .00497	4.11194 .01010	3.56706 .02502	3.04666 .04988	2.69622 .07490	2.42161 .09977

Table VI (*Continued*)

N	n	Nominal levels						
		.001	.005	.010	.025	.050	.075	.100
16	2		3.05074 .00833	2.75626 .01667	2.52916 .02500	2.16221 .05000	1.85475 .07500	1.68870 .10000
16	3	4.04101 .00179	3.62074 .00536	3.32627 .01071	2.92538 .02500	2.48073 .05000	2.19773 .07500	1.95915 .10000
16	4	4.61102 .00110	4.04765 .00495	3.72249 .00989	3.22977 .02473	2.75626 .05000	2.43148 .07473	2.17353 .10000
16	5	5.00724 .00092	4.32319 .00504	3.98769 .01007	3.45827 .02450	2.93686 .04991	2.60645 .07532	2.31826 .10005
16	6	5.14043 .00100	4.52116 .00500	4.15688 .00999	3.59087 .02498	3.06730 .05007	2.70310 .07505	2.42076 .09990
16	7	5.35686 .00096	4.64213 .00498	4.25149 .00997	3.67665 .02491	3.15249 .04991	2.78023 .07491	2.48073 .10009
16	8	5.43415 .00093	4.66692 .00497	4.29208 .00995	3.72249 .02510	3.15989 .05004	2.78711 .07498	2.51184 .10000
17	2		3.11272 .00735	2.82340 .01471	2.60133 .02206	2.12617 .05147	1.93824 .07353	1.64892 .10294
17	3	4.14218 .00147	3.73218 .00441	3.40791 .01029	2.96674 .02500	2.49423 .05000	2.23343 .07500	1.98018 .10000
17	4	4.76164 .00084	4.08212 .00504	3.77412 .01008	3.26458 .02479	2.78845 .05000	2.45044 .07479	2.19904 .10000
17	5	5.05683 .00097	4.39358 .00501	4.06610 .01002	3.48079 .02505	2.96939 .05026	2.63925 .07498	2.35314 .10003
17	6	5.28609 .00097	4.61244 .00501	4.24438 .01002	3.65907 .02497	3.11859 .05018	2.75667 .07498	2.45799 .10003
17	7	5.43207 .00098	4.76164 .00509	4.37731 .01003	3.77412 .02509	3.21998 .05003	2.83136 .07502	2.54046 .10022
17	8	5.55887 .00099	4.83475 .00502	4.44432 .00995	3.81811 .02497	3.26458 .05006	2.87438 .07499	2.57839 .09996
18	2		3.17045 .00654	2.88576 .01307	2.48483 .02614	2.17087 .05229	1.91385 .07190	1.73052 .09804
18	3	4.23617 .00123	3.73389 .00490	3.38734 .00980	2.96271 .02451	2.52935 .05025	2.23210 .07475	1.98324 .10049
18	4	4.73776 .00098	4.16737 .00490	3.80269 .01013	3.30938 .02484	2.80709 .05000	2.48057 .07484	2.20014 .10000
18	5	5.10870 .00105	4.45135 .00502	4.11178 .01004	3.54498 .02498	3.01899 .05007	2.67245 .07493	2.38675 .10002
18	6	5.39268 .00102	4.69253 .00501	4.33682 .01002	3.73818 .02494	3.17544 .04977	2.79520 .07498	2.50494 .09998
18	7	5.59835 .00101	4.87656 .00503	4.47006 .00996	3.84912 .02501	3.27683 .04999	2.89270 .07497	2.59046 .09999
18	8	5.69210 .00101	4.96540 .00503	4.55442 .01001	3.92906 .02502	3.34539 .04989	2.95019 .07507	2.63889 .09998

Table VI (*Continued*)

N	n	Nominal levels						
		.001	.005	.010	.025	.050	.075	.100
18	9	5.74909 .00103	5.01000 .00500	4.59445 .00995	3.95964 .02497	3.37553 .05004	2.97459 .07497	2.66430 .09994
19	2		3.22442 .00585	2.94393 .01170	2.55109 .02339	2.10822 .05263	1.92765 .07602	1.71376 .09942
19	3	4.32387 .00103	3.77213 .00516	3.43695 .01032	2.97241 .02477	2.53302 .04954	2.23895 .07534	2.00206 .10010
19	4	4.81689 .00103	4.19477 .00490	3.84654 .01006	3.33840 .02503	2.82201 .05005	2.50107 .07508	2.24906 .09933
19	5	5.20974 .00103	4.51192 .00499	4.17377 .00998	3.58302 .02503	3.06191 .04997	2.68278 .07499	2.40133 .10002
19	6	5.49533 .00100	4.79101 .00501	4.38736 .00999	3.78127 .02499	3.22605 .04968	2.83192 .07500	2.54229 .10010
19	7	5.75744 .00099	4.96296 .00500	4.56824 .01000	3.92941 .02503	3.34991 .04999	2.95133 .07498	2.64190 .10000
19	8	5.88817 .00102	5.08552 .00499	4.66943 .01002	4.02630 .02498	3.42021 .05000	3.01593 .07500	2.69260 .10000
19	9	5.92939 .00100	5.16350 .00499	4.73856 .01000	4.06469 .02500	3.45902 .05000	3.05311 .07501	2.73140 .09984
20	2		3.27508 .00526	2.99842 .01053	2.53855 .02632	2.18241 .04737	1.92947 .07368	1.72124 .10000
20	3	4.40603 .00088	3.74381 .00526	3.46204 .00965	2.98688 .02456	2.55009 .05000	2.25381 .07456	2.01365 .10000
20	4	4.85436 .00103	4.21907 .00495	3.90365 .00991	3.34688 .02497	2.85646 .04995	2.52577 .07492	2.25928 .09990
20	5	5.26501 .00097	4.57299 .00503	4.21646 .00993	3.61462 .02503	3.07968 .04999	2.72551 .07521	2.43454 .09997
20	6	5.59374 .00101	4.85971 .00501	4.46879 .00993	3.83100 .02500	3.26535 .05000	2.87423 .07492	2.57831 .09997
20	7	5.85268 .00101	5.06230 .00501	4.65498 .01001	3.99335 .02501	3.39273 .05003	2.99240 .07497	2.67789 .10000
20	8	6.01040 .00100	5.19980 .00500	4.77263 .00999	4.10043 .02501	3.48583 .05000	3.07408 .07500	2.74755 .10001
20	9	6.12214 .00099	5.28279 .00500	4.84855 .01000	4.16594 .02499	3.54228 .04999	3.12039 .07499	2.79469 .09997
20	10	6.14379 .00100	5.32106 .00500	4.87605 .01000	4.18573 .02501	3.55708 .04997	3.13824 .07499	2.80623 .10001

Table VII* (7.2.5): *Critical Frequencies of the Less Frequently Occurring Algebraic Sign for the Two-Tailed Sign Test When P = .5*

Largest value of r for which $P(r \text{ or fewer } +\text{'s}) + P(r \text{ or fewer } -\text{'s}) \leq \alpha$ or for which $P(r \text{ or fewer } +\text{'s}) \leq \alpha/2$

n	Two-tailed probability, α					
	.001	.01	.02	.05	.10	.50
1	—	—	—	—	—	—
2	—	—	—	—	—	0
3	—	—	—	—	—	0
4	—	—	—	—	—	0
5	—	—	—	—	0	1
6	—	—	—	0	0	1
7	—	—	0	0	0	2
8	—	0	0	0	1	2
9	—	0	0	1	1	2
10	—	0	0	1	1	3
11	0	0	1	1	2	3
12	0	1	1	2	2	4
13	0	1	1	2	3	4
14	0	1	2	2	3	5
15	1	2	2	3	3	5
16	1	2	2	3	4	6
17	1	2	3	4	4	6
18	1	3	3	4	5	7
19	2	3	4	4	5	7
20	2	3	4	5	5	7
21	2	4	4	5	6	8
22	3	4	5	5	6	8
23	3	4	5	6	7	9
24	3	5	5	6	7	9
25	4	5	6	7	7	10
26	4	6	6	7	8	10
27	4	6	7	7	8	11
28	5	6	7	8	9	11
29	5	7	7	8	9	12
30	5	7	8	9	10	12
31	6	7	8	9	10	13
32	6	8	8	9	10	13
33	6	8	9	10	11	14
34	7	9	9	10	11	14
35	7	9	10	11	12	15
36	7	9	10	11	12	15
37	8	10	10	12	13	15
38	8	10	11	12	13	16
39	8	11	11	12	13	16
40	9	11	12	13	14	17
41	9	11	12	13	14	17
42	10	12	13	14	15	18
43	10	12	13	14	15	18
44	10	13	13	15	16	19
45	11	13	14	15	16	19

n	Two-tailed probability, α					
	.001	.01	.02	.05	.10	.50
46	11	13	14	15	16	20
47	11	14	15	16	17	20
48	12	14	15	16	17	21
49	12	15	15	17	18	21
50	13	15	16	17	18	22
51	13	15	16	18	19	22
52	13	16	17	18	19	23
53	14	16	17	18	20	23
54	14	17	18	19	20	24
55	14	17	18	19	20	24
56	15	17	18	20	21	24
57	15	18	19	20	21	25
58	16	18	19	21	22	25
59	16	19	20	21	22	26
60	16	19	20	21	23	26
61	17	20	20	22	23	27
62	17	20	21	22	24	27
63	18	20	21	23	24	28
64	18	21	22	23	24	28
65	18	21	22	24	25	29
66	19	22	23	24	25	29
67	19	22	23	25	26	30
68	20	22	23	25	26	30
69	20	23	24	25	27	31
70	20	23	24	26	27	31
71	21	24	25	26	28	32
72	21	24	25	27	28	32
73	22	25	26	27	28	33
74	22	25	26	28	29	33
75	22	25	26	28	29	34
76	23	26	27	28	30	34
77	23	26	27	29	30	35
78	24	27	28	29	31	35
79	24	27	28	30	31	36
80	24	28	29	30	32	36
81	25	28	29	31	32	36
82	25	28	30	31	33	37
83	26	29	30	32	33	37
84	26	29	30	32	33	38
85	26	30	31	32	34	38
86	27	30	31	33	34	39
87	27	31	32	33	35	39
88	28	31	32	34	35	40
89	28	31	33	34	36	40
90	29	32	33	35	36	41

*Body of table is reproduced, with changes only in notation, from Table 1 in W. J. MacKinnon's "Table for Both the Sign Test and Distribution-Free Confidence Intervals of the Median for Sample Sizes to 1,000," *Journal of the American Statistical Association*, **59** (1964), 935–956, with permission of the author and editor.

Table VII (*Continued*)

n	Two-tailed probability, α					
	.001	.01	.02	.05	.10	.50
91	29	32	33	35	37	41
92	29	33	34	36	37	42
93	30	33	34	36	38	42
94	30	34	35	37	38	43
95	31	34	35	37	38	43
96	31	34	36	37	39	44
97	31	35	36	38	39	44
98	32	35	37	38	40	45
99	32	36	37	39	40	45
100	33	36	37	39	41	46
101	33	37	38	40	41	46
102	34	37	38	40	42	47
103	34	37	39	41	42	47
104	34	38	39	41	43	48
105	35	38	40	41	43	48
106	35	39	40	42	44	49
107	36	39	41	42	44	49
108	36	40	41	43	44	49
109	36	40	41	43	45	50
110	37	41	42	44	45	50
111	37	41	42	44	46	51
112	38	41	43	45	46	51
113	38	42	43	45	47	52
114	39	42	44	46	47	52
115	39	43	44	46	48	53
116	39	43	45	46	48	53
117	40	44	45	47	49	54
118	40	44	45	47	49	54
119	41	45	46	48	50	55
120	41	45	46	48	50	55
121	42	45	47	49	50	56
122	42	46	47	49	51	56
123	42	46	48	50	51	57
124	43	47	48	50	52	57
125	43	47	49	51	52	58
126	44	48	49	51	53	58
127	44	48	49	51	53	59
128	45	48	50	52	54	59
129	45	49	50	52	54	60
130	45	49	51	53	55	60
131	46	50	51	53	55	61
132	46	50	52	54	56	61
133	47	51	52	54	56	62
134	47	51	53	55	56	62
135	48	52	53	55	57	63
136	48	52	53	56	57	63
137	48	52	54	56	58	64
138	49	53	54	57	58	64
139	49	53	55	57	59	65
140	50	54	55	57	59	65
141	50	54	56	58	60	65
142	51	55	56	58	60	66
143	51	55	57	59	61	66
144	51	56	57	59	61	67
145	52	56	58	60	62	67
146	52	56	58	60	62	68
147	53	57	58	61	63	68
148	53	57	59	61	63	69
149	54	58	59	62	63	69
150	54	58	60	62	64	70
151	54	59	60	62	64	70
152	55	59	61	63	65	71
153	55	60	61	63	65	71
154	56	60	62	64	66	72
155	56	61	62	64	66	72
156	57	61	63	65	67	73
157	57	61	63	65	67	73
158	57	62	63	66	68	74
159	58	62	64	66	68	74
160	58	63	64	67	69	75
161	59	63	65	67	69	75
162	59	64	65	68	70	76
163	60	64	66	68	70	76
164	60	65	66	68	70	77
165	60	65	67	69	71	77
166	61	65	67	69	71	78
167	61	66	68	70	72	78
168	62	66	68	70	72	79
169	62	67	68	71	73	79
170	63	67	69	71	73	80
171	63	68	69	72	74	80
172	64	68	70	72	74	81
173	64	69	70	73	75	81
174	64	69	71	73	75	82
175	65	70	71	74	76	82
176	65	70	72	74	76	83
177	66	70	72	74	77	83
178	66	71	73	75	77	83
179	67	71	73	75	78	84
180	67	72	73	76	78	84

Table VII (*Continued*)

n	Two-tailed probability, α					
	.001	.01	.02	.05	.10	.50
181	67	72	74	76	78	85
182	68	73	74	77	79	85
183	68	73	75	77	79	86
184	69	74	75	78	80	86
185	69	74	76	78	80	87
186	70	74	76	79	81	87
187	70	75	77	79	81	88
188	71	75	77	80	82	88
189	71	76	78	80	82	89
190	71	76	78	81	83	89
191	72	77	78	81	83	90
192	72	77	79	81	84	90
193	73	78	79	82	84	91
194	73	78	80	82	85	91
195	74	79	80	83	85	92
196	74	79	81	83	85	92
197	75	79	81	84	86	93
198	75	80	82	84	86	93
199	75	80	82	85	87	94
200	76	81	83	85	87	94
201	76	81	83	86	88	95
202	77	82	83	86	88	95
203	77	82	84	87	89	96
204	78	83	84	87	89	96
205	78	83	85	87	90	97
206	78	84	85	88	90	97
207	79	84	86	88	91	98
208	79	84	86	89	91	98
209	80	85	87	89	92	99
210	80	85	87	90	92	99
211	81	86	88	90	93	100
212	81	86	88	91	93	100
213	82	87	89	91	94	101
214	82	87	89	92	94	101
215	82	88	89	92	94	102
216	83	88	90	93	95	102
217	83	89	90	93	95	103
218	84	89	91	94	96	103
219	84	89	91	94	96	104
220	85	90	92	94	97	104
221	85	90	92	95	97	104
222	86	91	93	95	98	105
223	86	91	93	96	98	105
224	86	92	94	96	99	106
225	87	92	94	97	99	106
226	87	93	95	97	100	107
227	88	93	95	98	100	107
228	88	94	95	98	101	108
229	89	94	96	99	101	108
230	89	95	96	99	102	109
231	90	95	97	100	102	109
232	90	95	97	100	102	110
233	90	96	98	101	103	110
234	91	96	98	101	103	111
235	91	97	99	101	104	111
236	92	97	99	102	104	112
237	92	98	100	102	105	112
238	93	98	100	103	105	113
239	93	99	101	103	106	113
240	94	99	101	104	106	114
241	94	100	101	104	107	114
242	94	100	102	105	107	115
243	95	100	102	105	108	115
244	95	101	103	106	108	116
245	96	101	103	106	109	116
246	96	102	104	107	109	117
247	97	102	104	107	110	117
248	97	103	105	108	110	118
249	98	103	105	108	111	118
250	98	104	106	109	111	119
251	99	104	106	109	111	119
252	99	105	107	109	112	120
253	99	105	107	110	112	120
254	100	106	107	110	113	121
255	100	106	108	111	113	121
256	101	106	108	111	114	122
257	101	107	109	112	114	122
258	102	107	109	112	115	123
259	102	108	110	113	115	123
260	103	108	110	113	116	124
261	103	109	111	114	116	124
262	103	109	111	114	117	125
263	104	110	112	115	117	125
264	104	110	112	115	118	126
265	105	111	113	116	118	126
266	105	111	113	116	119	126
267	106	111	114	117	119	127
268	106	112	114	117	120	127
269	107	112	114	117	120	128
270	107	113	115	118	120	128

Table VII (*Continued*)

n	Two-tailed probability, α					
	.001	.01	.02	.05	.10	.50
271	107	113	115	118	121	129
272	108	114	116	119	121	129
273	108	114	116	119	122	130
274	109	115	117	120	122	130
275	109	115	117	120	123	131
276	110	116	118	121	123	131
277	110	116	118	121	124	132
278	111	117	119	122	124	132
279	111	117	119	122	125	133
280	112	117	120	123	125	133
281	112	118	120	123	126	134
282	112	118	120	124	126	134
283	113	119	121	124	127	135
284	113	119	121	124	127	135
285	114	120	122	125	128	136
286	114	120	122	125	128	136
287	115	121	123	126	129	137
288	115	121	123	126	129	137
289	116	122	124	127	130	138
290	116	122	124	127	130	138
291	117	123	125	128	130	139
292	117	123	125	128	131	139
293	117	123	126	129	131	140
294	118	124	126	129	132	140
295	118	124	127	130	132	141
296	119	125	127	130	133	141
297	119	125	127	131	133	142
298	120	126	128	131	134	142
299	120	126	128	132	134	143
300	121	127	129	132	135	143
301	121	127	129	133	135	144
302	121	128	130	133	136	144
303	122	128	130	133	136	145
304	122	129	131	134	137	145
305	123	129	131	134	137	146
306	123	130	132	135	138	146
307	124	130	132	135	138	147
308	124	130	133	136	139	147
309	125	131	133	136	139	148
310	125	131	134	137	140	148
311	126	132	134	137	140	149
312	126	132	134	138	140	149
313	126	133	135	138	141	150
314	127	133	135	139	141	150
315	127	134	136	139	142	151
316	128	134	136	140	142	151
317	128	135	137	140	143	151
318	129	135	137	141	143	152
319	129	136	138	141	144	152
320	130	136	138	141	144	153
321	130	136	139	142	145	153
322	131	137	139	142	145	154
323	131	137	140	143	146	154
324	131	138	140	143	146	155
325	132	138	141	144	147	155
326	132	139	141	144	147	156
327	133	139	141	145	148	156
328	133	140	142	145	148	157
329	134	140	142	146	149	157
330	134	141	143	146	149	158
331	135	141	143	147	150	158
332	135	142	144	147	150	159
333	136	142	144	148	150	159
334	136	142	145	148	151	160
335	136	143	145	149	151	160
336	137	143	146	149	152	161
337	137	144	146	150	152	161
338	138	144	147	150	153	162
339	138	145	147	150	153	162
340	139	145	148	151	154	163
341	139	146	148	151	154	163
342	140	146	149	152	155	164
343	140	147	149	152	155	164
344	141	147	149	153	156	165
345	141	148	150	153	156	165
346	141	148	150	154	157	166
347	142	149	151	154	157	166
348	142	149	151	155	158	167
349	143	149	152	155	158	167
350	143	150	152	156	159	168
351	144	150	153	156	159	168
352	144	151	153	157	160	169
353	145	151	154	157	160	169
354	145	152	154	158	161	170
355	146	152	155	158	161	170
356	146	153	155	159	161	171
357	146	153	156	159	162	171
358	147	154	156	159	162	172
359	147	154	156	160	163	172
360	148	155	157	160	163	173

Table VII (*Continued*)

n	Two-tailed probability, α					
	.001	.01	.02	.05	.10	.50
361	148	155	157	161	164	173
362	149	156	158	161	164	174
363	149	156	158	162	165	174
364	150	156	159	162	165	175
365	150	157	159	163	166	175
366	151	157	160	163	166	176
367	151	158	160	164	167	176
368	152	158	161	164	167	177
369	152	159	161	165	168	177
370	152	159	162	165	168	178
371	153	160	162	166	169	178
372	153	160	163	166	169	178
373	154	161	163	167	170	179
374	154	161	164	167	170	179
375	155	162	164	168	171	180
376	155	162	164	168	171	180
377	156	163	165	168	172	181
378	156	163	165	169	172	181
379	157	163	166	169	172	182
380	157	164	166	170	173	182
381	157	164	167	170	173	183
382	158	165	167	171	174	183
383	158	165	168	171	174	184
384	159	166	168	172	175	184
385	159	166	169	172	175	185
386	160	167	169	173	176	185
387	160	167	170	173	176	186
388	161	168	170	174	177	186
389	161	168	171	174	177	187
390	162	169	171	175	178	187
391	162	169	172	175	178	188
392	162	170	172	176	179	188
393	163	170	172	176	179	189
394	163	170	173	177	180	189
395	164	171	173	177	180	190
396	164	171	174	178	181	190
397	165	172	174	178	181	191
398	165	172	175	178	182	191
399	166	173	175	179	182	192
400	166	173	176	179	183	192
401	167	174	176	180	183	193
402	167	174	177	180	184	193
403	168	175	177	181	184	194
404	168	175	178	181	184	194
405	168	176	178	182	185	195
406	169	176	179	182	185	195
407	169	177	179	183	186	196
408	170	177	180	183	186	196
409	170	177	180	184	187	197
410	171	178	180	184	187	197
411	171	178	181	185	188	198
412	172	179	181	185	188	198
413	172	179	182	186	189	199
414	173	180	182	186	189	199
415	173	180	183	187	190	200
416	174	181	183	187	190	200
417	174	181	184	187	191	201
418	174	182	184	188	191	201
419	175	182	185	188	192	202
420	175	183	185	189	192	202
421	176	183	186	189	193	203
422	176	184	186	190	193	203
423	177	184	187	190	194	204
424	177	185	187	191	194	204
425	178	185	188	191	195	205
426	178	185	188	192	195	205
427	179	186	188	192	196	206
428	179	186	189	193	196	206
429	179	187	189	193	196	207
430	180	187	190	194	197	207
431	180	188	190	194	197	207
432	181	188	191	195	198	208
433	181	189	191	195	198	208
434	182	189	192	196	199	209
435	182	190	192	196	199	209
436	183	190	193	197	200	210
437	183	191	193	197	200	210
438	184	191	194	198	201	211
439	184	192	194	198	201	211
440	185	192	195	198	202	212
441	185	192	195	199	202	212
442	185	193	196	199	203	213
443	186	193	196	200	203	213
444	186	194	197	200	204	214
445	187	194	197	201	204	214
446	187	195	197	201	205	215
447	188	195	198	202	205	215
448	188	196	198	202	206	216
449	189	196	199	203	206	216
450	189	197	199	203	207	217

Table VII (*Continued*)

n	.001	.01	.02	.05	.10	.50	n	.001	.01	.02	.05	.10	.50
	Two-tailed probability, α							Two-tailed probability, α					
451	190	197	200	204	207	217	496	210	218	221	225	229	239
452	190	198	200	204	208	218	497	211	219	222	226	229	240
453	191	198	201	205	208	218	498	211	219	222	226	230	240
454	191	199	201	205	208	219	499	212	220	223	227	230	241
455	191	199	202	206	209	219	500	212	220	223	227	231	241
456	192	200	202	206	209	220							
457	192	200	203	207	210	220	501	213	221	223	228	231	242
458	193	200	203	207	210	221	502	213	221	224	228	232	242
459	193	201	204	208	211	221	503	214	222	224	229	232	243
460	194	201	204	208	211	222	504	214	222	225	229	233	243
							505	215	223	225	229	233	244
461	194	202	205	208	212	222	506	215	223	226	230	234	244
462	195	202	205	209	212	223	507	216	224	226	230	234	245
463	195	203	205	209	213	223	508	216	224	227	231	234	245
464	196	203	206	210	213	224	509	216	224	227	231	235	246
465	196	204	206	210	214	224	510	217	225	228	232	235	246
466	197	204	207	211	214	225							
467	197	205	207	211	215	225	511	217	225	228	232	236	247
468	197	205	208	212	215	226	512	218	226	229	233	236	247
469	198	206	208	212	216	226	513	218	226	229	233	237	248
470	198	206	209	213	216	227	514	219	227	230	234	237	248
							515	219	227	230	234	238	249
471	199	207	209	213	217	227	516	220	228	231	235	238	249
472	199	207	210	214	217	228	517	220	228	231	235	239	250
473	200	208	210	214	218	228	518	221	229	232	236	239	250
474	200	208	211	215	218	229	519	221	229	232	236	240	251
475	201	208	211	215	219	229	520	222	230	232	237	240	251
476	201	209	212	216	219	230							
477	202	209	212	216	220	230	521	222	230	233	237	241	252
478	202	210	213	217	220	231	522	222	231	233	238	241	252
479	203	210	213	217	221	231	523	223	231	234	238	242	253
480	203	211	214	218	221	232	524	223	232	234	239	242	253
							525	224	232	235	239	243	254
481	203	211	214	218	221	232	526	224	232	235	240	243	254
482	204	212	214	218	222	233	527	225	233	236	240	244	255
483	204	212	215	219	222	233	528	225	233	236	240	244	255
484	205	213	215	219	223	234	529	226	234	237	241	245	256
485	205	213	216	220	223	234	530	226	234	237	241	245	256
486	206	214	216	220	224	235							
487	206	214	217	221	224	235	531	227	235	238	242	246	257
488	207	215	217	221	225	236	532	227	235	238	242	246	257
489	207	215	218	222	225	236	533	228	236	239	243	247	258
490	208	216	218	222	226	237	534	228	236	239	243	247	258
							535	229	237	240	244	247	259
491	208	216	219	223	226	237	536	229	237	240	244	248	259
492	209	216	219	223	227	238	537	229	238	241	245	248	260
493	209	217	220	224	227	238	538	230	238	241	245	249	260
494	209	217	220	224	228	239	539	230	239	242	246	249	261
495	210	218	221	225	228	239	540	231	239	242	246	250	261

Table VII (*Continued*)

n	Two-tailed probability, α					
	.001	.01	.02	.05	.10	.50
541	231	240	242	247	250	262
542	232	240	243	247	251	262
543	232	241	243	248	251	263
544	233	241	244	248	252	263
545	233	241	244	249	252	264
546	234	242	245	249	253	264
547	234	242	245	250	253	265
548	235	243	246	250	254	265
549	235	243	246	251	254	266
550	235	244	247	251	255	266
551	236	244	247	252	255	267
552	236	245	248	252	256	267
553	237	245	248	252	256	268
554	237	246	249	253	257	268
555	238	246	249	253	257	269
556	238	247	250	254	258	269
557	239	247	250	254	258	270
558	239	248	251	255	259	270
559	240	248	251	255	259	271
560	240	249	251	256	260	271
561	241	249	252	256	260	272
562	241	249	252	257	261	272
563	242	250	253	257	261	272
564	242	250	253	258	261	273
565	242	251	254	258	262	273
566	243	251	254	259	262	274
567	243	252	255	259	263	274
568	244	252	255	260	263	275
569	244	253	256	260	264	275
570	245	253	256	261	264	276
571	245	254	257	261	265	276
572	246	254	257	262	265	277
573	246	255	258	262	266	277
574	247	255	258	263	266	278
575	247	256	259	263	267	278
576	248	256	259	263	267	279
577	248	257	260	264	268	279
578	249	257	260	264	268	280
579	249	258	261	265	269	280
580	249	258	261	265	269	281
581	250	258	261	266	270	281
582	250	259	262	266	270	282
583	251	259	262	267	271	282
584	251	260	263	267	271	283
585	252	260	263	268	272	283

n	Two-tailed probability, α					
	.001	.01	.02	.05	.10	.50
586	252	261	264	268	272	284
587	253	261	264	269	273	284
588	253	262	265	269	273	285
589	254	262	265	270	274	285
590	254	263	266	270	274	286
591	255	263	266	271	275	286
592	255	264	267	271	275	287
593	255	264	267	272	275	287
594	256	265	268	272	276	288
595	256	265	268	273	276	288
596	257	266	269	273	277	289
597	257	266	269	274	277	289
598	258	267	270	274	278	290
599	258	267	270	275	278	290
600	259	267	271	275	279	291
601	259	268	271	275	279	291
602	260	268	271	276	280	292
603	260	269	272	276	280	292
604	261	269	272	277	281	293
605	261	270	273	277	281	293
606	262	270	273	278	282	294
607	262	271	274	278	282	294
608	262	271	274	279	283	295
609	263	272	275	279	283	295
610	263	272	275	280	284	296
611	264	273	276	280	284	296
612	264	273	276	281	285	297
613	265	274	277	281	285	297
614	265	274	277	282	286	298
615	266	275	278	282	286	298
616	266	275	278	283	287	299
617	267	276	279	283	287	299
618	267	276	279	284	288	300
619	268	276	280	284	288	300
620	268	277	280	285	289	301
621	269	277	281	285	289	301
622	269	278	281	286	289	302
623	269	278	281	286	290	302
624	270	279	282	287	290	303
625	270	279	282	287	291	303
626	271	280	283	287	291	304
627	271	280	283	288	292	304
628	272	281	284	288	292	305
629	272	281	284	289	293	305
630	273	282	285	289	293	306

Table VII (*Continued*)

n	Two-tailed probability, α					
	.001	.01	.02	.05	.10	.50
631	273	282	285	290	294	306
632	274	283	286	290	294	307
633	274	283	286	291	295	307
634	275	284	287	291	295	308
635	275	284	287	292	296	308
636	276	285	288	292	296	308
637	276	285	288	293	297	309
638	276	285	289	293	297	309
639	277	286	289	294	298	310
640	277	286	290	294	298	310
641	278	287	290	295	299	311
642	278	287	291	295	299	311
643	279	288	291	296	300	312
644	279	288	291	296	300	312
645	280	289	292	297	301	313
646	280	289	292	297	301	313
647	281	290	293	298	302	314
648	281	290	293	298	302	314
649	282	291	294	299	303	315
650	282	291	294	299	303	315
651	283	292	295	300	304	316
652	283	292	295	300	304	316
653	284	293	296	300	304	317
654	284	293	296	301	305	317
655	284	294	297	301	305	318
656	285	294	297	302	306	318
657	285	295	298	302	306	319
658	286	295	298	303	307	319
659	286	295	299	303	307	320
660	287	296	299	304	308	320
661	287	296	300	304	308	321
662	288	297	300	305	309	321
663	288	297	301	305	309	322
664	289	298	301	306	310	322
665	289	298	302	306	310	323
666	290	299	302	307	311	323
667	290	299	302	307	311	324
668	291	300	303	308	312	324
669	291	300	303	308	312	325
670	291	301	304	309	313	325
671	292	301	304	309	313	326
672	292	302	305	310	314	326
673	293	302	305	310	314	327
674	293	303	306	311	315	327
675	294	303	306	311	315	328

n	Two-tailed probability, α					
	.001	.01	.02	.05	.10	.50
676	294	304	307	312	316	328
677	295	304	307	312	316	329
678	295	304	308	312	317	329
679	296	305	308	313	317	330
680	296	305	309	313	318	330
681	297	306	309	314	318	331
682	297	306	310	314	319	331
683	298	307	310	315	319	332
684	298	307	311	315	319	332
685	298	308	311	316	320	333
686	299	308	312	316	320	333
687	299	309	312	317	321	334
688	300	309	313	317	321	334
689	300	310	313	318	322	335
690	301	310	313	318	322	335
691	301	311	314	319	323	336
692	302	311	314	319	323	336
693	302	312	315	320	324	337
694	303	312	315	320	324	337
695	303	313	316	321	325	338
696	304	313	316	321	325	338
697	304	314	317	322	326	339
698	305	314	317	322	326	339
699	305	314	318	323	327	340
700	306	315	318	323	327	340
701	306	315	319	324	328	341
702	306	316	319	324	328	341
703	307	316	320	325	329	342
704	307	317	320	325	329	342
705	308	317	321	325	330	343
706	308	318	321	326	330	343
707	309	318	322	326	331	344
708	309	319	322	327	331	344
709	310	319	323	327	332	345
710	310	320	323	328	332	345
711	311	320	324	328	333	346
712	311	321	324	329	333	346
713	312	321	324	329	334	346
714	312	322	325	330	334	347
715	313	322	325	330	335	347
716	313	323	326	331	335	348
717	313	323	326	331	335	348
718	314	324	327	332	336	349
719	314	324	327	332	336	349
720	315	324	328	333	337	350

Table VII (*Continued*)

n	.001	.01	.02	.05	.10	.50
	Two-tailed probability, α					
721	315	325	328	333	337	350
722	316	325	329	334	338	351
723	316	326	329	334	338	351
724	317	326	330	335	339	352
725	317	327	330	335	339	352
726	318	327	331	336	340	353
727	318	328	331	336	340	353
728	319	328	332	337	341	354
729	319	329	332	337	341	354
730	320	329	333	338	342	355
731	320	330	333	338	342	355
732	321	330	334	338	343	356
733	321	331	334	339	343	356
734	321	331	335	339	344	357
735	322	332	335	340	344	357
736	322	332	335	340	345	358
737	323	333	336	341	345	358
738	323	333	336	341	346	359
739	324	334	337	342	346	359
740	324	334	337	342	347	360
741	325	334	338	343	347	360
742	325	335	338	343	348	361
743	326	335	339	344	348	361
744	326	336	339	344	349	362
745	327	336	340	345	349	362
746	327	337	340	345	350	363
747	328	337	341	346	350	363
748	328	338	341	346	351	364
749	329	338	342	347	351	364
750	329	339	342	347	351	365
751	329	339	343	348	352	365
752	330	340	343	348	352	366
753	330	340	344	349	353	366
754	331	341	344	349	353	367
755	331	341	345	350	354	367
756	332	342	345	350	354	368
757	332	342	346	351	355	368
758	333	343	346	351	355	369
759	333	343	346	352	356	369
760	334	344	347	352	356	370
761	334	344	347	352	357	370
762	335	344	348	353	357	371
763	335	345	348	353	358	371
764	336	345	349	354	358	372
765	336	346	349	354	359	372
766	337	346	350	355	359	373
767	337	347	350	355	360	373
768	337	347	351	356	360	374
769	338	348	351	356	361	374
770	338	348	352	357	361	375
771	339	349	352	357	362	375
772	339	349	353	358	362	376
773	340	350	353	358	363	376
774	340	350	354	359	363	377
775	341	351	354	359	364	377
776	341	351	355	360	364	378
777	342	352	355	360	365	378
778	342	352	356	361	365	379
779	343	353	356	361	366	379
780	343	353	357	362	366	380
781	344	354	357	362	367	380
782	344	354	357	363	367	381
783	345	354	358	363	367	381
784	345	355	358	364	368	382
785	345	355	359	364	368	382
786	346	356	359	365	369	383
787	346	356	360	365	369	383
788	347	357	360	365	370	384
789	347	357	361	366	370	384
790	348	358	361	366	371	385
791	348	358	362	367	371	385
792	349	359	362	367	372	386
793	349	359	363	368	372	386
794	350	360	363	368	373	386
795	350	360	364	369	373	387
796	351	361	364	369	374	387
797	351	361	365	370	374	388
798	352	362	365	370	375	388
799	352	362	366	371	375	389
800	353	363	366	371	376	389
801	353	363	367	372	376	390
802	353	364	367	372	377	390
803	354	364	368	373	377	391
804	354	365	368	373	378	391
805	355	365	369	374	378	392
806	355	365	369	374	379	392
807	356	366	369	375	379	393
808	356	366	370	375	380	393
809	357	367	370	376	380	394
810	357	367	371	376	381	394

Table VII (*Continued*)

n	Two-tailed probability, α						n	Two-tailed probability, α					
	.001	.01	.02	.05	.10	.50		.001	.01	.02	.05	.10	.50
811	358	368	371	377	381	395	856	379	389	393	398	403	417
812	358	368	372	377	382	395	857	379	390	393	399	403	418
813	359	369	372	378	382	396	858	380	390	394	399	404	418
814	359	369	373	378	383	396	859	380	391	394	400	404	419
815	360	370	373	379	383	397	860	381	391	395	400	405	419
816	360	370	374	379	384	397							
817	361	371	374	379	384	398	861	381	392	395	401	405	420
818	361	371	375	380	384	398	862	382	392	396	401	406	420
819	361	372	375	380	385	399	863	382	393	396	402	406	421
820	362	372	376	381	385	399	864	383	393	397	402	407	421
							865	383	394	397	403	407	422
821	362	373	376	381	386	400	866	384	394	398	403	408	422
822	363	373	377	382	386	400	867	384	395	398	404	408	423
823	363	374	377	382	387	401	868	385	395	399	404	409	423
824	364	374	378	383	387	401	869	385	396	399	405	409	424
825	364	375	378	383	388	402	870	386	396	400	405	410	424
826	365	375	379	384	388	402							
827	365	375	379	384	389	403	871	386	397	400	406	410	425
828	366	376	380	385	389	403	872	386	397	401	406	411	425
829	366	376	380	385	390	404	873	387	397	401	407	411	426
830	367	377	381	386	390	404	874	387	398	402	407	412	426
							875	388	398	402	408	412	427
831	367	377	381	386	391	405	876	388	399	403	408	413	427
832	368	378	381	387	391	405	877	389	399	403	408	413	428
833	368	378	382	387	392	406	878	389	400	404	409	414	428
834	369	379	382	388	392	406	879	390	400	404	409	414	429
835	369	379	383	388	393	407	880	390	401	405	410	415	429
836	369	380	383	389	393	407							
837	370	380	384	389	394	408	881	391	401	405	410	415	429
838	370	381	384	390	394	408	882	391	402	405	411	416	430
839	371	381	385	390	395	409	883	392	402	406	411	416	430
840	371	382	385	391	395	409	884	392	403	406	412	417	431
							885	393	403	407	412	417	431
841	372	382	386	391	396	410	886	393	404	407	413	418	432
842	372	383	386	392	396	410	887	394	404	408	413	418	432
843	373	383	387	392	397	411	888	394	405	408	414	418	433
844	373	384	387	393	397	411	889	394	405	409	414	419	433
845	374	384	388	393	398	412	890	395	406	409	415	419	434
846	374	385	388	394	398	412							
847	375	385	389	394	399	413	891	395	406	410	415	420	434
848	375	386	389	394	399	413	892	396	407	410	416	420	435
849	376	386	390	395	400	414	893	396	407	411	416	421	435
850	376	386	390	395	400	414	894	397	408	411	417	421	436
							895	397	408	412	417	422	436
851	377	387	391	396	401	415	896	398	408	412	418	422	437
852	377	387	391	396	401	415	897	398	409	413	418	423	437
853	377	388	392	397	401	416	898	399	409	413	419	423	438
854	378	388	392	397	402	416	899	399	410	414	419	424	438
855	378	389	393	398	402	417	900	400	410	414	420	424	439

Table VII (*Continued*)

n	Two-tailed probability, α						n	Two-tailed probability, α					
	.001	.01	.02	.05	.10	.50		.001	.01	.02	.05	.10	.50
901	400	411	415	420	425	439	951	424	435	439	444	449	464
902	401	411	415	421	425	440	952	424	435	439	445	450	465
903	401	412	416	421	426	440	953	425	436	440	445	450	465
904	402	412	416	422	426	441	954	425	436	440	446	451	466
905	402	413	417	422	427	441	955	426	437	441	446	451	466
906	403	413	417	423	427	442	956	426	437	441	447	452	467
907	403	414	417	423	428	442	957	427	438	442	447	452	467
908	403	414	418	423	428	443	958	427	438	442	448	453	468
909	404	415	418	424	429	443	959	428	439	442	448	453	468
910	404	415	419	424	429	444	960	428	439	443	449	454	469
911	405	416	419	425	430	444	961	429	440	443	449	454	469
912	405	416	420	425	430	445	962	249	440	444	450	454	470
913	406	417	420	426	431	445	963	429	441	444	450	455	470
914	406	417	421	426	431	446	964	430	441	445	451	455	471
915	407	418	421	427	432	446	965	430	442	445	451	456	471
916	407	418	422	427	432	447	966	431	442	446	452	456	472
917	408	419	422	428	433	447	967	431	442	446	452	457	472
918	408	419	423	428	433	448	968	432	443	447	453	457	473
919	409	419	423	429	434	448	969	432	443	447	453	458	473
920	409	420	424	429	434	449	970	433	444	448	453	458	473
921	410	420	424	430	435	449	971	433	444	448	454	459	474
922	410	421	425	430	435	450	972	434	445	449	454	459	474
923	411	421	425	431	436	450	973	434	445	449	455	460	475
924	411	422	426	431	436	451	974	435	446	450	455	460	475
925	412	422	426	432	436	451	975	435	446	450	456	461	476
926	412	423	427	432	437	452	976	436	447	451	456	461	476
927	412	423	427	433	437	452	977	436	447	451	457	462	477
928	413	424	428	433	438	453	978	437	448	452	457	462	477
929	413	424	428	434	438	453	979	437	448	452	458	463	478
930	414	425	429	434	439	454	980	438	449	453	458	463	478
931	414	425	429	435	439	454	981	438	449	453	459	464	479
932	415	426	430	435	440	455	982	438	450	454	459	464	479
933	415	426	430	436	440	455	983	439	450	454	460	465	480
934	416	427	430	436	441	456	984	439	451	455	460	465	480
935	416	427	431	437	441	456	985	440	451	455	461	466	481
936	417	428	431	437	442	457	986	440	452	455	461	466	481
937	417	428	432	438	442	457	987	441	452	456	462	467	482
938	418	429	432	438	443	458	988	441	453	456	462	467	482
939	418	429	433	438	443	458	989	442	453	457	463	468	483
940	419	430	433	439	444	459	990	442	453	457	463	468	483
941	419	430	434	439	444	459	991	443	454	458	464	469	484
942	420	430	434	440	445	460	992	443	454	458	464	469	484
943	420	431	435	440	445	460	993	444	455	459	465	470	485
944	420	431	435	441	446	461	994	444	455	459	465	470	485
945	421	432	436	441	446	461	995	445	456	460	466	471	486
946	421	432	436	442	447	462	996	445	456	460	466	471	486
947	422	433	437	442	447	462	997	446	457	461	467	472	487
948	422	433	437	443	448	463	998	446	457	461	467	472	487
949	423	434	438	443	448	463	999	447	458	462	468	473	488
950	423	434	438	444	449	464	1000	447	458	462	468	473	488

Table VIII*† (8.2.5): *Critical Lower-Tail Values of b for Fisher's Exact Test for* 2×2 *Tables, Followed by Exact Probability Level for Test*

Largest value of b, given *fixed* values of A, B, and a, for the ratio a/A to be just significantly larger than b/B in the following 2×2 table

a	$A - a$	A
b	$B - b$	B
$a + b$	$(A + B) - (a + b)$	$A + B$

$A \geq B$
$a/A \geq b/B$

Thus largest value of b' for which

$$P(b \leq b' \mid A, B, a) = P\left\{\left(\frac{a}{A} - \frac{b}{B}\right) \geq \left(\frac{a}{A} - \frac{b'}{B}\right)\right\} \leq \alpha$$

followed by $P(b \leq b' \mid A, B, a + b')$

	a	Probability 0.05	0.025	0.01	0.005
$A = 3$ $B = 3$	3	**0** ·050	—	—	—
$A = 4$ $B = 4$	4	**0** ·014	**0** ·014	—	—
3	4	**0** ·029	—	—	—
$A = 5$ $B = 5$	5	**1** ·024	**1** ·024	**0** ·004	**0** ·004
	4	**0** ·024	**0** ·024	—	—
4	5	**1** ·048	**0** ·008	**0** ·008	—
	4	**0** ·040	—	—	—
3	5	**0** ·018	**0** ·018	—	—
2	5	**0** ·048	—	—	—
$A = 6$ $B = 6$	6	**2** ·030	**1** ·008	**1** ·008	**0** ·001
	5	**1** ·040	**0** ·008	**0** ·008	—
	4	**0** ·030	—	—	—
5	6	**1** ·015+	**1** ·015+	**0** ·002	**0** ·002
	5	**0** ·013	**0** ·013	—	—
	4	**0** ·045+	—	—	—
4	6	**1** ·033	**0** ·005−	**0** ·005−	**0** ·005−
	5	**0** ·024	**0** ·024	—	—
3	6	**0** ·012	**0** ·012	—	—
	5	**0** ·048	—	—	—
2	6	**0** ·036	—	—	—

*Body of table is an abridgement of D. J. Finney, R. Latscha, B. M. Bennett, and P. Hsu, (Introduction by E. S. Pearson), *Tables for Testing Significance in a* 2×2 *Contingency Table*, London and New York: Cambridge University Press, 1963, with permission of the authors and publishers.

†The table shows: (1) in bold type, for given A, B, and a, the value of $b(< a)$ which is just significant at the probability level quoted (single-tail test); (2) in small type, for given A, B, and $a + b$, the exact probability (if there is independence) that b is equal to or less than the integer shown in bold type.

Table VIII (*Continued*)

	a	Probability			
		0.05	**0.025**	**0.01**	**0.005**
$A = 7 \quad B = 7$	7	**3** ·035−	**2** ·010+	**1** ·002	**1** ·002
	6	**1** ·015−	**1** ·015−	**0** ·002	**0** ·002
	5	**0** ·010+	**0** ·010+	—	—
	4	**0** ·035−	—	—	—
6	7	**2** ·021	**2** ·021	**1** ·005−	**1** ·005−
	6	**1** ·025+	**0** ·004	**0** ·004	**0** ·004
	5	**0** ·016	**0** ·016	—	—
	4	**0** ·049	—	—	—
5	7	**2** ·045+	**1** ·010+	**0** ·001	**0** ·001
	6	**1** ·045+	**0** ·008	**0** ·008	—
	5	**0** ·027	—	—	—
4	7	**1** ·024	**1** ·024	**0** ·003	**0** ·003
	6	**0** ·015+	**0** ·015+	—	—
	5	**0** ·045+	—	—	—
3	7	**0** ·008	**0** ·008	**0** ·008	—
	6	**0** ·033	—	—	—
2	7	**0** ·028	—	—	—
$A = 8 \quad B = 8$	8	**4** ·038	**3** ·013	**2** ·003	**2** ·003
	7	**2** ·020	**2** ·020	**1** ·005+	**0** ·001
	6	**1** ·020	**1** ·020	**0** ·003	**0** ·003
	5	**0** ·013	**0** ·013	—	—
	4	**0** ·038	—	—	—
7	8	**3** ·026	**2** ·007	**2** ·007	**1** ·001
	7	**2** ·035−	**1** ·009	**1** ·009	**0** ·001
	6	**1** ·032	**0** ·006	**0** ·006	—
	5	**0** ·019	**0** ·019	—	—
6	8	**2** ·015−	**2** ·015−	**1** ·003	**1** ·003
	7	**1** ·016	**1** ·016	**0** ·002	**0** ·002
	6	**0** ·009	**0** ·009	**0** ·009	—
	5	**0** ·028	—	—	—
5	8	**2** ·035−	**1** ·007	**1** ·007	**0** ·001
	7	**1** ·032	**0** ·005−	**0** ·005−	**0** ·005−
	6	**0** ·016	**0** ·016	—	—
	5	**0** ·044	—	—	—
4	8	**1** ·018	**1** ·018	**0** ·002	**0** ·002
	7	**0** ·010+	**0** ·010+	—	—
	6	**0** ·030	—	—	—
3	8	**0** ·006	**0** ·006	**0** ·006	—
	7	**0** ·024	**0** ·024	—	—
2	8	**0** ·022	**0** ·022	—	—

Table VIII (*Continued*)

	a	Probability			
		0.05	0.025	0.01	0.005
$A = 9$ $B = 9$	9	**5** ·041	**4** ·015−	**3** ·005−	**3** ·005−
	8	**3** ·025−	**3** ·025−	**2** ·008	**1** ·002
	7	**2** ·028	**1** ·008	**1** ·008	**0** ·001
	6	**1** ·025−	**1** ·025−	**0** ·005−	**0** ·005−
	5	**0** ·015−	**0** ·015−	—	—
	4	**0** ·041	—	—	—
8	9	**4** ·029	**3** ·009	**3** ·009	**2** ·002
	8	**3** ·043	**2** ·013	**1** ·003	**1** ·003
	7	**2** ·044	**1** ·012	**0** ·002	**0** ·002
	6	**1** ·036	**0** ·007	**0** ·007	—
	5	**0** ·020	**0** ·020	—	—
7	9	**3** ·019	**3** ·019	**2** ·005−	**2** ·005−
	8	**2** ·024	**2** ·024	**1** ·006	**0** ·001
	7	**1** ·020	**1** ·020	**0** ·003	**0** ·003
	6	**0** ·010+	**0** ·010+	—	—
	5	**0** ·029	—	—	—
6	9	**3** ·044	**2** ·011	**1** ·002	**1** ·002
	8	**2** ·047	**1** ·011	**0** ·001	**0** ·001
	7	**1** ·035−	**0** ·006	**0** ·006	—
	6	**0** ·017	**0** ·017	—	—
	5	**0** ·042	—	—	—
5	9	**2** ·027	**1** ·005−	**1** ·005−	**1** ·005−
	8	**1** ·023	**1** ·023	**0** ·003	**0** ·003
	7	**0** ·010+	**0** ·010+	—	—
	6	**0** ·028	—	—	—
4	9	**1** ·014	**1** ·014	**0** ·001	**0** ·001
	8	**0** ·007	**0** ·007	**0** ·007	—
	7	**0** ·021	**0** ·021	—	—
	6	**0** ·049	—	—	—
3	9	**1** ·045+	**0** ·005−	**0** ·005−	**0** ·005−
	8	**0** ·018	**0** ·018	—	—
	7	**0** ·045+	—	—	—
2	9	**0** ·018	**0** ·018	—	—
$A = 10$ $B = 10$	10	**6** ·043	**5** ·016	**4** ·005+	**3** ·002
	9	**4** ·029	**3** ·010−	**3** ·010−	**2** ·003
	8	**3** ·035−	**2** ·012	**1** ·003	**1** ·003
	7	**2** ·035−	**1** ·010−	**1** ·010−	**0** ·002
	6	**1** ·029	**0** ·005+	**0** ·005+	—
	5	**0** ·016	**0** ·016	—	—
	4	**0** ·043	—	—	—
9	10	**5** ·033	**4** ·011	**3** ·003	**3** ·003
	9	**4** ·050−	**3** ·017	**2** ·005−	**2** ·005−
	8	**2** ·019	**2** ·019	**1** ·004	**1** ·004
	7	**1** ·015−	**1** ·015−	**0** ·002	**0** ·002
	6	**1** ·040	**0** ·008	**0** ·008	—
	5	**0** ·022	**0** ·022	—	—

Table VIII (*Continued*)

	a	Probability			
		0.05	**0.025**	**0.01**	**0.005**
$A = 10$ $B = 8$	10	4 ·023	4 ·023	3 ·007	2 ·002
	9	3 ·032	2 ·009	2 ·009	1 ·002
	8	2 ·031	1 ·008	1 ·008	0 ·001
	7	1 ·023	1 ·023	0 ·004	0 ·004
	6	0 ·011	0 ·011	—	—
	5	0 ·029	—	—	—
7	10	3 ·015−	3 ·015−	2 ·003	2 ·003
	9	2 ·018	2 ·018	1 ·004	1 ·004
	8	1 ·013	1 ·013	0 ·002	0 ·002
	7	1 ·036	0 ·006	0 ·006	—
	6	0 ·017	0 ·017	—	—
	5	0 ·041	—	—	—
6	10	3 ·036	2 ·008	2 ·008	1 ·001
	9	2 ·036	1 ·008	1 ·008	0 ·001
	8	1 ·024	1 ·024	0 ·003	0 ·003
	7	0 ·010+	0 ·010+	—	—
	6	0 ·026	—	—	—
5	10	2 ·022	2 ·022	1 ·004	1 ·004
	9	1 ·017	1 ·017	0 ·002	0 ·002
	8	1 ·047	0 ·007	0 ·007	—
	7	0 ·019	0 ·019	—	—
	6	0 ·042	—	—	—
4	10	1 ·011	1 ·011	0 ·001	0 ·001
	9	1 ·041	0 ·005−	0 ·005−	0 ·005−
	8	0 ·015−	0 ·015−	—	—
	7	0 ·035−	—	—	—
3	10	1 ·038	0 ·003	0 ·003	0 ·003
	9	0 ·014	0 ·014	—	—
	8	0 ·035−	—	—	—
2	10	0 ·015+	0 ·015+	—	—
	9	0 ·045+	—	—	—
$A = 11$ $B = 11$	11	7 ·045+	6 ·018	5 ·006	4 ·002
	10	5 ·032	4 ·012	3 ·004	3 ·004
	9	4 ·040	3 ·015−	2 ·004	2 ·004
	8	3 ·043	2 ·015−	1 ·004	1 ·004
	7	2 ·040	1 ·012	0 ·002	0 ·002
	6	1 ·032	0 ·006	0 ·006	—
	5	0 ·018	0 ·018	—	—
	4	0 ·045+	—	—	—
10	11	6 ·035+	5 ·012	4 ·004	4 ·004
	10	4 ·021	4 ·021	3 ·007	2 ·002
	9	3 ·024	3 ·024	2 ·007	1 ·002
	8	2 ·023	2 ·023	1 ·006	0 ·001
	7	1 ·017	1 ·017	0 ·003	0 ·003
	6	1 ·043	0 ·009	0 ·009	—
	5	0 ·023	0 ·023	—	—

Table VIII (*Continued*)

	a	Probability			
		0.05	**0.025**	**0.01**	**0.005**
$A = 11$ $B = 9$	11	**5** ·026	**4** ·008	**4** ·008	**3** ·002
	10	**4** ·038	**3** ·012	**2** ·003	**2** ·003
	9	**3** ·040	**2** ·012	**1** ·003	**1** ·003
	8	**2** ·035−	**1** ·009	**1** ·009	**0** ·001
	7	**1** ·025−	**1** ·025−	**0** ·004	**0** ·004
	6	**0** ·012	**0** ·012	—	—
	5	**0** ·030	—	—	—
8	11	**4** ·018	**4** ·018	**3** ·005−	**3** ·005−
	10	**3** ·024	**3** ·024	**2** ·006	**1** ·001
	9	**2** ·022	**2** ·022	**1** ·005−	**1** ·005−
	8	**1** ·015−	**1** ·015−	**0** ·002	**0** ·002
	7	**1** ·037	**0** ·007	**0** ·007	—
	6	**0** ·017	**0** ·017	—	—
	5	**0** ·040	—	—	—
7	11	**4** ·043	**3** ·011	**2** ·002	**2** ·002
	10	**3** ·047	**2** ·013	**1** ·002	**1** ·002
	9	**2** ·039	**1** ·009	**1** ·009	**0** ·001
	8	**1** ·025−	**1** ·025−	**0** ·004	**0** ·004
	7	**0** ·010+	**0** ·010+	—	—
	6	**0** ·025−	**0** ·025−	—	—
6	11	**3** ·029	**2** ·006	**2** ·006	**1** ·001
	10	**2** ·028	**1** ·005+	**1** ·005+	**0** ·001
	9	**1** ·018	**1** ·018	**0** ·002	**0** ·002
	8	**1** ·043	**0** ·007	**0** ·007	—
	7	**0** ·017	**0** ·017	—	—
	6	**0** ·037	—	—	—
5	11	**2** ·018	**2** ·018	**1** ·003	**1** ·003
	10	**1** ·013	**1** ·013	**0** ·001	**0** ·001
	9	**1** ·036	**0** ·005−	**0** ·005−	**0** ·005−
	8	**0** ·013	**0** ·013	—	—
	7	**0** ·029	—	—	—
4	11	**1** ·009	**1** ·009	**1** ·009	**0** ·001
	10	**1** ·033	**0** ·004	**0** ·004	**0** ·004
	9	**0** ·011	**0** ·011	—	—
	8	**0** ·026	—	—	—
3	11	**1** ·033	**0** ·003	**0** ·003	**0** ·003
	10	**0** ·011	**0** ·011	—	—
	9	**0** ·027	—	—	—
2	11	**0** ·013	**0** ·013	—	—
	10	**0** ·038	—	—	—

Table VIII (*Continued*)

	a	Probability			
		0.05	**0.025**	**0.01**	**0.005**
$A = 12$ $B = 12$	12	**8** ·047	**7** ·019	**6** ·007	**5** ·002
	11	**6** ·034	**5** ·014	**4** ·005−	**4** ·005−
	10	**5** ·045−	**4** ·018	**3** ·006	**2** ·002
	9	**4** ·050−	**3** ·020	**2** ·006	**1** ·001
	8	**3** ·050−	**2** ·018	**1** ·005−	**1** ·005−
	7	**2** ·045−	**1** ·014	**0** ·002	**0** ·002
	6	**1** ·034	**0** ·007	**0** ·007	—
	5	**0** ·019	**0** ·019	—	—
	4	**0** ·047	—	—	—
11	12	**7** ·037	**6** ·014	**5** ·005−	**5** ·005−
	11	**5** ·024	**5** ·024	**4** ·008	**3** ·002
	10	**4** ·029	**3** ·010+	**2** ·003	**2** ·003
	9	**3** ·030	**2** ·009	**2** ·009	**1** ·002
	8	**2** ·026	**1** ·007	**1** ·007	**0** ·001
	7	**1** ·019	**1** ·019	**0** ·003	**0** ·003
	6	**1** ·045−	**0** ·009	**0** ·009	—
	5	**0** ·024	**0** ·024	—	—
10	12	**6** ·029	**5** ·010−	**5** ·010−	**4** ·003
	11	**5** ·043	**4** ·015+	**3** ·005−	**3** ·005−
	10	**4** ·048	**3** ·017	**2** ·005−	**2** ·005−
	9	**3** ·046	**2** ·015−	**1** ·004	**1** ·004
	8	**2** ·038	**1** ·010+	**0** ·002	**0** ·002
	7	**1** ·026	**0** ·005−	**0** ·005−	**0** ·005−
	6	**0** ·012	**0** ·012	—	—
	5	**0** ·030	—	—	—
9	12	**5** ·021	**5** ·021	**4** ·006	**3** ·002
	11	**4** ·029	**3** ·009	**3** ·009	**2** ·002
	10	**3** ·029	**2** ·008	**2** ·008	**1** ·002
	9	**2** ·024	**2** ·024	**1** ·006	**0** ·001
	8	**1** ·016	**1** ·016	**0** ·002	**0** ·002
	7	**1** ·037	**0** ·007	**0** ·007	—
	6	**0** ·017	**0** ·017	—	—
	5	**0** ·039	—	—	—
8	12	**5** ·049	**4** ·014	**3** ·004	**3** ·004
	11	**3** ·018	**3** ·018	**2** ·004	**2** ·004
	10	**2** ·015+	**2** ·015+	**1** ·003	**1** ·003
	9	**2** ·040	**1** ·010−	**1** ·010−	**0** ·001
	8	**1** ·025−	**1** ·025−	**0** ·004	**0** ·004
	7	**0** ·010+	**0** ·010+	—	—
	6	**0** ·024	**0** ·024	—	—
7	12	**4** ·036	**3** ·009	**3** ·009	**2** ·002
	11	**3** ·038	**2** ·010−	**2** ·010−	**1** ·002
	10	**2** ·029	**1** ·006	**1** ·006	**0** ·001
	9	**1** ·017	**1** ·017	**0** ·002	**0** ·002
	8	**1** ·040	**0** ·007	**0** ·007	—
	7	**0** ·016	**0** ·016	—	—
	6	**0** ·034	—	—	—

Table VIII (*Continued*)

	a	Probability			
		0.05	0.025	0.01	0.005
$A = 12$ $B = 6$	12	3 ·025−	3 ·025−	2 ·005−	2 ·005−
	11	2 ·022	2 ·022	1 ·004	1 ·004
	10	1 ·013	1 ·013	0 ·002	0 ·002
	9	1 ·032	0 ·005−	0 ·005−	0 ·005−
	8	0 ·011	0 ·011	—	—
	7	0 ·025−	0 ·025−	—	—
	6	0 ·050−	—	—	—
5	12	2 ·015−	2 ·015−	1 ·002	1 ·002
	11	1 ·010−	1 ·010−	1 ·010−	0 ·001
	10	1 ·028	0 ·003	0 ·003	0 ·003
	9	0 ·009	0 ·009	0 ·009	—
	8	0 ·020	0 ·020	—	—
	7	0 ·041	—	—	—
4	12	2 ·050	1 ·007	1 ·007	0 ·001
	11	1 ·027	0 ·003	0 ·003	0 ·003
	10	0 ·008	0 ·008	0 ·008	—
	9	0 ·019	0 ·019	—	—
	8	0 ·038	—	—	—
3	12	1 ·029	0 ·002	0 ·002	0 ·002
	11	0 ·009	0 ·009	0 ·009	—
	10	0 ·022	0 ·022	—	—
	9	0 ·044	—	—	—
2	12	0 ·011	0 ·011	—	—
	11	0 ·033	—	—	—
$A = 13$ $B = 13$	13	9 ·048	8 ·020	7 ·007	6 ·003
	12	7 ·037	6 ·015+	5 ·006	4 ·002
	11	6 ·048	5 ·021	4 ·008	3 ·002
	10	4 ·024	4 ·024	3 ·008	2 ·002
	9	3 ·024	3 ·024	2 ·008	1 ·002
	8	2 ·021	2 ·021	1 ·006	0 ·001
	7	2 ·048	1 ·015+	0 ·003	0 ·003
	6	1 ·037	0 ·007	0 ·007	—
	5	0 ·020	0 ·020	—	—
	4	0 ·048	—	—	—
12	13	8 ·039	7 ·015−	6 ·005+	5 ·002
	12	6 ·027	5 ·010−	5 ·010−	4 ·003
	11	5 ·033	4 ·013	3 ·004	3 ·004
	10	4 ·036	3 ·013	2 ·004	2 ·004
	9	3 ·034	2 ·011	1 ·003	1 ·003
	8	2 ·029	1 ·008	1 ·008	0 ·001
	7	1 ·020	1 ·020	0 ·004	0 ·004
	6	1 ·046	0 ·010−	0 ·010−	—
	5	0 ·024	0 ·024	—	—

Table VIII (*Continued*)

	a	Probability			
		0.05	**0·025**	**0.01**	**0.005**
$A = 13$ $B = 11$	13	7 ·031	6 ·011	5 ·003	5 ·003
	12	6 ·048	5 ·018	4 ·006	3 ·002
	11	4 ·021	4 ·021	3 ·007	2 ·002
	10	3 ·021	3 ·021	2 ·006	1 ·001
	9	3 ·050−	2 ·017	1 ·004	1 ·004
	8	2 ·040	1 ·011	0 ·002	0 ·002
	7	1 ·027	0 ·005−	0 ·005−	0 ·005−
	6	0 ·013	0 ·013	—	—
	5	0 ·030	—	—	—
10	13	6 ·024	6 ·024	5 ·007	4 ·002
	12	5 ·035−	4 ·012	3 ·003	3 ·003
	11	4 ·037	3 ·012	2 ·003	2 ·003
	10	3 ·033	2 ·010+	1 ·002	1 ·002
	9	2 ·026	1 ·006	1 ·006	0 ·001
	8	1 ·017	1 ·017	0 ·003	0 ·003
	7	1 ·038	0 ·007	0 ·007	—
	6	0 ·017	0 ·017	—	—
	5	0 ·038	—	—	—
9	13	5 ·017	5 ·017	4 ·005−	4 ·005−
	12	4 ·023	4 ·023	3 ·007	2 ·001
	11	3 ·022	3 ·022	2 ·006	1 ·001
	10	2 ·017	2 ·017	1 ·004	1 ·004
	9	2 ·040	1 ·010+	0 ·001	0 ·001
	8	1 ·025−	1 ·025−	0 ·004	0 ·004
	7	0 ·010+	0 ·010+	—	—
	6	0 ·023	0 ·023	—	—
	5	0 ·049	—	—	—
8	13	5 ·042	4 ·012	3 ·003	3 ·003
	12	4 ·047	3 ·014	2 ·003	2 ·003
	11	3 ·041	2 ·011	1 ·002	1 ·002
	10	2 ·029	1 ·007	1 ·007	0 ·001
	9	1 ·017	1 ·017	0 ·002	0 ·002
	8	1 ·037	0 ·006	0 ·006	—
	7	0 ·015−	0 ·015−	—	—
	6	0 ·032	—	—	—
7	13	4 ·031	3 ·007	3 ·007	2 ·001
	12	3 ·031	2 ·007	2 ·007	1 ·001
	11	2 ·022	2 ·022	1 ·004	1 ·004
	10	1 ·012	1 ·012	0 ·002	0 ·002
	9	1 ·029	0 ·004	0 ·004	0 ·004
	8	0 ·010+	0 ·010+	—	—
	7	0 ·022	0 ·022	—	—
	6	0 ·044	—	—	—

Table VIII (*Continued*)

	a	Probability			
		0.05	0.025	0.01	0.005
$A = 13$ $B = 6$	13	3 ·021	3 ·021	2 ·004	2 ·004
	12	2 ·017	2 ·017	1 ·003	1 ·003
	11	2 ·046	1 ·010−	1 ·010−	0 ·001
	10	1 ·024	1 ·024	0 ·003	0 ·003
	9	1 ·050−	0 ·008	0 ·008	—
	8	0 ·017	0 ·017	—	—
	7	0 ·034	—	—	—
5	13	2 ·012	2 ·012	1 ·002	1 ·002
	12	2 ·044	1 ·008	1 ·008	0 ·001
	11	1 ·022	1 ·022	0 ·002	0 ·002
	10	1 ·047	0 ·007	0 ·007	—
	9	0 ·015−	0 ·015−	—	—
	8	0 ·029	—	—	—
4	13	2 ·044	1 ·006	1 ·006	0 ·000
	12	1 ·022	1 ·022	0 ·002	0 ·002
	11	0 ·006	0 ·006	0 ·006	—
	10	0 ·015−	0 ·015−	—	—
	9	0 ·029	—	—	—
3	13	1 ·025	1 ·025	0 ·002	0 ·002
	12	0 ·007	0 ·007	0 ·007	—
	11	0 ·018	0 ·018	—	—
	10	0 ·036	—	—	—
2	13	0 ·010−	0 ·010−	0 ·010−	—
	12	0 ·029	—	—	—
$A = 14$ $B = 14$	14	10 ·049	9 ·020	8 ·008	7 ·003
	13	8 ·038	7 ·016	6 ·006	5 ·002
	12	6 ·023	6 ·023	5 ·009	4 ·003
	11	5 ·027	4 ·011	3 ·004	3 ·004
	10	4 ·028	3 ·011	2 ·003	2 ·003
	9	3 ·027	2 ·009	2 ·009	1 ·002
	8	2 ·023	2 ·023	1 ·006	0 ·001
	7	1 ·016	1 ·016	0 ·003	0 ·003
	6	1 ·038	0 ·008	0 ·008	—
	5	0 ·020	0 ·020	—	—
	4	0 ·049	—	—	—
13	14	9 ·041	8 ·016	7 ·006	6 ·002
	13	7 ·029	6 ·011	5 ·004	5 ·004
	12	6 ·037	5 ·015+	4 ·005+	3 ·002
	11	5 ·041	4 ·017	3 ·006	2 ·001
	10	4 ·041	3 ·016	2 ·005−	2 ·005−
	9	3 ·038	2 ·013	1 ·003	1 ·003
	8	2 ·031	1 ·009	1 ·009	0 ·001
	7	1 ·021	1 ·021	0 ·004	0 ·004
	6	1 ·048	0 ·010+	—	—
	5	0 ·025−	0 ·025−	—	—

Table VIII (*Continued*)

	a	Probability			
		0.05	0.025	0.01	0.005
$A = 14$ $B = 12$	14	8 ·033	7 ·012	6 ·004	6 ·004
	13	6 ·021	6 ·021	5 ·007	4 ·002
	12	5 ·025+	4 ·009	4 ·009	3 ·003
	11	4 ·026	3 ·009	3 ·009	2 ·002
	10	3 ·024	3 ·024	2 ·007	1 ·002
	9	2 ·019	2 ·019	1 ·005−	1 ·005−
	8	2 ·042	1 ·012	0 ·002	0 ·002
	7	1 ·028	0 ·005+	0 ·005+	—
	6	0 ·013	0 ·013	—	—
	5	0 ·030	—	—	—
11	14	7 ·026	6 ·009	6 ·009	5 ·003
	13	6 ·039	5 ·014	4 ·004	4 ·004
	12	5 ·043	4 ·016	3 ·005−	3 ·005−
	11	4 ·042	3 ·015−	2 ·004	2 ·004
	10	3 ·036	2 ·011	1 ·003	1 ·003
	9	2 ·027	1 ·007	1 ·007	0 ·001
	8	1 ·017	1 ·017	0 ·003	0 ·003
	7	1 ·038	0 ·007	0 ·007	—
	6	0 ·017	0 ·017	—	—
	5	0 ·038	—	—	—
10	14	6 ·020	6 ·020	5 ·006	4 ·002
	13	5 ·028	4 ·009	4 ·009	3 ·002
	12	4 ·028	3 ·009	3 ·009	2 ·002
	11	3 ·024	3 ·024	2 ·007	1 ·001
	10	2 ·018	2 ·018	1 ·004	1 ·004
	9	2 ·040	1 ·011	0 ·002	0 ·002
	8	1 ·024	1 ·024	0 ·004	0 ·004
	7	0 ·010−	0 ·010−	0 ·010−	—
	6	0 ·022	0 ·022	—	—
	5	0 ·047	—	—	—
9	14	6 ·047	5 ·014	4 ·004	4 ·004
	13	4 ·018	4 ·018	3 ·005−	3 ·005−
	12	3 ·017	3 ·017	2 ·004	2 ·004
	11	3 ·042	2 ·012	1 ·002	1 ·002
	10	2 ·029	1 ·007	1 ·007	0 ·001
	9	1 ·017	1 ·017	0 ·002	0 ·002
	8	1 ·036	0 ·006	0 ·006	—
	7	0 ·014	0 ·014	—	—
	6	0 ·030	—	—	—
8	14	5 ·036	4 ·010−	4 ·010−	3 ·002
	13	4 ·039	3 ·011	2 ·002	2 ·002
	12	3 ·032	2 ·008	2 ·008	1 ·001
	11	2 ·022	2 ·022	1 ·005−	1 ·005−
	10	2 ·048	1 ·012	0 ·002	0 ·002
	9	1 ·026	0 ·004	0 ·004	0 ·004
	8	0 ·009	0 ·009	0 ·009	—
	7	0 ·020	0 ·020	—	—
	6	0 ·040	—	—	—

Table VIII (*Continued*)

	a	Probability			
		0.05	**0.025**	**0.01**	**0.005**
$A = 14$ $B = 7$	14	4 ·026	3 ·006	3 ·006	2 ·001
	13	3 ·025	2 ·006	2 ·006	1 ·001
	12	2 ·017	2 ·017	1 ·003	1 ·003
	11	2 ·041	1 ·009	1 ·009	0 ·001
	10	1 ·021	1 ·021	0 ·003	0 ·003
	9	1 ·043	0 ·007	0 ·007	—
	8	0 ·015−	0 ·015−	—	—
	7	0 ·030	—	—	—
6	14	3 ·018	3 ·018	2 ·003	2 ·003
	13	2 ·014	2 ·014	1 ·002	1 ·002
	12	2 ·037	1 ·007	1 ·007	0 ·001
	11	1 ·018	1 ·018	0 ·002	0 ·002
	10	1 ·038	0 ·005+	0 ·005+	—
	9	0 ·012	0 ·012	—	—
	8	0 ·024	0 ·024	—	—
	7	0 ·044	—	—	—
5	14	2 ·010+	2 ·010+	1 ·001	1 ·001
	13	2 ·037	1 ·006	1 ·006	0 ·001
	12	1 ·017	1· 017	0 ·002	0 ·002
	11	1 ·038	0 ·005−	0 ·005−	0 ·005−
	10	0 ·011	0 ·011	—	—
	9	0 ·022	0 ·022	—	—
	8	0 ·040	—	—	—
4	14	2 ·039	1 ·005−	1 ·005−	1 ·005−
	13	1 ·019	1 ·019	0 ·002	0 ·002
	12	1 ·044	0 ·005−	0 ·005−	0 ·005−
	11	0 ·011	0 ·011	—	—
	10	0 ·023	0 ·023	—	—
	9	0 ·041	—	—	—
3	14	1 ·022	1 ·022	0 ·001	0 ·001
	13	0 ·006	0 ·006	0 ·006	—
	12	0 ·015−	0 ·015−	—	—
	11	0 ·029	—	—	—
2	14	0 ·008	0 ·008	0 ·008	—
	13	0 ·025	0 ·025	—	—
	12	0 ·050	—	—	—
$A = 15$ $B = 15$	15	11 ·050−	10 ·021	9 ·008	8 ·003
	14	9 ·040	8 ·018	7 ·007	6 ·003
	13	7 ·025+	6 ·010+	5 ·004	5 ·004
	12	6 ·030	5 ·013	4 ·005−	4 ·005−
	11	5 ·033	4 ·013	3 ·005−	3 ·005−
	10	4 ·033	3 ·013	2 ·004	2 ·004
	9	3 ·030	2 ·010+	1 ·003	1 ·003
	8	2 ·025+	1 ·007	1 ·007	0 ·001
	7	1 ·018	1 ·018	0 ·003	0 ·003
	6	1 ·040	0 ·008	0 ·008	—
	5	0 ·021	0 ·021	—	—
	4	0 ·050−	—	—	—

Table VIII (*Continued*)

	a	Probability			
		0.05	**0.025**	**0.01**	**0.005**
$A = 15$ $B = 14$	15	10 ·042	9 ·017	8 ·006	7 ·002
	14	8 ·031	7 ·013	6 ·005−	6 ·005−
	13	7 ·041	6 ·017	5 ·007	4 ·002
	12	6 ·046	5 ·020	4 ·007	3 ·002
	11	5 ·048	4 ·020	3 ·007	2 ·002
	10	4 ·046	3 ·018	2 ·006	1 ·001
	9	3 ·041	2 ·014	1 ·004	1 ·004
	8	2 ·033	1 ·009	1 ·009	0 ·001
	7	1 ·022	1 ·022	0 ·004	0 ·004
	6	1 ·049	0 ·011	—	—
	5	0 ·025+	—	—	—
13	15	9 ·035−	8 ·013	7 ·005−	7 ·005−
	14	7 ·023	7 ·023	6 ·009	5 ·003
	13	6 ·029	5 ·011	4 ·004	4 ·004
	12	5 ·031	4 ·012	3 ·004	3 ·004
	11	4 ·030	3 ·011	2 ·003	2 ·003
	10	3 ·026	2 ·008	2 ·008	1 ·002
	9	2 ·020	2 ·020	1 ·005+	0 ·001
	8	2 ·043	1 ·013	0 ·002	0 ·002
	7	1 ·029	0 ·005+	0 ·005+	—
	6	0 ·013	0 ·013	—	—
	5	0 ·031	—	—	—
12	15	8 ·028	7 ·010−	7 ·010−	6 ·003
	14	7 ·043	6 ·016	5 ·006	4 ·002
	13	6 ·049	5 ·019	4 ·007	3 ·002
	12	5 ·049	4 ·019	3 ·006	2 ·002
	11	4 ·045+	3 ·017	2 ·005−	2 ·005−
	10	3 ·038	2 ·012	1 ·003	1 ·003
	9	2 ·028	1 ·007	1 ·007	0 ·001
	8	1 ·018	1 ·018	0 ·003	0 ·003
	7	1 ·038	0 ·007	0 ·007	—
	6	0 ·017	0 ·017	—	—
	5	0 ·037	—	—	—
11	15	7 ·022	7 ·022	6 ·007	5 ·002
	14	6 ·032	5 ·011	4 ·003	4· 003
	13	5 ·034	4 ·012	3 ·003	3 ·003
	12	4 ·032	3 ·010+	2 ·003	2 ·003
	11	3 ·026	2 ·008	2 ·008	1 ·002
	10	2 ·019	2 ·019	1 ·004	1 ·004
	9	2 ·040	1 ·011	0 ·002	0 ·002
	8	1 ·024	1 ·024	0 ·004	0 ·004
	7	1 ·049	0 ·010−	0 ·010−	—
	6	0 ·022	0 ·022	—	—
	5	0 ·046	—	—	—

Table VIII (*Continued*)

	a	Probability			
		0.05	**0.025**	**0.01**	**0.005**
$A = 15$ $B = 10$	15	**6** ·017	**6** ·017	**5** ·005−	**5** ·005−
	14	**5** ·023	**5** ·023	**4** ·007	**3** ·002
	13	**4** ·022	**4** ·022	**3** ·007	**2** ·001
	12	**3** ·018	**3** ·018	**2** ·005−	**2** ·005−
	11	**3** ·042	**2** ·013	**1** ·003	**1** ·003
	10	**2** ·029	**1** ·007	**1** ·007	**0** ·001
	9	**1** ·016	**1** ·016	**0** ·002	**0** ·002
	8	**1** ·034	**0** ·006	**0** ·006	—
	7	**0** ·013	**0** ·013	—	—
	6	**0** ·028	—	—	—
9	15	**6** ·042	**5** ·012	**4** ·003	**4** ·003
	14	**5** ·047	**4** ·015−	**3** ·004	**3** ·004
	13	**4** ·042	**3** ·013	**2** ·003	**2** ·003
	12	**3** ·032	**2** ·009	**2** ·009	**1** ·002
	11	**2** ·021	**2** ·021	**1** ·005−	**1** ·005−
	10	**2** ·045−	**1** ·011	**0** ·002	**0** ·002
	9	**1** ·024	**1** ·024	**0** ·004	**0** ·004
	8	**1** ·048	**0** ·009	**0** ·009	—
	7	**0** ·019	**0** ·019	—	—
	6	**0** ·037	—	—	—
8	15	**5** ·032	**4** ·008	**4** ·008	**3** ·002
	14	**4** ·033	**3** ·009	**3** ·009	**2** ·002
	13	**3** ·026	**2** ·006	**2** ·006	**1** ·001
	12	**2** ·017	**2** ·017	**1** ·003	**1** ·003
	11	**2** ·037	**1** ·008	**1** ·008	**0** ·001
	10	**1** ·019	**1** ·019	**0** ·003	**0** ·003
	9	**1** ·038	**0** ·006	**0** ·006	—
	8	**0** ·013	**0** ·013	—	—
	7	**0** ·026	—	—	—
	6	**0** ·050−	—	—	—
7	15	**4** ·023	**4** ·023	**3** ·005−	**3** ·005−
	14	**3** ·021	**3** ·021	**2** ·004	**2** ·004
	13	**2** ·014	**2** ·014	**1** ·002	**1** ·002
	12	**2** ·032	**1** ·007	**1** ·007	**0** ·001
	11	**1** ·015+	**1** ·015+	**0** ·002	**0** ·002
	10	**1** ·032	**0** ·005−	**0** ·005−	**0** ·005−
	9	**0** ·010+	**0** ·010+	—	—
	8	**0** ·020	**0** ·020	—	—
	7	**0** ·038	—	—	—
6	15	**3** ·015+	**3** ·015+	**2** ·003	**2** ·003
	14	**2** ·011	**2** ·011	**1** ·002	**1** ·002
	13	**2** ·031	**1** ·006	**1** ·006	**0** ·001
	12	**1** ·014	**1** ·014	**0** ·002	**0** ·002
	11	**1** ·029	**0** ·004	**0** ·004	**0** ·004
	10	**0** ·009	**0** ·009	**0** ·009	—
	9	**0** ·017	**0** ·017	—	—
	8	**0** ·032	—	—	—

Table VIII (*Continued*)

	a	Probability			
		0.05	**0.025**	**0.01**	**0.005**
$A = 15$ $B = 5$	15	2 ·009	2 ·009	2 ·009	1 ·001
	14	2 ·032	1 ·005−	1 ·005−	1 ·005−
	13	1 ·014	1 ·014	0 ·001	0 ·001
	12	1 ·031	0 ·004	0 ·004	0 ·004
	11	0 ·008	0 ·008	0 ·008	—
	10	0 ·016	0 ·016	—	—
	9	0 ·030	—	—	—
4	15	2 ·035+	1 ·004	1 ·004	1 ·004
	14	1 ·016	1 ·016	0 ·001	0 ·001
	13	1 ·037	0 ·004	0 ·004	0 ·004
	12	0 ·009	0 ·009	0 ·009	—
	11	0 ·018	0 ·018	—	—
	10	0 ·033	—	—	—
3	15	1 ·020	1 ·020	0 ·001	0 ·001
	14	0 ·005−	0 ·005−	0 ·005−	0 ·005−
	13	0 ·012	0 ·012	—	—
	12	0 ·025−	0 ·025−	—	—
	11	0 ·043	—	—	—
2	15	0 ·007	0 ·007	0 ·007	—
	14	0 ·022	0 ·022	—	—
	13	0 ·044	—	—	—
$A = 16$ $B = 16$	16	11 ·022	11 ·022	10 ·009	9 ·003
	15	10 ·041	9 ·019	8 ·008	7 ·003
	14	8 ·027	7 ·012	6 ·005−	6 ·005−
	13	7 ·033	6 ·015−	5 ·006	4 ·002
	12	6 ·037	5 ·016	4 ·006	3 ·002
	11	5 ·038	4 ·016	3 ·006	2 ·002
	10	4 ·037	3 ·015−	2 ·005−	2 ·005−
	9	3 ·033	2 ·012	1 ·003	1 ·003
	8	2 ·027	1 ·008	1 ·008	0 ·001
	7	1 ·019	1 ·019	0 ·003	0 ·003
	6	1 ·041	0 ·009	0 ·009	—
	5	0 ·022	0 ·022	—	—
$A = 17$ $B = 17$	17	12 ·022	12 ·022	11 ·009	10 ·004
	16	11 ·043	10 ·020	9 ·008	8 ·003
	15	9 ·029	8 ·013	7 ·005+	6 ·002
	14	8 ·035+	7 ·016	6 ·007	5 ·002
	13	7 ·040	6 ·018	5 ·007	4 ·003
	12	6 ·042	5 ·019	4 ·007	3 ·002
	11	5 ·042	4 ·018	3 ·007	2 ·002
	10	4 ·040	3 ·016	2 ·005+	1 ·001
	9	3 ·035+	2 ·013	1 ·003	1 ·003
	8	2 ·029	1 ·008	1 ·008	0 ·001
	7	1 ·020	1 ·020	0 ·004	0 ·004
	6	1 ·043	0 ·009	0 ·009	—
	5	0 ·022	0 ·022	—	—

Table VIII (*Continued*)

	a	Probability			
		0.05	0.025	0.01	0.005
$A = 18$ $B = 18$	18	13 ·023	13 ·023	12 ·010−	11 ·004
	17	12 ·044	11 ·020	10 ·009	9 ·004
	16	10 ·030	9 ·014	8 ·006	7 ·002
	15	9 ·038	8 ·018	7 ·008	6 ·003
	14	8 ·043	7 ·020	6 ·009	5 ·003
	13	7 ·046	6 ·022	5 ·009	4 ·003
	12	6 ·047	5 ·022	4 ·009	3 ·003
	11	5 ·046	4 ·020	3 ·008	2 ·002
	10	4 ·043	3 ·018	2 ·006	1 ·001
	9	3 ·038	2 ·014	1 ·004	1 ·004
	8	2 ·030	1 ·009	1 ·009	0 ·001
	7	1 ·020	1 ·020	0 ·004	0 ·004
	6	1 ·044	0 ·010−	0 ·010−	—
	5	0 ·023	0 ·023	—	—
$A = 19$ $B = 19$	19	14 ·023	14 ·023	13 ·010−	12 ·004
	18	13 ·045−	12 ·021	11 ·009	10 ·004
	17	11 ·031	10 ·015−	9 ·006	8 ·003
	16	10 ·039	9 ·019	8 ·009	7 ·003
	15	9 ·046	8 ·022	6 ·004	6 ·004
	14	8 ·050−	7 ·024	5 ·004	5 ·004
	13	6 ·025+	5 ·011	4 ·004	4 ·004
	12	5 ·024	5 ·024	3 ·003	3 ·003
	11	5 ·050−	4 ·022	3 ·009	2 ·003
	10	4 ·046	3 ·019	2 ·006	1 ·002
	9	3 ·039	2 ·015−	1 ·004	1 ·004
	8	2 ·031	1 ·009	1 ·009	0 ·002
	7	1 ·021	1 ·021	0 ·004	0 ·004
	6	1 ·045−	0 ·010−	0 ·010−	—
	5	0 ·023	0 ·023	—	—
$A = 20$ $B = 20$	20	15 ·024	15 ·024	13 ·004	13 ·004
	19	14 ·046	13 ·022	12 ·010−	11 ·004
	18	12 ·032	11 ·015+	10 ·007	9 ·003
	17	11 ·041	10 ·020	9 ·009	8 ·004
	16	10 ·048	9 ·024	7 ·005−	7 ·005−
	15	8 ·027	7 ·012	6 ·005+	5 ·002
	14	7 ·028	6 ·013	5 ·005+	4 ·002
	13	6 ·028	5 ·012	4 ·005−	4 ·005−
	12	5 ·027	4 ·011	3 ·004	3 ·004
	11	4 ·024	4 ·024	3 ·009	2 ·003
	10	4 ·048	3 ·020	2 ·007	1 ·002
	9	3 ·041	2 ·015+	1 ·004	1 ·004
	8	2 ·032	1 ·010−	1 ·010−	0 ·002
	7	1 ·022	1 ·022	0 ·004	0 ·004
	6	1 ·046	0 ·010+	—	—
	5	0 ·024	0 ·024	—	—
$A = 21$ $B = 21$	21	16 ·0239	16 ·0239	14 ·0043	14 ·0043
	20	15 ·0465	14 ·0224	12 ·0044	12 ·0044
	19	13 ·0335	12 ·0162	11 ·0074	10 ·0032
	18	12 ·0427	11 ·0215	9 ·0044	9 ·0044

Table VIII (*Continued*)

	a	Probability			
		0.05	0.025	0.01	0.005
$A = 21$ $B = 21$	17	10 ·0258	9 ·0123	8 ·0054	7 ·0022
	16	9 ·0289	8 ·0139	7 ·0061	6 ·0024
	15	8 ·0308	7 ·0147	6 ·0063	5 ·0024
	14	7 ·0314	6 ·0147	5 ·0061	4 ·0022
	13	6 ·0308	5 ·0139	4 ·0054	3 ·0018
	12	5 ·0289	4 ·0123	3 ·0044	3 ·0044
	11	4 ·0258	3 ·0101	2 ·0032	2 ·0032
	10	3 ·0215	3 ·0215	2 ·0074	1 ·0018
	9	3 ·0427	2 ·0162	1 ·0044	1 ·0044
	8	2 ·0335	1 ·0102	0 ·0017	0 ·0017
	7	1 ·0224	1 ·0224	0 ·0043	0 ·0043
	6	1 ·0465	0 ·0103	—	—
	5	0 ·0239	0 ·0239	—	—
$A = 22$ $B = 22$	22	17 ·0242	17 ·0242	15 ·0045	15 ·0045
	21	16 ·0473	15 ·0230	13 ·0047	13 ·0047
	20	14 ·0344	13 ·0169	12 ·0078	11 ·0034
	19	13 ·0442	12 ·0226	10 ·0049	10 ·0049
	18	11 ·0273	10 ·0134	9 ·0061	8 ·0025
	17	10 ·0309	9 ·0153	8 ·0069	7 ·0029
	16	9 ·0333	8 ·0164	7 ·0074	6 ·0030
	15	8 ·0345	7 ·0168	6 ·0074	5 ·0029
	14	7 ·0345	6 ·0164	5 ·0069	4 ·0025
	13	6 ·0333	5 ·0153	4 ·0061	3 ·0020
	12	5 ·0309	4 ·0134	3 ·0049	3 ·0049
	11	4 ·0273	3 ·0108	2 ·0034	2 ·0034
	10	3 ·0226	3 ·0226	2 ·0078	1 ·0019
	9	3 ·0442	2 ·0169	1 ·0047	1 ·0047
	8	2 ·0344	1 ·0106	0 ·0018	0 ·0018
	7	1 ·0230	1 ·0230	0 ·0045	0 ·0045
	6	1 ·0473	0 ·0106	—	—
	5	0 ·0242	0 ·0242	—	—
$A = 23$ $B = 23$	23	18 ·0245	18 ·0245	16 ·0046	16 ·0046
	22	17 ·0480	16 ·0235	14 ·0049	14 ·0049
	21	15 ·0353	14 ·0176	13 ·0083	12 ·0037
	20	14 ·0455	13 ·0236	11 ·0053	10 ·0023
	19	12 ·0287	11 ·0143	10 ·0067	9 ·0029
	18	11 ·0327	10 ·0166	9 ·0078	8 ·0034
	17	10 ·0356	9 ·0181	8 ·0085	7 ·0036
	16	9 ·0373	8 ·0188	7 ·0087	6 ·0036
	15	8 ·0379	7 ·0188	6 ·0085	5 ·0034
	14	7 ·0373	6 ·0181	5 ·0078	4 ·0029
	13	6 ·0356	5 ·0166	4 ·0067	3 ·0023
	12	5 ·0327	4 ·0143	3 ·0053	2 ·0015
	11	4 ·0287	3 ·0115	2 ·0037	2 ·0037
	10	3 ·0236	3 ·0236	2 ·0083	1 ·0021
	9	3 ·0455	2 ·0176	1 ·0049	1 ·0049
	8	2 ·0353	1 ·0110	0 ·0019	0 ·0019
	7	1 ·0235	1 ·0235	0 ·0046	0 ·0046
	6	1 ·0480	0 ·0108	—	—
	5	0 ·0245	0 ·0245	—	—

Table VIII (*Continued*)

	a	Probability			
		0.05	0.025	0.01	0.005
$A = 24$ $B = 24$	24	19 ·0248	19 ·0248	17 ·0047	17 ·0047
	23	18 ·0486	17 ·0240	15 ·0051	14 ·0022
	22	16 ·0361	15 ·0182	14 ·0087	13 ·0039
	21	15 ·0467	14 ·0245	12 ·0057	11 ·0025
	20	13 ·0299	12 ·0152	11 ·0073	10 ·0032
	19	12 ·0344	11 ·0178	10 ·0086	9 ·0038
	18	11 ·0377	10 ·0196	9 ·0095	8 ·0042
	17	10 ·0399	9 ·0207	8 ·0099	7 ·0043
	16	9 ·0410	8 ·0211	7 ·0099	6 ·0042
	15	8 ·0410	7 ·0207	6 ·0095	5 ·0038
	14	7 ·0399	6 ·0196	5 ·0086	4 ·0032
	13	6 ·0377	5 ·0178	4 ·0073	3 ·0025
	12	5 ·0344	4 ·0152	3 ·0057	2 ·0017
	11	4 ·0299	3 ·0122	2 ·0039	2 ·0039
	10	3 ·0245	3 ·0245	2 ·0087	1 ·0022
	9	3 ·0467	2 ·0182	1 ·0051	0 ·0008
	8	2 ·0361	1 ·0113	0 ·0019	0 ·0019
	7	1 ·0240	1 ·0240	0 ·0047	0 ·0047
	6	1 ·0486	0 ·0110	—	—
	5	0 ·0248	0 ·0248	—	—
$A = 25$ $B = 25$	25	20 ·0251	19 ·0111	18 ·0048	18 ·0048
	24	19 ·0491	18 ·0244	16 ·0053	15 ·0023
	23	17 ·0369	16 ·0187	15 ·0090	14 ·0041
	22	16 ·0477	14 ·0127	13 ·0061	12 ·0027
	21	14 ·0311	13 ·0161	12 ·0078	11 ·0036
	20	13 ·0359	12 ·0189	11 ·0093	10 ·0043
	19	12 ·0396	11 ·0210	9 ·0048	9 ·0048
	18	11 ·0423	10 ·0225	8 ·0051	7 ·0021
	17	10 ·0438	9 ·0232	7 ·0051	6 ·0020
	16	9 ·0444	8 ·0232	6 ·0048	6 ·0048
	15	8 ·0438	7 ·0225	5 ·0043	5 ·0043
	14	7 ·0423	6 ·0210	5 ·0093	4 ·0036
	13	6 ·0396	5 ·0189	4 ·0078	3 ·0027
	12	5 ·0359	4 ·0161	3 ·0061	2 ·0018
	11	4 ·0311	3 ·0127	2 ·0041	2 ·0041
	10	3 ·0253	2 ·0090	2 ·0090	1 ·0023
	9	3 ·0477	2 ·0187	1 ·0053	0 ·0008
	8	2 ·0369	1 ·0116	0 ·0020	0 ·0020
	7	1 ·0244	1 ·0244	0 ·0048	0 ·0048
	6	1 ·0491	0 ·0111	—	—
	5	0 ·0251	—	—	—

Table VIII (*Continued*)

	a	Probability			
		0.05	**0.025**	**0.01**	**0.005**
$A = 26$ $B = 26$	**26**	**21** ·0253	**20** ·0113	**19** ·0049	**19** ·0049
	25	**20** ·0497	**19** ·0248	**17** ·0055	**16** ·0024
	24	**18** ·0376	**17** ·0192	**16** ·0094	**15** ·0044
	23	**17** ·0487	**15** ·0133	**14** ·0064	**13** ·0029
	22	**15** ·0322	**14** ·0169	**13** ·0084	**12** ·0039
	21	**14** ·0373	**13** ·0199	**11** ·0047	**11** ·0047
	20	**13** ·0414	**12** ·0224	**10** ·0054	**9** ·0024
	19	**12** ·0444	**11** ·0242	**9** ·0058	**8** ·0025
	18	**11** ·0465	**9** ·0127	**8** ·0059	**7** ·0025
	17	**10** ·0475	**8** ·0127	**7** ·0058	**6** ·0024
	16	**9** ·0475	**7** ·0123	**6** ·0054	**5** ·0021
	15	**8** ·0465	**7** ·0242	**5** ·0047	**5** ·0047
	14	**7** ·0444	**6** ·0224	**4** ·0039	**4** ·0039
	13	**6** ·0414	**5** ·0199	**4** ·0084	**3** ·0029
	12	**5** ·0373	**4** ·0169	**3** ·0064	**2** ·0019
	11	**4** ·0322	**3** ·0133	**2** ·0044	**2** ·0044
	10	**3** ·0261	**2** ·0094	**2** ·0094	**1** ·0024
	9	**3** ·0487	**2** ·0192	**1** ·0055	**0** ·0008
	8	**2** ·0376	**1** ·0119	**0** ·0021	**0** ·0021
	7	**1** ·0248	**1** ·0248	**0** ·0049	**0** ·0049
	6	**1** ·0497	**0** ·0113	—	—
	5	**0** ·0253	—	—	—
$A = 27$ $B = 27$	**27**	**22** ·0255	**21** ·0115	**20** ·0050	**19** ·0021
	26	**20** ·0252	**19** ·0122	**18** ·0056	**17** ·0025
	25	**19** ·0382	**18** ·0197	**17** ·0097	**16** ·0046
	24	**18** ·0497	**16** ·0138	**15** ·0068	**14** ·0031
	23	**16** ·0332	**15** ·0176	**14** ·0088	**13** ·0042
	22	**15** ·0386	**14** ·0209	**12** ·0052	**11** ·0023
	21	**14** ·0430	**13** ·0236	**11** ·0059	**10** ·0027
	20	**13** ·0465	**11** ·0133	**10** ·0065	**9** ·0029
	19	**12** ·0489	**10** ·0140	**9** ·0067	**8** ·0030
	18	**10** ·0278	**9** ·0143	**8** ·0067	**7** ·0029
	17	**9** ·0278	**8** ·0140	**7** ·0065	**6** ·0027
	16	**8** ·0271	**7** ·0133	**6** ·0059	**5** ·0023
	15	**8** ·0489	**6** ·0122	**5** ·0052	**4** ·0019
	14	**7** ·0465	**6** ·0236	**4** ·0042	**4** ·0042
	13	**6** ·0430	**5** ·0209	**4** ·0088	**3** ·0031
	12	**5** ·0386	**4** ·0176	**3** ·0068	**2** ·0021
	11	**4** ·0332	**3** ·0138	**2** ·0046	**2** ·0046
	10	**3** ·0268	**2** ·0097	**2** ·0097	**1** ·0025
	9	**3** ·0497	**2** ·0197	**1** ·0056	**0** ·0009
	8	**2** ·0382	**1** ·0122	**0** ·0021	**0** ·0021
	7	**1** ·0252	**0** ·0050	**0** ·0050	—
	6	**0** ·0115	**0** ·0115	—	—
	5	**0** ·0255	—	—	—

Table VIII (*Continued*)

	a	Probability			
		0.05	0.025	0.01	0.005
$A = 28$ $B = 28$	28	23 ·0257	22 ·0116	21 ·0051	20 ·0022
	27	21 ·0255	20 ·0124	19 ·0058	18 ·0026
	26	20 ·0388	19 ·0201	17 ·0048	17 ·0048
	25	18 ·0275	17 ·0143	16 ·0071	15 ·0033
	24	17 ·0341	16 ·0183	15 ·0093	14 ·0045
	23	16 ·0398	15 ·0218	13 ·0056	12 ·0026
	22	15 ·0446	14 ·0248	12 ·0065	11 ·0030
	21	14 ·0483	12 ·0144	11 ·0071	10 ·0033
	20	12 ·0289	11 ·0153	10 ·0076	9 ·0035
	19	11 ·0299	10 ·0157	9 ·0077	8 ·0035
	18	10 ·0302	9 ·0157	8 ·0076	7 ·0033
	17	9 ·0299	8 ·0153	7 ·0071	6 ·0030
	16	8 ·0289	7 ·0144	6 ·0065	5 ·0026
	15	7 ·0272	6 ·0131	5 ·0056	4 ·0021
	14	7 ·0483	6 ·0248	4 ·0045	4 ·0045
	13	6 ·0446	5 ·0218	4 ·0093	3 ·0033
	12	5 ·0398	4 ·0183	3 ·0071	2 ·0022
	11	4 ·0341	3 ·0143	2 ·0048	2 ·0048
	10	3 ·0275	2 ·0100	1 ·0026	1 ·0026
	9	2 ·0201	2 ·0201	1 ·0058	0 ·0009
	8	2 ·0388	1 ·0124	0 ·0022	0 ·0022
	7	1 ·0255	0 ·0051	0 ·0051	—
	6	0 ·0116	0 ·0116	—	—
	5	0 ·0257	—	—	—
$A = 29$ $B = 29$	29	24 ·0259	23 ·0117	22 ·0052	21 ·0022
	28	22 ·0259	21 ·0126	20 ·0059	19 ·0027
	27	21 ·0393	20 ·0206	18 ·0049	18 ·0049
	26	19 ·0281	18 ·0147	17 ·0074	16 ·0035
	25	18 ·0350	17 ·0189	16 ·0098	15 ·0048
	24	17 ·0410	16 ·0227	14 ·0060	13 ·0028
	23	16 ·0460	14 ·0138	13 ·0070	12 ·0033
	22	14 ·0285	13 ·0154	12 ·0078	11 ·0037
	21	13 ·0305	12 ·0165	11 ·0084	10 ·0040
	20	12 ·0318	11 ·0172	10 ·0086	9 ·0040
	19	11 ·0325	10 ·0174	9 ·0086	8 ·0040
	18	10 ·0325	9 ·0172	8 ·0084	7 ·0037
	17	9 ·0318	8 ·0165	7 ·0078	6 ·0033
	16	8 ·0305	7 ·0154	6 ·0070	5 ·0028
	15	7 ·0285	6 ·0138	5 ·0060	4 ·0022
	14	6 ·0259	5 ·0120	4 ·0048	4 ·0048
	13	6 ·0460	5 ·0227	4 ·0098	3 ·0035
	12	5 ·0410	4 ·0189	3 ·0074	2 ·0023
	11	4 ·0350	3 ·0147	2 ·0049	2 ·0049
	10	3 ·0281	2 ·0103	1 ·0027	1 ·0027
	9	2 ·0206	2 ·0206	1 ·0059	0 ·0009
	8	2 ·0393	1 ·0126	0 ·0022	0 ·0022
	7	1 ·0259	0 ·0052	0 ·0052	—
	6	0 ·0117	0 ·0117	—	—
	5	0 ·0259	—	—	—

Table VIII (*Continued*)

	a	Probability			
		0.05	**0.025**	**0.01**	**0.005**
$A = 30$ $B = 30$	**30**	**25** ·0261	**24** ·0119	**23** ·0053	**22** ·0023
	29	**23** ·0262	**22** ·0128	**21** ·0061	**20** ·0028
	28	**22** ·0399	**21** ·0210	**19** ·0051	**18** ·0024
	27	**20** ·0287	**19** ·0152	**18** ·0077	**17** ·0037
	26	**19** ·0358	**18** ·0195	**16** ·0051	**15** ·0024
	25	**18** ·0420	**17** ·0235	**15** ·0063	**14** ·0031
	24	**17** ·0473	**15** ·0146	**14** ·0075	**13** ·0036
	23	**15** ·0298	**14** ·0163	**13** ·0084	**12** ·0041
	22	**14** ·0320	**13** ·0176	**12** ·0091	**11** ·0044
	21	**13** ·0336	**12** ·0185	**11** ·0095	**10** ·0046
	20	**12** ·0346	**11** ·0189	**10** ·0097	**9** ·0046
	19	**11** ·0349	**10** ·0189	**9** ·0095	**8** ·0044
	18	**10** ·0346	**9** ·0185	**8** ·0091	**7** ·0041
	17	**9** ·0336	**8** ·0176	**7** ·0084	**6** ·0036
	16	**8** ·0320	**7** ·0163	**6** ·0075	**5** ·0031
	15	**7** ·0298	**6** ·0146	**5** ·0063	**4** ·0024
	14	**6** ·0269	**5** ·0125	**4** ·0051	**3** ·0017
	13	**6** ·0473	**5** ·0235	**3** ·0037	**3** ·0037
	12	**5** ·0420	**4** ·0195	**3** ·0077	**2** ·0024
	11	**4** ·0358	**3** ·0152	**2** ·0051	**1** ·0012
	10	**3** ·0287	**2** ·0106	**1** ·0028	**1** ·0028
	9	**2** ·0210	**2** ·0210	**1** ·0061	**0** ·0010
	8	**2** ·0399	**1** ·0128	**0** ·0023	**0** ·0023
	7	**1** ·0262	**0** ·0053	**0** ·0053	—
	6	**0** ·0119	**0** ·0119	—	—
	5	**0** ·0261	—	—	—

Table IX* (11.2.5): *Critical Values of U for Total-Number-of-Runs Test*

Largest value of U' for which $P(U \leq U') \leq \alpha$

	n_1																		
	2	3	4	5	6	7	8	9	10	11	12	13	14	15	16	17	18	19	20
n_2	$\alpha = .05$																		
4			2																
5		2	2	3															
6		2	3	3	3														
7		2	3	3	4	4													
8	2	2	3	3	4	4	5												
9	2	2	3	4	4	5	5	6											
10	2	3	3	4	5	5	6	6	6										
11	2	3	3	4	5	5	6	6	7	7									
12	2	3	4	4	5	6	6	7	7	8	8								
13	2	3	4	4	5	6	6	7	8	8	9	9							
14	2	3	4	5	5	6	7	7	8	8	9	9	10						
15	2	3	4	5	6	6	7	8	8	9	9	10	10	11					
16	2	3	4	5	6	6	7	8	8	9	10	10	11	11	11				
17	2	3	4	5	6	7	7	8	9	9	10	10	11	11	12	12			
18	2	3	4	5	6	7	8	8	9	10	10	11	11	12	12	13	13		
19	2	3	4	5	6	7	8	8	9	10	10	11	12	12	13	13	14	14	
20	2	3	4	5	6	7	8	9	9	10	11	11	12	12	13	13	14	14	15
n_2	$\alpha = .01$																		
5				2															
6			2	2	2														
7			2	2	3	3													
8			2	2	3	3	4												
9		2	2	3	3	4	4	4											
10		2	2	3	3	4	4	5	5										
11		2	2	3	4	4	5	5	5	6									
12		2	3	3	4	4	5	5	6	6	7								
13		2	3	3	4	5	5	6	6	6	7	7							
14		2	3	3	4	5	5	6	6	7	7	8	8						
15		2	3	4	4	5	5	6	7	7	8	8	8	9					
16		2	3	4	4	5	6	6	7	7	8	8	9	9	10				
17		2	3	4	5	5	6	7	7	8	8	9	9	10	10	10			
18		2	3	4	5	5	6	7	7	8	8	9	9	10	10	11	11		
19	2	2	3	4	5	6	6	7	8	8	9	9	10	10	11	11	12	12	
20	2	2	3	4	5	6	6	7	8	8	9	10	10	11	11	11	12	12	13

*Adapted from Table II in Frieda S. Swed and C. Eisenhart's "Tables for Testing Randomness of Grouping in a Sequence of Alternatives," *Annals of Mathematical Statistics*, **14** (1943), 66–87, with permission of the authors and editor.

Table X* (12.2.5): *Cumulative Probability Distribution for r, the Total Number of Runs Up and Down*

$P(r \leq r')$

Number of runs r'	Number of observations, n																								
	1	2	3	4	5	6	7	8	9	10	11	12	13	14	15	16	17	18	19	20	21	22	23	24	25
1	......	1.0000	.3333	.0833	.0167	.0028	.0004	.0000	.0000	.0000	.0000	.0000	.0000	.0000	.0000	.0000	.0000	.0000	.0000	.0000	.0000	.0000	.0000	.0000	.0000
2	......	......	1.0000	.5833	.2500	.0861	.0250	.0063	.0014	.0003	.0001	.0000	.0000	.0000	.0000	.0000	.0000	.0000	.0000	.0000	.0000	.0000	.0000	.0000	.0000
3	......	......	......	1.0000	.7333	.4139	.1909	.0749	.0257	.0079	.0022	.0005	.0001	.0000	.0000	.0000	.0000	.0000	.0000	.0000	.0000	.0000	.0000	.0000	.0000
4	......	......	......	......	1.0000	.8306	.5583	.3124	.1500	.0633	.0239	.0082	.0026	.0007	.0002	.0001	.0000	.0000	.0000	.0000	.0000	.0000	.0000	.0000	.0000
5	......	......	......	......	......	1.0000	.8921	.6750	.4347	.2427	.1196	.0529	.0213	.0079	.0027	.0009	.0003	.0001	.0000	.0000	.0000	.0000	.0000	.0000	.0000
6	......	......	......	......	......	......	1.0000	.9313	.7653	.5476	.3438	.1918	.0964	.0441	.0186	.0072	.0026	.0009	.0003	.0001	.0000	.0000	.0000	.0000	.0000
7	......	......	......	......	......	......	......	1.0000	.9563	.8329	.6460	.4453	.2749	.1534	.0782	.0367	.0160	.0065	.0025	.0009	.0003	.0001	.0000	.0000	.0000
8	......	......	......	......	......	......	......	......	1.0000	.9722	.8823	.7280	.5413	.3633	.2216	.1238	.0638	.0306	.0137	.0058	.0023	.0009	.0003	.0001	.0000
9	......	......	......	......	......	......	......	......	......	1.0000	.9823	.9179	.7942	.6278	.4520	.2975	.1799	.1006	.0523	.0255	.0117	.0050	.0021	.0008	.0003
10	......	......	......	......	......	......	......	......	......	......	1.0000	.9887	.9432	.8464	.7030	.5369	.3770	.2443	.1467	.0821	.0431	.0213	.0099	.0044	.0018
11	......	......	......	......	......	......	......	......	......	......	......	1.0000	.9928	.9609	.8866	.7665	.6150	.4568	.3144	.2012	.1202	.0674	.0356	.0177	.0084
12	......	......	......	......	......	......	......	......	......	......	......	......	1.0000	.9954	9733	.9172	.8188	.6848	.5337	.3873	.2622	.1661	.0988	.0554	.0294
13	......	......	......	......	......	......	......	......	......	......	......	......	......	1.0000	.9971	.9818	.9400	.8611	.7454	.6055	.4603	.3276	.2188	.1374	.0815
14	......	......	......	......	......	......	......	......	......	......	......	......	......	......	1.0000	.9981	.9877	.9569	.8945	.7969	.6707	.5312	.3953	.2768	.1827
15	......	......	......	......	......	......	......	......	......	......	......	......	......	......	......	1.0000	.9988	.9917	.9692	.9207	.8398	.7286	.5980	.4631	.3384
16	......	......	......	......	......	......	......	......	......	......	......	......	......	......	......	......	1.0000	.9992	.9944	.9782	.9409	.8749	.7789	.6595	.5292
17	......	......	......	......	......	......	......	......	......	......	......	......	......	......	......	......	......	1.0000	.9995	.9962	.9846	.9563	.9032	.8217	.7148
18	......	......	......	......	......	......	......	......	......	......	......	......	......	......	......	......	......	......	1.0000	.9997	.9975	.9892	.9679	.9258	.8577
19	......	......	......	......	......	......	......	......	......	......	......	......	......	......	......	......	......	......	......	1.0000	.9998	.9983	.9924	.9765	.9436
20	......	......	......	......	......	......	......	......	......	......	......	......	......	......	......	......	......	......	......	......	1.0000	.9999	.9989	.9947	.9830
21	......	......	......	......	......	......	......	......	......	......	......	......	......	......	......	......	......	......	......	......	......	1.0000	.9999	.9993	.9963
22	......	......	......	......	......	......	......	......	......	......	......	......	......	......	......	......	......	......	......	......	......	......	1.0000	1.0000	.9995
23	......	......	......	......	......	......	......	......	......	......	......	......	......	......	......	......	......	......	......	......	......	......	......	1.0000	1.0000
24	......	......	......	......	......	......	......	......	......	......	......	......	......	......	......	......	......	......	......	......	......	......	......	......	1.0000

*Body of table is reproduced, with changes only in notation, from Table 1 in E. S. Edgington's "Probability Table for Number of Runs of Signs of First Differences in Ordered Series," *Journal of the American Statistical Association*, **56** (1961), 156–159, with permission of the author and editor.

Table XI* (13.1.5): *Critical Upper-Tail Values of S for Kendall's Rank-Order Correlation Test*

Smallest value of S' for which $P(S \geq S') \leq \alpha$

n	$\alpha = .005$	$\alpha = .010$	$\alpha = .025$	$\alpha = .050$	$\alpha = .100$
4	8	8	8	6	6
5	12	10	10	8	8
6	15	13	13	11	9
7	19	17	15	13	11
8	22	20	18	16	12
9	26	24	20	18	14
10	29	27	23	21	17
11	33	31	27	23	19
12	38	36	30	26	20
13	44	40	34	28	24
14	47	43	37	33	25
15	53	49	41	35	29
16	58	52	46	38	30
17	64	58	50	42	34
18	69	63	53	45	37
19	75	67	57	49	39
20	80	72	62	52	42
21	86	78	66	56	44
22	91	83	71	61	47
23	99	89	75	65	51
24	104	94	80	68	54
25	110	100	86	72	58
26	117	107	91	77	61
27	125	113	95	81	63
28	130	118	100	86	68
29	138	126	106	90	70
30	145	131	111	95	75
31	151	137	117	99	77
32	160	144	122	104	82
33	166	152	128	108	86
34	175	157	133	113	89
35	181	165	139	117	93
36	190	172	146	122	96
37	198	178	152	128	100
38	205	185	157	133	105
39	213	193	163	139	109
40	222	200	170	144	112

*Body of table is reproduced from Table III in L. Kaarsemaker and A. van Wijngaarden's "Tables for Use in Rank Correlation," *Statistica Neerlandica*, 7 (1953), 41–54 (reproduced as Report R73 of the Computation Department of the Mathematical Centre, Amsterdam), with permission of the authors, the Mathematical Centre, and the editor of *Statistica Neerlandica*.

Table XII* (13.3.5): *Critical Values of k for Smirnov's Maximum Deviation Tests for Identical Populations*

Smallest value of k for which $P(D^+ \geq k/n) \leq \alpha$ [or for which $P(D \geq k/n) \leq 2\alpha$], followed by $P(D^+ \geq k/n)$ [which equals $\frac{1}{2}P(D \geq k/n)$]

$n = m$	Significance level, α					
	.05	.025	.01	.005	.001	.0005
3	3 (.05000)	—	—	—	—	—
4	4 (.01429)	4 (.01429)	—	—	—	—
5	4 (.03968)	5 (.00397)	5 (.00397)	5 (.00397)	—	—
6	5 (.01299)	5 (.01299)	6 (.00108)	6 (.00108)	—	—
7	5 (.02652)	6 (.00408)	6 (.00408)	6 (.00408)	7 (.00029)	7 (.00029)
8	5 (.04351)	6 (.00932)	6 (.00932)	7 (.00124)	8 (.00008)	8 (.00008)
9	6 (.01678)	6 (.01678)	7 (.00315)	7 (.00315)	8 (.00037)	8 (.00037)
10	6 (.02622)	7 (.00617)	7 (.00617)	8 (.00103)	9 (.00011)	9 (.00011)
11	6 (.03733)	7 (.01037)	8 (.00218)	8 (.00218)	9 (.00033)	9 (.00033)
12	6 (.04977)	7 (.01572)	8 (.00393)	8 (.00393)	9 (.00075)	10 (.00010)
13	7 (.02214)	7 (.02214)	8 (.00633)	9 (.00144)	10 (.00025)	10 (.00025)
14	7 (.02952)	8 (.00939)	8 (.00939)	9 (.00245)	10 (.00051)	11 (.00008)
15	7 (.03773)	8 (.01312)	9 (.00383)	9 (.00383)	10 (.00092)	11 (.00018)
16	7 (.04666)	8 (.01750)	9 (.00560)	10 (.00151)	11 (.00034)	11 (.00034)
17	8 (.02248)	8 (.02248)	9 (.00778)	10 (.00231)	11 (.00058)	12 (.00012)
18	8 (.02801)	9 (.01037)	10 (.00333)	10 (.00333)	11 (.00092)	12 (.00022)
19	8 (.03405)	9 (.01338)	10 (.00461)	10 (.00461)	12 (.00036)	12 (.00036)

*Adapted from Table 3 in Z. W. Birnbaum and R. A. Hall's "Small Sample Distributions for Multi-Sample Statistics of the Smirnov Type," *Annals of Mathematical Statistics*, **31** (1960), 710–720, with permission of the authors and editor.

Table XII (*Continued*)

$n = m$	Significance Level, α					
	.05	.025	.01	.005	.001	.0005
20	8	9	10	11	12	13
	(.04053)	(.01677)	(.00615)	(.00198)	(.00056)	(.00014)
21	8	9	10	11	12	13
	(.04741)	(.02054)	(.00795)	(.00273)	(.00083)	(.00022)
22	9	9	11	11	13	13
	(.02467)	(.02467)	(.00365)	(.00365)	(.00034)	(.00034)
23	9	10	11	11	13	13
	(.02914)	(.01236)	(.00473)	(.00473)	(.00050)	(.00050)
24	9	10	11	12	13	14
	(.03390)	(.01496)	(.00598)	(.00216)	(.00070)	(.00020)
25	9	10	11	12	13	14
	(.03895)	(.01781)	(.00742)	(.00281)	(.00096)	(.00030)
26	9	10	11	12	14	14
	(.04425)	(.02090)	(.00904)	(.00357)	(.00042)	(.00042)
27	9	10	12	12	14	15
	(.04978)	(.02422)	(.00445)	(.00445)	(.00057)	(.00018)
28	10	11	12	13	14	15
	(.02776)	(.01281)	(.00545)	(.00213)	(.00076)	(.00025)
29	10	11	12	13	14	15
	(.03151)	(.01497)	(.00657)	(.00266)	(.00099)	(.00034)
30	10	11	12	13	15	15
	(.03544)	(.01729)	(.00782)	(.00327)	(.00045)	(.00045)
31	10	11	12	13	15	16
	(.03956)	(.01978)	(.00920)	(.00397)	(.00059)	(.00020)
32	10	11	13	13	15	16
	(.04384)	(.02243)	(.00476)	(.00476)	(.00075)	(.00027)
33	10	12	13	14	15	16
	(.04828)	(.01234)	(.00563)	(.00240)	(.00095)	(.00035)
34	11	12	13	14	16	16
	(.02819)	(.01409)	(.00660)	(.00289)	(.00045)	(.00045)
35	11	12	13	14	16	17
	(.03127)	(.01597)	(.00765)	(.00344)	(.00057)	(.00021)
36	11	12	13	14	16	17
	(.03450)	(.01797)	(.00880)	(.00405)	(.00071)	(.00027)
37	11	12	14	14	16	17
	(.03784)	(.02008)	(.00472)	(.00472)	(.00087)	(.00034)
38	11	12	14	15	17	17
	(.04130)	(.02230)	(.00547)	(.00248)	(.00042)	(.00042)
39	11	12	14	15	17	18
	(.04487)	(.02463)	(.00628)	(.00291)	(.00052)	(.00020)
40	11	13	14	15	17	18
	(.04854)	(.01430)	(.00715)	(.00338)	(.00064)	(.00025)

Table XIII* (13.5.5): *Critical Values of K^+ for Kolmogorov-Smirnov Maximum Deviation Tests for Goodness of Fit*

Values of K' for which $P(K^+ \geq K') = \alpha$ [or for which $P(K \geq K') \cong 2\alpha$ when $\alpha \leq .05$]

n	$\alpha = .10$	$\alpha = .05$	$\alpha = .025$	$\alpha = .01$	$\alpha = .005$
1	.90000	.95000	.97500	.99000	.99500
2	.68377	.77639	.84189	.90000	.92929
3	.56481	.63604	.70760	.78456	.82900
4	.49265	.56522	.62394	.68887	.73424
5	.44698	.50945	.56328	.62718	.66853
6	.41037	.46799	.51926	.57741	.61661
7	.38148	.43607	.48342	.53844	.57581
8	.35831	.40962	.45427	.50654	.54179
9	.33910	.38746	.43001	.47960	.51332
10	.32260	.36866	.40925	.45662	.48893
11	.30829	.35242	.39122	.43670	.46770
12	.29577	.33815	.37543	.41918	.44905
13	.28470	.32549	.36143	.40362	.43247
14	.27481	.31417	.34890	.38970	.41762
15	.26588	.30397	.33760	.37713	.40420
16	.25778	.29472	.32733	.36571	.39201
17	.25039	.28627	.31796	.35528	.38086
18	.24360	.27851	.30936	.34569	.37062
19	.23735	.27136	.30143	.33685	.36117
20	.23156	.26473	.29408	.32866	.35241
21	.22617	.25858	.28724	.32104	.34427
22	.22115	.25283	.28087	.31394	.33666
23	.21645	.24746	.27490	.30728	.32954
24	.21205	.24242	.26931	.30104	.32286
25	.20790	.23768	.26404	.29516	.31657
26	.20399	.23320	.25907	.28962	.31064
27	.20030	.22898	.25438	.28438	.30502
28	.19680	.22497	.24993	.27942	.29971
29	.19348	.22117	.24571	.27471	.29466
30	.19032	.21756	.24170	.27023	.28987
31	.18732	.21412	.23788	.26596	.28530
32	.18445	.21085	.23424	.26189	.28094
33	.18171	.20771	.23076	.25801	.27677
34	.17909	.20472	.22743	.25429	.27279
35	.17659	.20185	.22425	.25073	.26897
36	.17418	.19910	.22119	.24732	.26532
37	.17188	.19646	.21826	.24404	.26180
38	.16966	.19392	.21544	.24089	.25843
39	.16753	.19148	.21273	.23786	.25518
40	.16547	.18913	.21012	.23494	.25205

*Body of table is reproduced, with changes only in notation, from Table 1 in L. H. Miller's "Table of Percentage Points of Kolmogorov Statistics," *Journal of the American Statistical Association*, **51** (1956), 111–121, with permission of the author and editor.

Table XIII (*Continued*)

n	$\alpha = .10$	$\alpha = .05$	$\alpha = .025$	$\alpha = .01$	$\alpha = .005$
41	.16349	.18687	.20760	.23213	.24904
42	.16158	.18468	.20517	.22941	.24613
43	.15974	.18257	.20283	.22679	.24332
44	.15796	.18053	.20056	.22426	.24060
45	.15623	.17856	.19837	.22181	.23798
46	.15457	.17665	.19625	.21944	.23544
47	.15295	.17481	.19420	.21715	.23298
48	.15139	.17302	.19221	.21493	.23059
49	.14987	.17128	.19028	.21277	.22828
50	.14840	.16959	.18841	.21068	.22604
51	.14697	.16796	.18659	.20864	.22386
52	.14558	.16637	.18482	.20667	.22174
53	.14423	.16483	.18311	.20475	.21968
54	.14292	.16332	.18144	.20289	.21768
55	.14164	.16186	.17981	.20107	.21574
56	.14040	.16044	.17823	.19930	.21384
57	.13919	.15906	.17669	.19758	.21199
58	.13801	.15771	.17519	.19590	.21019
59	.13686	.15639	.17373	.19427	.20844
60	.13573	.15511	.17231	.19267	.20673
61	.13464	.15385	.17091	.19112	.20506
62	.13357	.15263	.16956	.18960	.20343
63	.13253	.15144	.16823	.18812	.20184
64	.13151	.15027	.16693	.18667	.20029
65	.13052	.14913	.16567	.18525	.19877
66	.12954	.14802	.16443	.18387	.19729
67	.12859	.14693	.16322	.18252	.19584
68	.12766	.14587	.16204	.18119	.19442
69	.12675	.14483	.16088	.17990	.19303
70	.12586	.14381	.15975	.17863	.19167
71	.12499	.14281	.15864	.17739	.19034
72	.12413	.14183	.15755	.17618	.18903
73	.12329	.14087	.15649	.17498	.18776
74	.12247	.13993	.15544	.17382	.18650
75	.12167	.13901	.15442	.17268	.18528
76	.12088	.13811	.15342	.17155	.18408
77	.12011	.13723	.15244	.17045	.18290
78	.11935	.13636	.15147	.16938	.18174
79	.11860	.13551	.15052	.16832	.18060
80	.11787	.13467	.14960	.16728	.17949
81	.11716	.13385	.14868	.16626	.17840
82	.11645	.13305	.14779	.16526	.17732
83	.11576	.13226	.14691	.16428	.17627
84	.11508	.13148	.14605	.16331	.17523
85	.11442	.13072	.14520	.16236	.17421

Table XIII (*Continued*)

n	$\alpha = .10$	$\alpha = .05$	$\alpha = .025$	$\alpha = .01$	$\alpha = .005$
86	.11376	.12997	.14437	.16143	.17321
87	.11311	.12923	.14355	.16051	.17223
88	.11248	.12850	.14274	.15961	.17126
89	.11186	.12779	.14195	.15873	.17031
90	.11125	.12709	.14117	.15786	.16938
91	.11064	.12640	.14040	.15700	.16846
92	.11005	.12572	.13965	.15616	.16755
93	.10947	.12506	.13891	.15533	.16666
94	.10889	.12440	.13818	.15451	.16579
95	.10833	.12375	.13746	.15371	.16493
96	.10777	.12312	.13675	.15291	.16408
97	.10722	.12249	.13606	.15214	.16324
98	.10668	.12187	.13537	.15137	.16242
99	.10615	.12126	.13469	.15061	.16161
100	.10563	.12067	.13403	.14987	.16081

Table XIV* (13.7.5): *Critical Upper-Tail Values of E for David's Empty-Cell Test for Goodness of Fit*

Smallest value of E' for which $P(E \geq E') \leq \alpha$, followed by $P(E \geq E')$

$\alpha = .05$

C	n = 5	6	7	8	9	10	11	12	13	14	15	16
2		1	1	1	1	1	1	1	1	1	1	1
		(.0313)	(.0156)	(.0078)	(.0039)	(.0020)	(.0010)	(.0005)	(.0002)	(.0001)	(.0001)	(.0000)
3	2	2	2	2	2	2	1	1	1	1	1	1
	(.0124)	(.0041)	(.0014)	(.0005)	(.0002)	(.0001)	(.0347)	(.0231)	(.0154)	(.0103)	(.0069)	(.0046)
4	3	3	2	2	2	2	2	2	2	2	2	1
	(.0039)	(.0010)	(.0463)	(.0233)	(.0117)	(.0059)	(.0029)	(.0015)	(.0007)	(.0004)	(.0002)	(.0400)
5	4	3	3	3	3	3	2	2	2	2	2	2
	(.0016)	(.0400)	(.0162)	(.0065)	(.0026)	(.0010)	(.0354)	(.0214)	(.0129)	(.0078)	(.0047)	(.0028)
6	5	4	4	4	3	3	3	3	3	2	2	2
	(.0008)	(.0200)	(.0068)	(.0023)	(.0368)	(.0188)	(.0095)	(.0048)	(.0028)	(.0489)	(.0330)	(.0222)
7	5	5	5	4	4	4	4	3	3	3	3	3
	(.0379)	(.0111)	(.0033)	(.0361)	(.0160)	(.0070)	(.0031)	(.0384)	(.0225)	(.0131)	(.0075)	(.0044)
8	6	6	6	5	5	5	4	4	4	4	3	3
	(.0259)	(.0067)	(.0017)	(.0197)	(.0076)	(.0029)	(.0296)	(.0154)	(.0078)	(.0040)	(.0446)	(.0272)
9	7	7	6	6	6	5	5	5	4	4	4	4
	(.0184)	(.0042)	(.0327)	(.0115)	(.0040)	(.0309)	(.0145)	(.0067)	(.0477)	(.0279)	(.0161)	(.0092)
10	8	8	7	7	6	6	6	5	5	5	5	4
	(.0136)	(.0028)	(.0222)	(.0070)	(.0413)	(.0179)	(.0076)	(.0449)	(.0240)	(.0126)	(.0066)	(.0448)

*Body of table is reproduced, with changes only in notation, from Tables la and 1b in M. Csorgo and I. Guttman's "On the Empty Cell Test," *Technometrics*, **4** (1962), 235–247, with permission of the authors and editor.

Table XIV (*Continued*)

$\alpha = .05$

C	n											
	17	18	19	20	21	22	23	24	25	26	27	28
2	1 (.0000)	1 (.0000)	1 (.0000)	1 (.0000)	1 (.0000)	1 (.0000)	1 (.0000)	1 (.0000)	1 (.0000)	1 (.0000)	1 (.0000)	1 (.0000)
3	1 (.0031)	1 (.0020)	1 (.0014)	1 (.0009)	1 (.0006)	1 (.0004)	1 (.0003)	1 (.0002)	1 (.0001)	1 (.0001)	1 (.0001)	1 (.0000)
4	1 (.0300)	1 (.0225)	1 (.0169)	1 (.0127)	1 (.0095)	1 (.0071)	1 (.0053)	1 (.0040)	1 (.0030)	1 (.0023)	1 (.0017)	1 (.0013)
5	2 (.0017)	2 (.0010)	2 (.0006)	2 (.0004)	1 (.0459)	1 (.0368)	1 (.0294)	1 (.0236)	1 (.0189)	1 (.0151)	1 (.0121)	1 (.0097)
6	2 (.0149)	2 (.0100)	2 (.0067)	2 (.0045)	2 (.0030)	2 (.0020)	2 (.0013)	2 (.0009)	2 (.0006)	2 (.0004)	1 (.0434)	1 (.0362)
7	3 (.0025)	2 (.0462)	2 (.0335)	2 (.0241)	2 (.0173)	2 (.0125)	2 (.0089)	2 (.0064)	2 (.0046)	2 (.0033)	2 (.0024)	2 (.0017)
8	3 (.0174)	3 (.0111)	3 (.0070)	3 (.0044)	3 (.0028)	2 (.0464)	2 (.0352)	2 (.0267)	2 (.0202)	2 (.0153)	2 (.0115)	2 (.0086)
9	4 (.0052)	3 (.0476)	3 (.0327)	3 (.0224)	3 (.0152)	3 (.0103)	3 (.0070)	3 (.0046)	3 (.0032)	2 (.0479)	2 (.0377)	2 (.0297)
10	4 (.0282)	4 (.0176)	4 (.0109)	4 (.0067)	4 (.0041)	3 (.0390)	3 (.0280)	3 (.0201)	3 (.0143)	3 (.0102)	3 (.0073)	3 (.0051)

Table XIV (*Continued*)

$\alpha = .05$

C	n											
	29	30	31	32	33	34	35	36	37	38	39	40
2	1	1	1	1	1	1	1	1	1	1	1	1
	(.0000)	(.0000)	(.0000)	(.0000)	(.0000)	(.0000)	(.0000)	(.0000)	(.0000)	(.0000)	(.0000)	(.0000)
3	1	1	1	1	1	1	1	1	1	1	1	1
	(.0000)	(.0000)	(.0000)	(.0000)	(.0000)	(.0000)	(.0000)	(.0000)	(.0000)	(.0000)	(.0000)	(.0000)
4	1	1	1	1	1	1	1	1	1	1	1	1
	(.0010)	(.0007)	(.0005)	(.0004)	(.0003)	(.0002)	(.0002)	(.0001)	(.0001)	(.0001)	(.0001)	(.0000)
5	1	1	1	1	1	1	1	1	1	1	1	1
	(.0077)	(.0062)	(.0050)	(.0040)	(.0032)	(.0025)	(.0020)	(.0016)	(.0013)	(.0010)	(.0008)	(.0007)
6	1	1	1	1	1	1	1	1	1	1	1	1
	(.0302)	(.0252)	(.0210)	(.0175)	(.0146)	(.0122)	(.0101)	(.0085)	(.0071)	(.0059)	(.0049)	(.0041)
7	2	2	2	2	1	1	1	1	1	1	1	1
	(.0012)	(.0009)	(.0006)	(.0004)	(.0429)	(.0368)	(.0315)	(.0271)	(.0233)	(.0199)	(.0171)	(.0147)
8	2	2	2	2	2	2	2	2	2	2	1	1
	(.0065)	(.0049)	(.0037)	(.0028)	(.0021)	(.0016)	(.0012)	(.0009)	(.0007)	(.0005)	(.0434)	(.0380)
9	2	2	2	2	2	2	2	2	2	2	2	2
	(.0233)	(.0183)	(.0143)	(.0112)	(.0087)	(.0068)	(.0053)	(.0042)	(.0032)	(.0025)	(.0020)	(.0015)
10	3	3	2	2	2	2	2	2	2	2	2	2
	(.0036)	(.0026)	(.0408)	(.0330)	(.0267)	(.0215)	(.0174)	(.0140)	(.0112)	(.0090)	(.0073)	(.0058)

Table XIV (*Continued*)

$\alpha = .05$

C	n = 41	42	43	44	45	46	47	48	49	50
2	1 (.0000)	1 (.0000)	1 (.0000)	1 (.0000)	1 (.0000)	1 (.0000)	1 (.0000)	1 (.0000)	1 (.0000)	1 (.0000)
3	1 (.0000)	1 (.0000)	1 (.0000)	1 (.0000)	1 (.0000)	1 (.0000)	1 (.0000)	1 (.0000)	1 (.0000)	1 (.0000)
4	1 (.0000)	1 (.0000)	1 (.0000)	1 (.0000)	1 (.0000)	1 (.0000)	1 (.0000)	1 (.0000)	1 (.0000)	1 (.0000)
5	1 (.0005)	1 (.0004)	1 (.0003)	1 (.0003)	1 (.0002)	1 (.0002)	1 (.0001)	1 (.0000)	1 (.0000)	1 (.0000)
6	1 (.0034)	1 (.0028)	1 (.0024)	1 (.0020)	1 (.0016)	1 (.0012)	1 (.0008)	1 (.0004)	1 (.0000)	1 (.0000)
7	1 (.0126)	1 (.0108)	1 (.0092)	1 (.0079)	1 (.0067)	1 (.0054)	1 (.0042)	1 (.0030)	1 (.0017)	1 (.0004)
8	1 (.0333)	1 (.0292)	1 (.0255)	1 (.0224)	1 (.0196)	1 (.0171)	1 (.0150)	1 (.0131)	1 (.0115)	1 (.0101)
9	2 (.0012)	2 (.0009)	2 (.0007)	2 (.0006)	1 (.0445)	1 (.0396)	1 (.0352)	1 (.0313)	1 (.0279)	1 (.0248)
10	2 (.0047)	2 (.0038)	2 (.0030)	2 (.0024)	2 (.0019)	2 (.0016)	2 (.0013)	2 (.0010)	2 (.0007)	2 (.0004)

Table XIV (*Continued*)

$\alpha = .01$

C	n											
	5	6	7	8	9	10	11	12	13	14	15	16
2				1	1	1	1	1	1	1	1	1
				(.0078)	(.0039)	(.0020)	(.0010)	(.0005)	(.0002)	(.0001)	(.0001)	(.0000)
3		2	2	2	2	2	2	2	2	2	1	1
		(.0041)	(.0014)	(.0005)	(.0002)	(.0001)	(.0000)	(.0000)	(.0000)	(.0000)	(.0069)	(.0046)
4	3	3	3	3	3	2	2	2	2	2	2	2
	(.0039)	(.0010)	(.0002)	(.0001)	(.0000)	(.0059)	(.0029)	(.0015)	(.0007)	(.0004)	(.0002)	(.0001)
5	4	4	4	3	3	3	3	3	3	2	2	2
	(.0016)	(.0003)	(.0001)	(.0065)	(.0026)	(.0010)	(.0004)	(.0002)	(.0001)	(.0078)	(.0047)	(.0028)
6	5	5	4	4	4	4	3	3	3	3	3	3
	(.0008)	(.0001)	(.0068)	(.0023)	(.0008)	(.0003)	(.0095)	(.0048)	(.0024)	(.0012)	(.0006)	(.0003)
7	6	6	5	5	5	4	4	4	4	4	3	3
	(.0004)	(.0001)	(.0033)	(.0009)	(.0003)	(.0070)	(.0031)	(.0013)	(.0006)	(.0002)	(.0075)	(.0044)
8	7	6	6	6	5	5	5	5	4	4	4	4
	(.0002)	(.0067)	(.0017)	(.0004)	(.0076)	(.0029)	(.0011)	(.0004)	(.0078)	(.0040)	(.0020)	(.0010)
9	8	7	7	7	6	6	6	5	5	5	5	4
	(.0002)	(.0042)	(.0010)	(.0002)	(.0040)	(.0014)	(.0005)	(.0067)	(.0031)	(.0014)	(.0006)	(.0092)
10	9	8	8	7	7	7	6	6	6	6	5	5
	(.0001)	(.0028)	(.0006)	(.0070)	(.0022)	(.0007)	(.0076)	(.0031)	(.0013)	(.0005)	(.0066)	(.0034)

Table XIV (*Continued*)

$\alpha = .01$

C	n = 17	18	19	20	21	22	23	24	25	26	27	28
2	1 (.0000)	1 (.0000)	1 (.0000)	1 (.0000)	1 (.0000)	1 (.0000)	1 (.0000)	1 (.0000)	1 (.0000)	1 (.0000)	1 (.0000)	1 (.0000)
3	1 (.0031)	1 (.0020)	1 (.0014)	1 (.0009)	1 (.0006)	1 (.0004)	1 (.0003)	1 (.0002)	1 (.0001)	1 (.0001)	1 (.0001)	1 (.0000)
4	2 (.0001)	2 (.0000)	2 (.0000)	2 (.0000)	1 (.0095)	1 (.0071)	1 (.0053)	1 (.0040)	1 (.0030)	1 (.0025)	1 (.0017)	1 (.0013)
5	2 (.0017)	2 (.0010)	2 (.0006)	2 (.0004)	2 (.0002)	2 (.0001)	2 (.0001)	2 (.0001)	2 (.0000)	2 (.0000)	2 (.0000)	1 (.0097)
6	3 (.0002)	3 (.0001)	2 (.0067)	2 (.0045)	2 (.0030)	2 (.0020)	2 (.0013)	2 (.0009)	2 (.0006)	2 (.0004)	2 (.0003)	2 (.0002)
7	3 (.0025)	3 (.0015)	3 (.0008)	3 (.0005)	3 (.0003)	3 (.0002)	2 (.0089)	2 (.0064)	2 (.0046)	2 (.0033)	2 (.0024)	2 (.0017)
8	4 (.0005)	4 (.0003)	3 (.0070)	3 (.0044)	3 (.0028)	3 (.0018)	3 (.0011)	3 (.0007)	3 (.0004)	3 (.0003)	3 (.0002)	2 (.0086)
9	4 (.0052)	4 (.0030)	4 (.0017)	4 (.0009)	4 (.0005)	4 (.0003)	3 (.0070)	3 (.0046)	3 (.0032)	3 (.0021)	3 (.0014)	3 (.0010)
10	5 (.0017)	5 (.0009)	5 (.0005)	4 (.0067)	4 (.0041)	4 (.0025)	4 (.0015)	4 (.0009)	4 (.0006)	4 (.0003)	3 (.0073)	3 (.0051)

Table XIV (*Continued*)

$\alpha = .01$

C	n = 29	30	31	32	33	34	35	36	37	38	39	40
2	1	1	1	1	1	1	1	1	1	1	1	1
	(.0000)	(.0000)	(.0000)	(.0000)	(.0000)	(.0000)	(.0000)	(.0000)	(.0000)	(.0000)	(.0000)	(.0000)
3	1	1	1	1	1	1	1	1	1	1	1	1
	(.0000)	(.0000)	(.0000)	(.0000)	(.0000)	(.0000)	(.0000)	(.0000)	(.0000)	(.0000)	(.0000)	(.0000)
4	1	1	1	1	1	1	1	1	1	1	1	1
	(.0010)	(.0007)	(.0005)	(.0004)	(.0003)	(.0002)	(.0002)	(.0001)	(.0001)	(.0001)	(.0001)	(.0000)
5	1	1	1	1	1	1	1	1	1	1	1	1
	(.0077)	(.0062)	(.0050)	(.0040)	(.0032)	(.0025)	(.0020)	(.0016)	(.0013)	(.0010)	(.0008)	(.0007)
6	2	2	2	2	2	2	2	1	1	1	1	1
	(.0001)	(.0001)	(.0001)	(.0000)	(.0000)	(.0000)	(.0000)	(.0085)	(.0071)	(.0059)	(.0049)	(.0041)
7	2	2	2	2	2	2	2	2	2	2	2	2
	(.0012)	(.0009)	(.0006)	(.0004)	(.0003)	(.0002)	(.0002)	(.0001)	(.0001)	(.0001)	(.0000)	(.0000)
8	2	2	2	2	2	2	2	2	2	2	2	2
	(.0065)	(.0049)	(.0037)	(.0028)	(.0021)	(.0016)	(.0012)	(.0009)	(.0007)	(.0005)	(.0004)	(.0003)
9	3	3	3	3	2	2	2	2	2	2	2	2
	(.0006)	(.0004)	(.0003)	(.0002)	(.0087)	(.0068)	(.0053)	(.0042)	(.0032)	(.0025)	(.0020)	(.0015)
10	3	3	3	3	3	3	3	3	3	2	2	2
	(.0036)	(.0026)	(.0018)	(.0013)	(.0009)	(.0006)	(.0004)	(.0003)	(.0002)	(.0090)	(.0073)	(.0058)

Table XIV (*Continued*)

$\alpha = .01$

C	n = 41	42	43	44	45	46	47	48	49	50
2	1 (.0000)	1 (.0000)	1 (.0000)	1 (.0000)	1 (.0000)	1 (.0000)	1 (.0000)	1 (.0000)	1 (.0000)	1 (.0000)
3	1 (.0000)	1 (.0000)	1 (.0000)	1 (.0000)	1 (.0000)	1 (.0000)	1 (.0000)	1 (.0000)	1 (.0000)	1 (.0000)
4	1 (.0000)	1 (.0000)	1 (.0000)	1 (.0000)	1 (.0000)	1 (.0000)	1 (.0000)	1 (.0000)	1 (.0000)	1 (.0000)
5	1 (.0005)	1 (.0004)	1 (.0003)	1 (.0003)	1 (.0002)	1 (.0002)	1 (.0001)	1 (.0000)	1 (.0000)	1 (.0000)
6	1 (.0034)	1 (.0028)	1 (.0024)	1 (.0020)	1 (.0016)	1 (.0012)	1 (.0008)	1 (.0004)	1 (.0000)	1 (.0000)
7	2 (.0000)	2 (.0000)	1 (.0092)	1 (.0079)	1 (.0067)	1 (.0054)	1 (.0042)	1 (.0030)	1 (.0017)	1 (.0004)
8	2 (.0002)	2 (.0002)	2 (.0001)	2 (.0001)	2 (.0001)	2 (.0001)	2 (.0000)	2 (.0000)	2 (.0000)	2 (.0000)
9	2 (.0012)	2 (.0009)	2 (.0007)	2 (.0006)	2 (.0004)	2 (.0003)	2 (.0003)	2 (.0002)	2 (.0002)	2 (.0001)
10	2 (.0047)	2 (.0038)	2 (.0030)	2 (.0024)	2 (.0019)	2 (.0016)	2 (.0013)	2 (.0010)	2 (.0007)	2 (.0004)

Table XV*: *Cumulative Probability Points in the Standard Normal Distribution*
Z = Random normal variate with zero mean and unit variance
$P = P(Z \leq Z'); Q = 1 - P = P(Z \geq Z') = P(Z \leq -Z')$

P	Z'	Q	P	Z'	Q
0.50	0.00000	0.50	0.925	1.43953	0.075
0.51	0.02507	0.49	0.930	1.47579	0.070
0.52	0.05015	0.48	0.935	1.51410	0.065
0.53	0.07527	0.47	0.940	1.55477	0.060
0.54	0.10043	0.46	0.945	1.59819	0.055
0.55	0.12566	0.45	0.950	1.64485	0.050
0.56	0.15097	0.44	0.955	1.69540	0.045
0.57	0.17637	0.43	0.960	1.75069	0.040
0.58	0.20189	0.42	0.965	1.81191	0.035
0.59	0.22754	0.41	0.970	1.88079	0.030
0.60	0.25335	0.40	0.975	1.95996	0.025
0.61	0.27932	0.39	0.980	2.05375	0.020
0.62	0.30548	0.38	0.985	2.17009	0.015
0.63	0.33185	0.37	0.990	2.32635	0.010
0.64	0.35846	0.36			
0.65	0.38532	0.35	0.991	2.36562	0.009
0.66	0.41246	0.34	0.992	2.40892	0.008
0.67	0.43991	0.33	0.993	2.45726	0.007
0.68	0.46770	0.32	0.994	2.51214	0.006
0.69	0.49585	0.31	0.995	2.57583	0.005
0.70	0.52440	0.30	0.996	2.65207	0.004
0.71	0.55338	0.29	0.997	2.74778	0.003
0.72	0.58284	0.28	0.998	2.87816	0.002
0.73	0.61281	0.27	0.999	3.09023	0.001
0.74	0.64335	0.26			
0.75	0.67449	0.25	0.9991	3.12139	0.0009
0.76	0.70630	0.24	0.9992	3.15591	0.0008
0.77	0.73885	0.23	0.9993	3.19465	0.0007
0.78	0.77219	0.22	0.9994	3.23888	0.0006
0.79	0.80642	0.21	0.9995	3.29053	0.0005
0.80	0.84162	0.20	0.9996	3.35279	0.0004
0.81	0.87790	0.19	0.9997	3.43161	0.0003
0.82	0.91537	0.18	0.9998	3.54008	0.0002
0.83	0.95417	0.17	0.9999	3.71902	0.0001
0.84	0.99446	0.16			
0.85	1.03643	0.15	0.9999 5	3.89059	0.0000 5
0.86	1.08032	0.14	0.9999 9	4.26489	0.0000 1
0.87	1.12639	0.13	0.9999 95	4.41717	0.0000 05
0.88	1.17499	0.12	0.9999 99	4.75342	0.0000 01
0.89	1.22653	0.11	0.9999 995	4.89164	0.0000 005
0.900	1.28155	0.10	0.9999 999	5.19934	0.0000 001
0.905	1.31058	0.095	0.9999 9995	5.32672	0.0000 0005
0.910	1.34076	0.090	0.9999 9999	5.61200	0.0000 0001
0.915	1.37220	0.085	0.9999 9999 5	5.73073	0.0000 0000 5
0.920	1.40507	0.080	0.9999 9999 9	5.99781	0.0000 0000 1

*Reproduced, with changes only in notation, from a portion of Table 1.2 in D. B. Owen's *Handbook of Statistical Tables*, Reading, Mass.: Addison-Wesley Publishing Company, Inc., 1962, pp. 11–12, with permission of the author, the publisher, the Sandia Corporation, and the United States Atomic Energy Commission.

Table XVI*: *Critical Values of Chi-Square*

Values of χ^2_α, where $P(\chi^2 \geq \chi^2_\alpha) = \alpha$

d.f.	$\alpha = .05$	$\alpha = .01$	d.f.	$\alpha = .05$	$\alpha = .01$	d.f.	$\alpha = .05$	$\alpha = .01$
1	3.841	6.635	46	62.830	71.201	91	114.268	125.289
2	5.991	9.210	47	64.001	72.443	92	115.390	126.462
3	7.815	11.345	48	65.171	73.683	93	116.511	127.633
4	9.488	13.277	49	66.339	74.919	94	117.632	128.803
5	11.071	15.086	50	67.505	76.154	95	118.752	129.973
6	12.592	16.812	51	68.669	77.386	96	119.871	131.141
7	14.067	18.475	52	69.832	78.616	97	120.990	132.309
8	15.507	20.090	53	70.993	79.843	98	122.108	133.476
9	16.919	21.666	54	72.153	81.069	99	123.225	134.642
10	18.307	23.209	55	73.311	82.292	100	124.342	135.807
11	19.675	24.725	56	74.468	83.513	102	126.574	138.134
12	21.026	26.217	57	75.624	84.733	104	128.804	140.459
13	22.362	27.688	58	76.778	85.950	106	131.031	142.780
14	23.685	29.141	59	77.931	87.166	108	133.257	145.099
15	24.996	30.578	60	79.082	88.379	110	135.480	147.414
16	26.296	32.000	61	80.232	89.591	112	137.701	149.727
17	27.587	33.409	62	81.381	90.802	114	139.921	152.037
18	28.869	34.805	63	82.529	92.010	116	142.138	154.344
19	30.144	36.191	64	83.675	93.217	118	144.354	156.648
20	31.410	37.566	65	84.821	94.422	120	146.567	158.950
21	32.671	38.932	66	85.965	95.626	122	148.779	161.250
22	33.924	40.289	67	87.108	96.828	124	150.989	163.546
23	35.172	41.638	68	88.250	98.028	126	153.198	165.841
24	36.415	42.980	69	89.391	99.228	128	155.405	168.133
25	37.652	44.314	70	90.531	100.425	130	157.610	170.423
26	38.885	45.642	71	91.670	101.621	132	159.814	172.711
27	40.113	46.963	72	92.808	102.816	134	162.016	174.996
28	41.337	48.278	73	93.945	104.010	136	164.216	177.280
29	42.557	49.588	74	95.081	105.202	138	166.415	179.561
30	43.773	50.892	75	96.217	106.393	140	168.613	181.840
31	44.985	52.191	76	97.351	107.583	142	170.809	184.118
32	46.194	53.486	77	98.484	108.771	144	173.004	186.393
33	47.400	54.776	78	99.617	109.958	146	175.198	188.666
34	48.602	56.061	79	100.749	111.144	148	177.390	190.938
35	49.802	57.342	80	101.879	112.329	150	179.581	193.208
36	50.998	58.619	81	103.010	113.512	200	233.994	249.445
37	52.192	59.892	82	104.139	114.695	250	287.882	304.940
38	53.384	61.162	83	105.267	115.876	300	341.395	359.906
39	54.572	62.428	84	106.395	117.057	400	447.632	468.724
40	55.758	63.691	85	107.522	118.236	500	553.127	576.493
41	56.942	64.950	86	108.648	119.414	600	658.094	683.516
42	58.124	66.206	87	109.773	120.591	700	762.661	789.974
43	59.304	67.459	88	110.898	121.767	800	866.911	895.984
44	60.481	68.710	89	112.022	122.942	900	970.904	1001.630
45	61.656	69.957	90	113.145	124.116	1000	1074.679	1106.969

*Reproduced, with changes only in notation, from a portion of Table 3.1 in D. B. Owen's *Handbook of Statistical Tables*, Reading Mass.: Addison-Wesley Publishing Company, Inc., 1962, pp. 49–55, with permission of the author, the publisher, the Sandia Corporation, and the United States Atomic Energy Commission.

AUTHOR INDEX

SUBJECT INDEX